Blue meridia
Peter Macthuisson

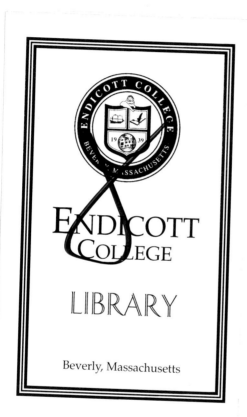

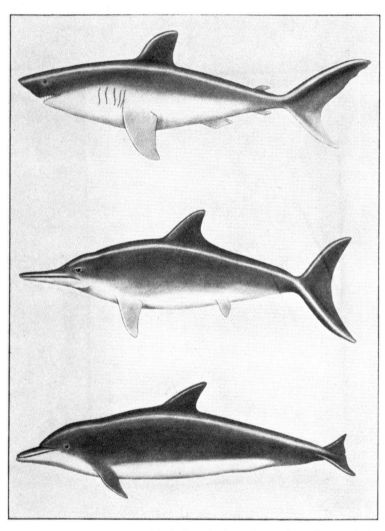

FRONTISPIECE. Convergence of body form in three classes of vertebrates: shark, a fish, above; ichthyosaur, a reptile; and porpoise, a mammal, below.

AQUATIC MAMMALS

Their Adaptations to Life in the Water

By

A. BRAZIER HOWELL

Lecturer in Comparative Anatomy
Department of Anatomy
The Johns Hopkins University

With fifty-four illustrations

DOVER PUBLICATIONS, INC.

NEW YORK

599
H

International Standard Book Number: 0-486-22593-3
Library of Congress Catalog Card Number: 79-116825

Manufactured in the United States of America
Dover Publications, Inc.
180 Varick Street
New York, N.Y. 10014

DEDICATED to those who are endeavoring to save from commercial extinction the great whales, the largest animals that have ever lived.

Foreword

F OR SOME years I have been investigating the functional anatomy of mammals specialized for particular modes of progression and endeavoring to correlate the facts discovered with their morphology and habits. I have already published several papers of a technical nature on the anatomy of aquatic mammals and the present book is the result of these studies. My endeavor is to present the subject in an essentially nontechnical manner, for it is believed that it is of sufficient interest to appeal to a reading public which is larger than that contained in the anatomical or even the purely mammalian field. Partly for this reason lengthy and tedious descriptions and impressive array of data have been omitted.

The subject of aquatic adaptation in mammals is somewhat peculiar in that there are numerous points which are difficult or impossible of thoroughly conclusive investigation. Habits are at times extremely hard to study and many of the specializations are of such a character that there are no intermediate stages with which they can properly be compared. As a result there has been advanced a truly appalling list of improperly founded theories in attempted explanation of these points. This, however, has perhaps been justified for the reason that only by this means can much progress be made. Many of the specializations are so unusual that in order finally to gain an understanding of them it is first necessary to have a variety of plausible theories from which to work.

. The present contribution is far from the final word on the subject. I believe that I have herewith shown that some of the theories so far tentatively accepted are really untenable, and I have advanced others that to me seem better founded, but some of my readers will justifiably hold contrary opinions, for in morphological interpretations there is invariably produced some chaff with the wheat and further discoveries will surely oblige some revision. In fact I am constantly revising my own beliefs as my knowledge of conditions increases. I have endeavored to evaluate every hypothesis that seemed to hold merit. In attempting to arrive at the proper explanation of some detail I have frequently advanced two or more essentially diverse theories, possibly without in my own mind accepting either of them, but discussing them pro and con in order that the reader may understand the various angles of the case. It is thus desired to emphasize the fact that the present report is more purely philosophical in character than would be justifiable in almost any other branch of mammalogy. We really know so little regarding the multitude of unique aquatic specializations that were statements confined to proven facts the most interesting parts of this fascinating topic would be omitted. In formulating theories and endeavoring to interpret conditions I have taken frequent opportunity to discuss problems with my colleagues. Often have they made remarks and offered suggestions that have given me an entirely new angle from which to approach some particular detail. For this I am appropriately grateful, and although I have tried always to make acknowledgement at the proper place, still I have found it difficult to give stated credit where credit may really be due. I can and do, however, give full measure of appreciation to Remington Kellogg, whose unflagging interest in my undertaking and broad knowledge of whales, unstintedly shared, have proven quite indispensable.

Contents

Illustrations

xi

Introduction

In such an investigation as the following the writer is forcibly impressed with the soundness of that tenet expressed by Jennings (1924) when he said that "under the same conditions objects of different material behave diversely; under diverse conditions objects of the same material behave diversely", and that "neither the material constitution alone, nor the conditions alone, will account for any event whatever; it is always the combination that has to be considered".

In considering such a phenomenally specialized organism as the whale and studying it part by part one is also forcibly impressed by the staggering length of time that it must have taken to produce such a creature, and the intricacy of the evolutional processes that have shaped countless anatomical details to effect proper interaction and interrelationship of parts to a single harmonious end.

The evidences of evolution are most abundant upon every hand, and still we know so very little about them. Tentative theories are advanced and discussed, but the ways of evolution are so exceedingly deliberate that the lifetime of a single investigator is not sufficiently long for him to observe its natural processes. O. P. Hay (1928) has stated "A learned writer on mammals tells me he doubts that a single new species has developed since the first interglacial stage", which was "perhaps 400,000 years" ago. If this be the case I do not see how the whales have evolved from a terrestrial ancestor in less than one hundred million years, and two hundred million might be all too short. How, then, can we expect successfully to investigate the processes of evolution in the laboratory? This element of vast time should be stressed, and also, although we know little about the ways of evolution, we cannot help formulating tentative hypotheses of explanation.

The writer is strongly convinced that we cannot look to any one theory to explain evolution. The whale is very different indeed from the generalized type of mammal in a great many major details, and hundreds and even thousands of minor details, involving billions of cells. These items have been changed doubtless by numberless evolutional stimuli, some simple and others complex, involving the inherent tendencies and limitations of the organism, antagonistic influences, and all the intricacies of the relationship of an animal to its environment.

[1]

The reader will doubtless discover in the following pages that the writer subscribes to a modification of the Lamarckian theory, or theory of the inheritance of acquired characters, perhaps better expressed as inheritance based upon the use or disuse of parts. This theory has fallen into disfavor for the reason that no one has been able to demonstrate it, but that is no reason for ignoring its possibility. The failure to prove it doubtless lies in the probability that it operates too slowly to be discernable in the lifetime of one man. There have been futile experiments such as the one in which the tails of white mice were removed throughout several generations in the expectation of producing a race of bobtails. Naturally this was foredoomed to failure under any circumstances for account was not taken of the fact that amputation of the tail did not remove the mouse's *need* for a tail as a balancing organ; or that the germ plasm remained unaltered. But if we could cause white mice to dig industriously in the ground for several hours each and every day, and oblige all their descendants to do likewise with conscientiousness, then would there be produced, according to my belief, a race of mice with feet much more specialized for digging: not, emphatically in twenty or a hundred generations, but in twenty or a hundred *thousand* generations.

In a study of this kind one frequently encounters the puzzling condition illustrated by the following situation. In the rorquals the narrow part of the tail—the peduncle—has built up a sharply ridged keel, both above and below, of fibrous tissue which acts to reduce water resistance during swimming. The broad part of the tail—the flukes—in this animal has built up precisely the same sort of fibrous expansions, but for the purpose of *increasing* water resistance. Thus the same sort of structure has been developed in response to stimuli which appear diametrically opposite. This is entirely beyond our comprehension, and if the condition is referred to at all it is usually in the most cautious, evasive terms. All that can be told is that often a structure will *seem* to develop where there has clearly become a need for it without our being able to distinguish any really directive stimulus at all, in spite of the fact that logic dictates that a new modification cannot anticipate a new function.

In considering the specialization exhibited by aquatic mammals one must constantly bear in mind that the results seen depend upon the strength of the stimuli involved, the capacity of the animal to respond to them, the strength of possible antagonistic stimuli, and the length of time involved. The aquatic evolution of one particular type of mammal may be in a straight line with slight deviation, while in the case of another, perhaps handicapped by some bodily equipment less readily adaptable to aquatic requirements, there may be much trial and error, with the employment of temporary makeshifts. Thus one pair of limbs may become quite highly modified for propulsion, but yet incapable, because of some mechanical defect in the method of employment, of developing as high speed as can the altered tail, in which case the latter is apt to take over the duty of locomotion and the former fall into disuse. But this is so only up to a certain point, and the "law of irreversible evolution" is believed to prevent the redevelopment of a part once atrophy is very definitely under way. The sea-lion might develop its muscles so that the hind limbs would become the chief organ of propulsion, but whales have lost their hind limbs forever.

Thorough modification for an aquatic life by a mammal involves not only more numerous but more profound anatomical and physiological alteration than for any other sort of existence. The volant specialization of the bat has resulted in considerable change; but this has concerned only lengthening or rotation of bones, development of membranes, and not particularly profound muscular modification. In the whale, however, the evolution of bodily details has been truly astonishing, and to only a slightly lesser extent in sirenians. Thus in mammals the aquatic life is the only one that has resulted in the complete loss of the external portions of the hind limb, acquisition of supernumerary phalanges to the digits, migration of the nostrils to the top of the head, and entire inability to travel for at least a few yards when placed on land. The entire specialized development has been confined to the ends of securing food and of locomotion involving but a single pair of major movements. In some sorts the former specialization does not have much effect upon body form, so that as far as concerns external conformation the only major stimulus has been a simplified propulsive one. This being the case it should not cause undue astonishment that the most perfect degree of highly specialized convergence in bodily form to be found between different vertebrate classes is furnished by some species of shark, ichthyosaur, and whale (frontispiece).

But while the external form, and even some of the internal anatomy, of a whale may be said to be simplified, internal alteration from the normal mammalian sort is often so profound in both quantity and quality that one may be at a complete loss to interpret it correctly. And yet all organs of the terrestrial mammal are present, even though they may be vestigeal. The change from what we consider as the mammalian norm has been insensibly deliberate, and if we could but have the complete picture before us there would be no difficulty in its correct interpretation. The change which the abandonment of the atmospheric for the completely aquatic environment necessitated was prodigeous, but nevertheless it is probable that with few exceptions (as the spermaceti organ of the cachalot) there have been no fundamentally new details developed, but only the gradual change and specialization of old ones.

In discussing aquatic modification, and most other things, the majority of writers are prone to consider that there is one, or at most a few, stimuli operating to accomplish precisely the same end. In other words it seems to be a common belief that when mammals take to the water there will be a tendency for them all to develop along precisely the same lines to exactly the same end. This is true to only a definitely limited extent. A dozen, say, types of terrestrial mammals taking to the water may have twelve individually peculiar organizations, differing in their capabilities for variation, amenable to diverse stimuli of dissimilar strength, with twelve combinations of interdependence of parts. The convergence which they will ultimately show will be in two ideal types of body form and toward two (presumably) final methods of swimming. But these ends may be attained in considerably different manners, involving entirely different specialization of details, and one must be cautious in stating that the aquatic life will result in such an item as a shortened neck without a very careful analysis of the factors involved.

Almost all mammals can swim after a fashion, and even some of those which one would naturally expect to be the least capable of doing so, such as the sloths (*Bradypus* and *Choloepus*) can cross broad rivers. To merit the term aquatic, however, a vertebrate must be sufficiently at home in the water so that when near this element it will instinctively seek it for concealment or escape. It must not only be able to swim with adequate speed but must be able to dive with facility and remain submerged. In almost all cases there is necessary the further corollary that it habitually seeks its food in the water. The ability to propel itself through the water does not mean that it has necessarily developed organs that are appreciably more highly specialized for such use than are found in its nearest, wholly terrestrial relative, but it does mean that such a vertebrate must have evolved some sort of valvular mechanism for closing both nostrils and ears while submerged. In other words an aquatic vertebrate must have the ability to close all orifices leading within the body.

[4]

All aquatic mammals have, of course, evolved from terrestrial ones, although paleontology does not furnish us with evidence regarding the early stages of any aquatic genus, but many of the steps may be reconstructed with a feeling of considerable assurance.

The surf is usually too boisterous to appeal to a small or medium-sized mammal with a desire for an occasional swim, and shore waters offer too few places for concealment. Large carnivores might take directly to the sea, as the polar bear is doing, but it is probable that most aquatic mammals first ventured into fresh water, the smaller sorts into bogs and small streams, and the larger into swamps or rivers. Those of carnivorous (including insectivorous) propensities took this step almost invariably in search of prey which seemed to them easier of capture than terrestrial food, or else more palatable. At this stage carnivores probably would not seek the water as a means of escape from their enemies and their swimming must have been only superficial, consisting of a plunge after a fish or a short sally across a stream. An aquatic herbivore probably would start its career either by seeking the shelter of densely grown swamps as a diurnal refuge from its enemies, to secure aquatic plants which attracted it, or to escape the torment of insects by standing in the water. It would not take long, geologically speaking, for either the carnivore or herbivore to discover that the water offered a safe refuge from most of its enemies. But this must also be taken into consideration: an abundance of aquatic enemies, such as crocodiles, would tend to discourage the adoption of an aquatic existence and just this factor may have prevented the aquatic development of many a promising mammalian stock.

A mammal with somewhat palustrine tendencies might, therefore, likely have been forced into the water either by similar forms that competed with it for terrestrial food, or by enemies which pursued it— perhaps both. If in the water it encountered an abundance of enemies in the shape of large fish or hungry reptiles it might either be exterminated, forced back to face terrestrial enemies as being the lesser of two evils, or obliged to content itself with a borderline existence, foraging in shallow water amid protective vegetation. This latter has constituted the fate of most small, semi-aquatic insectivores and rodents. Some of them may eventually encounter an environment sufficiently favorable so that they can relinquish such a game of hide-and-seek and boldly take to the open water, but most of them will not, and the latter can never develop very highly specialized aquatic modifications.

It is probable that the original adoption of a habitat that was largely aquatic usually, if not almost invariably, meant for a terrestrial mammal a diminution in the severity of the competition to which it was accustomed, else it would not have taken to the water. And it is well known that a life wherein competition is reduced is not conducive to the rapid modification of bodily form. On the other hand, any such drastic change in the functions of the body as is experienced by a terrestrial mammal when it takes definitely to the water is highly conducive to evolutionary changes. To what effect these diametrically opposed tendencies have acted in the past upon the aquatic mammals with which we are now acquainted cannot be conjectured. But we do know, by such paleontological evidence as is at hand, that the course of development and change in our aquatic mammals has been an extremely slow and lengthy process, probably lasting, in those groups most highly adapted, throughout tens if not scores of millions of years.

Let us take the hypothetical case of a carnivore that eventually becomes exclusively marine. For our purpose this must be of an active type that prefers live food, and not such an omnivorous and slightly sluggish animal as the racoon (*Procyon*), which is slightly aquatic in its habits. Such an active carnivore likes fish and finds that it is easier to catch them in the shallows than to compete with other sorts of predators catching mice in the woods. He will spend an increasing amount of time in the water and throughout successive generations gain increasing confidence in swimming considerable distances. During this time he has naturally discovered that few of his terrestrial enemies will follow him into the water, and he therefore instinctively seeks this element when startled. But he still journeys overland from one stream to another, may hunt on shore occasionally when fish are temporarily shy, and sleeps and raises his family in a hole beneath a bank. While he is content to linger in a small river he can do little else and his bodily modifications will not only be correspondingly circumscribed, but the change throughout long ages will be slow.

Perhaps in a few million years this carnivore has gained such a facility in swimming that he finds narrow quarters irksome, and seeks the greater freedom of large rivers. He is now able to catch fish in fair chase, and can also escape from his enemies by his speed. The water is his home but although he frequently takes short naps while floating on the surface, he still seeks a hole in the bank, or if large and bulky, a sandbar, for a sound sleep, and his mate must seek the land to raise a family. At this time it makes little difference in the degree of his aquatic adaptation whether he inhabit large rivers or coastal waters. And heretofore all steps in his development have presumably been by slow stages. He himself soon has no further need for the land, for although he still likes to bask on a rock in the sun, it is not long before he can forego this luxury, does such a course seem expedient. But his young must pass their early life upon the land and if this necessity persists he can never become exclusively aquatic.

The latter is a fundamental factor in the evolution of a highly developed aquatic mammal—the requirements of its young. The newborn seal and sea-lion will drown if forced into the water. If the terrestrial or amphibious enemies of pinnipeds should multiply sufficiently to destroy more than the critical number of young needed to replace the breeding stock this order would become extinct, for they are not yet ready to desert the land entirely. In a few million years they may be ready to do so. At that time, if driven from their rookeries, they may have become sufficiently modified so that enough young might survive an aquatic birth for the perpetuation of the race. The whales and sirenians have successfully taken this step and forever severed their slightest connection with dry land.

The development of existing aquatic herbivores was probably considerably different, for this sort of mammal is under no necessity for rapid movement in order to secure sustenance. From a palustral habitat the ancestors of the sirenians doubtless subsided sluggishly into deep water that was free from large enemies, chiefly for the purpose of escaping troublesome terrestrial carnivores, and have since been under the necessity of doing little but move from one submerged pasture to another.

All the above resolves into the simple statement, first advanced by Kükenthal (1890) that the degree of aquatic specialization in mammals is corollary to the amount of connection retained with the land.

Chapter One

The Mechanics of Swimming

IN CONSIDERING the characteristic modes of locomotion of any organism there are other factors besides simple progression to be heeded. Perhaps the most important of these is posture, which in turn depends largely upon bodily form as well as many environmental conditions. The posture of an elephant and a mouse must be different, necessitating differences in the skeletal framework and consequently in the muscular equipment. Dissimilarity of muscular equipment involves corresponding diversification of the controlling nervous mechanism, according as a mammal may habitually trot, gallop, pace or hop. Muscular action that is certainly reflex, if not actually involuntary in such forms as can sleep while standing, is a necessary agent in maintaining posture in a terrestrial mammal, while in a thoroughly aquatic one, such as a whale, posture need not involve action of the muscles, and balancing actions while moving are entirely different from those in which the non-aquatic sort must indulge. Similarly there must be important modifications in the nervous equipment for determining the swimming actions of such a mammal as the seal as contrasted with that of the sea-lion.

The laws underlying the mechanics of swimming as employed by vertebrates are of almost hopeless complexity for the reason that the body is not rigid and the force is not applied at any one point, but more or less continuously over a greater or lesser area. It has taken many years to calculate the factors encountered by a rigid ship moving through the water at a given speed with all the driving force applied at one point. It has taken years for a multitude of highly trained technicians to discover the proper lines for a rigid aeroplane fuselage and wings. If the latter were propelled to the accompaniment of contortive wrigglings by the hinder end, or convulsive gyrations of the wings by means of a multitude of small engines (i.e. muscles) of unknowable horse-power delivering their power at vague points, one may visualize how little would be known about aerodynamics at the present time. Precisely this situation is encountered by one who would investigate the principles underlying the swimming of mammals, and

[9]

the higher the attainments of any qualified physicist with whom the questions are discussed the more emphatic is he in his indication that he will have nothing to do with it. It therefore seems entirely impossible to arrive at an exact conception of the mechanics of swimming by mammals, but certain generalities we know to be facts, and the evidence for certain others appears to be sufficiently strong to justify us in tentatively offering hypotheses.

Stream-line, as I understand it, is that indefinite term by which we designate the precise shape of a given mass in order that it will slip through the water (or air) with less friction or resistance than any other shape of equal mass. But it is perhaps requisite that this body be towed. If it be self-propelled then the propeller will introduce a complication. It therefore is self-evident that no vertebrate can be perfectly stream-line in form. Not only must it have appendageous means of propulsion, but it usually has a steering or equilibrating apparatus. Because of feeding or other requirements its head may not be of a shape best adapted for speedy passage through the water. Either to compensate for such external irregularities or because obligated by muscular or visceral requirements, the cross section of the body may depart in one direction or the other from the circle that is ideal in a stream-line form. This being the case, two fish may be equally efficient in body form and yet have a considerably different appearance. Both may be 90 per cent efficient, and yet those details which from one detract 10 per cent may be entirely different from those amounting to the 10 per cent in the other. Or another viewpoint may be accorded this fact by the statement that what proves stream-line for one mode of propulsion does not prove so for some other method of swimming. Not only that but if one animal swims with its entire caudal and lumbar regions involved, the degree of its departure from an ideal stream-line form will not be the same as if it swims by vibrations of the tip of the tail only.

Thus, in effect, it is found that when we speak of a body as stream-line we mean merely that it tapers gradually, without sharp angles or excrescences that would offer resistance as it passes through the water. But even though our knowledge of what is, or might be, the ideal of stream-line form in any particular case is so vague, the evidence furnished by the tendency of every essentially active form of aquatic life to assume certain definite body-shapes is so overwhelming that we know beyond question that this is one of the most fundamental of aquatic stimuli.

Terrestrial vertebrates may swim with much lost motion, thrashing about and straining with most of the muscles of the body. As soon as one of these has become even slightly aquatic, however, movement through the water is accomplished more easily and effectively, as it must be in order not to exhaust the swimmer.

By such of the vertebrates as spend much of their time in the water, aquatic propulsion would seem to be accomplished by three methods:

(1) Rhythmic undulatory or oscillating movements of the body proper.

(2) Rhythmic movements of the appendages.

(3) Expulsion of jets of water.

As far as I am aware, no vertebrate swims exclusively by expelling jets of water, but Breder (1926) and others have shown that many fish habitually employ the force of the water expelled through their gills as a material though secondary aid to some other primary means of propulsion. For our present purpose this method of swimming may be dismissed with no more than this brief mention.

Swimming by means of rhythmic movements of the body include oscillations either in a vertical or horizontal plane of the body proper and these movements are in consequence always transmitted to the tail, which, if the latter be of sufficient size, then acts as a primary means of locomotion. This includes all vertebrates which swim chiefly by means of the tail except such fishes as hold the body rigid while rapidly oscillating the tail *tip*.

Swimming by means of rhythmic movements of the appendages is employed by all vertebrates which propel themselves through the water by movements of their limbs, even though their bodies be wriggled during the process, by undulations of long dorsal or ventral fins, or of a vibratory tail tip.

In the case of such forms as are but slightly modified for an aquatic life it is frequently difficult to decide which is the primary means of propulsion through the water, as body, tail and limbs may all be used to some extent, but such doubtful cases are always mentioned in the text. Another difficulty in properly classifying swimming methods is introduced by the fact that some particular vertebrate may have experienced a sequence in its propulsive mechanism during its evolutionary modification, a more inefficient method being temporarily employed pending development of the final primary propulsive organ.

The means of aquatic locomotion mentioned above in (1) and (2), together with representative vertebrates employing them, may be arranged as follows.

MODES OF AQUATIC PROPULSION EMPLOYED BY VERTEBRATES

(1) Propulsion chiefly by means of oscillations of the body and base of tail.

(*a*) Body fusiform

Pisces (the majority of fish with body form as in mackerel, shark etc.)

Reptilia

Caudata (the majority of salamanders)

Crocodilia (crocodiles and alligators; chiefly)

Ichthyosauria (of the better known, short-tailed sorts)

Mammalia

Insectivora (probably only *Potomogale* and *Limnogale*)

Carnivora (Lutrinae—the river otters, chiefly)

Sirenia

Cetacea (exclusive of zeuglodonts of *Basilosaurus* type)

(*b*) Body not fusiform but largely anguilliform or eel-like

Pisces (eels and fish of this form)

Reptilia

Serpentes (all swimming snakes)

Amblyrhynchus (màrine iguanas)

Mosasaurs

Mammalia

Cetacea (only zeuglodonts of *Basilosaurus* type)

(2) Propulsion by means of appendageous movements.

(*a*) Propulsion by both fore and hind limbs

Reptilia

Chelonia (turtles of the mud-turtle type)

Plesiosaurus and doubtless others

Mammalia

Carnivora (*Thalarctos,* the polar bear)

Pinnipedia (Odobenidae)

Ungulata (Hippopotamidae)

Rodentia (Hydrochoeridae)

(*b*)   Propulsion by means of pectoral appendages

Pisces (many fish use the pectoral fins for slow propulsion; the skates and rays (Batoidea) exclusively)

Reptilia

Chelonia (the marine turtles)

Aves (penguins and such birds as "fly" under water)

Mammalia

Monotremata

Pinnipedia (Otariidae)

(*c*)   Propulsion by pelvic appendages

Reptilia (Salientia—frogs and toads)

Aves (all birds that swim with the feet)

Mammalia

Marsupialia (*Chironectes*)

Insectivora (practically all aquatic forms save *Potomogale* and *Limnogale*)

Carnivora (Enhydrinae)

Pinnipedia (Phocidae)

Rodentia (all aquatic forms chiefly, most of them exclusively)

(*d*)   Propulsion by undulations of a longitudinal fin

Pisces (an important, although possibly not the most effective, means of locomotion of such forms as *Amiatus* and *Gymnotus,* and exclusively of the Hippocampidae or sea horses)

(*e*)   Propulsion by vibration or undulation of the tail tip

Pisces (used by many fish at slow speeds, and by such as are incased in an unyielding body covering, as the Ostraciidae)

In some cases a certain vertebrate may swim by a combination of the above methods, and at times, especially in the case of certain extinct reptiles, it is extremely difficult to determine the precise method of swimming.

Although beset with several difficulties which are at present insurmountable, some of the principles underlying the aquatic progression of mammals may be discussed with confidence.

If one drop a cutworm into a dish of water the worm is unable to progress, or to do aught but bend first to one side and then to the other. The reason for this is that the center of gravity and center of

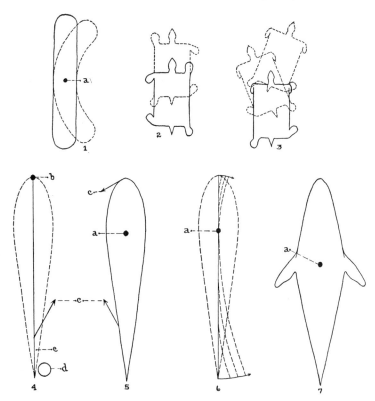

FIGURE 1. Diagrams illustrative of some swimming principles of vertebrates: (1) a cut worm or similar form; (2) mud turtle swimming freely; (3) mud turtle with fore legs bound down by adhesive tape; (4) fish with its snout held rigid; (5 and 6) fish swimming free; (7) the sea-lion principle: (a) pivot of movement; (b) fulcrum; (c) power arm; (d) weight to be moved (i.e. the water); (e) lever.

motion are at the middle of the body, and both the form and mass of the anterior and of the posterior parts are equal, and hence, any motion by one end is equalized by a corresponding motion of the other. Bending of the body in a curve hitches it first an infinitesimal distance to one side, and antagonistic bending by the body then hitches it to the other side, but all actions and all motions are equalized.[1]

As our cutworm cannot swim because there are two antagonistic propulsive strokes of equal force and two antagonistic body divisions of equal mass and shape, it is evident that in order to make a swimming organism out of it we would have to make certain interdependent alterations as follows: There would have to be provision for holding the forward part of the body relatively inert to furnish a base from which propulsive movements could be initiated. This could be accomplished by increasing the mass of the anterior part of the animal and altering its shape so that it offered more resistance to torsional stress. This would have the effect of displacing the center of mass of the entire animal forward of the middle. A relative increase in the mass of the anterior part would cause a relative decrease in that of the posterior portion, and the latter should also experience some flattening and broadening, to provide a greater area for operating against the water. The result may be likened to a man that is propelling a skiff by a single sculling-oar from the stern. But this sculling-oar may be very short and light, or very long and heavy. If the former, conditions are comparable to those encountered in fishes of the Ostraciidae type, in which the entire body is incased in an unyielding covering and locomotion is accomplished by vibration of the tail tip. Or if the sculling-oar be very long and heavy, conditions may better be compared to the mackerel, whale and similar marine types which propel themselves by oscillations of the entire body. Both these groups might conceivably have body forms of essentially the same degree of stream-line perfection, and the center of gravity would therefore tend to be located at the same point in both, although varying somewhat with the speed of movement. But the actual center or pivot of motion would be very different indeed. In the ostraciiform fish it would be located in the tail proper and far from the center of mass, while in the other sort it would be much farther forward and nearer the center of mass. In effect, a fish can hold its body rigid and move only the extreme tip of the tail in order to glide forward, or it can move the entire body; and this is what introduces difficult physics. Not only is the amount of force applied in swimming indeterminate, but the location of the exact area over which it operates is

[1] It must be understood in this connection that there is no assertion made of the inability of a short, worm-like body to swim through the water. Some larval forms can do so very well, but this ability depends upon some particular specialization which they have been able to develop.

unknown, and can never be known with certainty. Thus in some sorts of whales it seems that the entire body takes part in swimming movements, and that the center of motion accordingly migrates forward to the center of mass, while it has been observed that when some sorts of porpoises are swimming at speed the oscillations of the tail are through such a short arc and are so rapid that the animal appears to be making no movement at all. Which type of action is the more efficient cannot be stated.

If one should hold a fish by the snout and allow it to wriggle in the water from side to side its motions might in some respects be compared to the principles of a lever of the third order. The snout held immovable may be called the fulcrum while the water to be moved at either side of the tail is the weight (fig. 1) The entire posterior half, or even more, of the fish is the lever, and the muscles concerned in pulling the tail first to one side and then to the other constitute a double, compound power arm. When the fish is released and it darts away, the circumstances will have been altered as follows. The fulcrum shifts backward and becomes either a center of mass or a pivot of motion, according to conditions, and to the pair of caudal power arms is added a pair of anterior power arms (fig. 1), and consequently, a second lever. In fish of the usual type the anterior lever will always be much shorter than the posterior. The posterior one may also be so short that as a result the length of the anterior lever will be negligible (as in fish swimming by vibration of the tail tip), or the posterior may be of sufficient length so that it is continuous with the anterior lever—which is another way of saying that swimming will then be accomplished by throwing the entire body into a continuous curve.

For an interpretation of the motions followed by a fusiform fish (Breder's carangiform, and Abel's "torpedo principle") I follow Breder (1926), who has discussed the aquatic movements of this class with great thoroughness, and those who are interested in following the details at greater length may refer to him. If we take a fish of the form of a mackerel the swimming motions will theoretically be as follows: The contraction of the muscles of one side will throw the body into a curve, but because of the resistance which the flattened posterior part offers to the water this results in the head being thrown farther to one side than the tail. With the momentum thus acquired by the head and consequent inertia of the more massive anterior part of the body, the fish is enabled to swing the tail to the opposite side with considerable force (fig. 2). The pressure of the curved tail against the water initiates forward motion and this is maintained by alternations of the movements from side to side. The important points to note are that the body and tail do not oscillate about the snout, but that the anterior and posterior parts of the animal oscillate from a point in the body anterior to the middle. Whether this point is actually the center of gravity I do not know, for an aquatic vertebrate of this sort really has no center of gravity while it is in the water, so I prefer to employ the term "center of equilibrium" to designate this point. If the animal throw itself

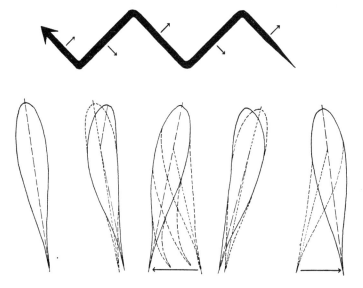

FIGURE 2. Diagrams illustrative of the swimming principles of an anguilli-form or eel-like body above, and of a fusiform body below (redrawn from Breder).

into an uninterrupted curve while swimming, then the center of equi-librium will constitute a pivot of motion, from which both ends are curved. If only the tail proper be involved in swimming motions (as in ostraciiform fish), then this curve of the caudal amphikinetic part (as of Breder) will begin at a point considerably posterior to the center of equilibrium, and as an accompaniment the curve of the cephalic amphikinetic part (always present in theory although at times reduced to an inappreciable amount) *must begin at a point correspondingly an-terior to the center of equilibrium.* The difficulty lies in defining the significance of the word "correspondingly" in the last sentence. The principle is clear but the formula cannot be worked out.

The above mode of swimming, termed the fusiform type because a spindle-shaped body is the best fitted to employ it, is followed by most fish, *Potomogale,* Sirenia, Phocidae (in seals the hind feet are employed the same as a tail), and Cetacea. Almost always there are, or finally will be, at least one rudder or pair of rudders or stabilators disassociated from the primary organ of propulsion, and not infrequently there is one or more accessory equilibrators (as a dorsal fin).

If the body and tail become disproportionately long for their diameter this mode of swimming will change to the anguilliform or eel-like type. The length, relative to diameter, must be sufficiently great for the animal to assume an S-shaped posture, when it will then swim by what constitutes, in reality, a duplication of the fusiform method of swimming, slipping through the water as a snake slips through the grass. This is of interest in the present connection chiefly for the reason that this was undoubtedly the method of swimming employed by long-tailed zeuglodonts of the *Basilosaurus* type. But there may be body shapes and swimming methods intermediate between the anguilliform and fusiform types. Some of the longer bodied, modern whales may be partially anguilliform in their motions, and the Weddell seal (*Leptonychotes*) certainly is remarkably serpentine in its terrestrial movements. Then too there were short-bodied and long, anguilliform-bodied zeuglodonts, and ichthyosaurs of similarly different conformation. Presumably the tendency should be for a shortening of an anguilliform vertebrate because the fusiform method of swimming is capable of higher speed, and the shorter body encounters less water friction.

Propulsion by the fore limbs, and by the hind limbs in other manner from that in which the hair seals (Phocidae) use these members, involves different principles. The most specialized instance of propulsion by the fore limbs among Mammalia is encountered in the sea-lions and fur seals (Otariidae). The principle is much the same as that of a row-boat. The latter should be fairly long for its width, for if it be short and tubby it will tend to wabble rather than progress in a straight line. It is evident also that the oars should operate from near the center of the craft, and that the latter should taper from the middle toward either end. This description fits the sea-lion, which apparently can steer with ease by means of either the head and neck or the hinder end. The description does not fit the platypus, however, which also swims chiefly by means of the pectoral limbs. In the latter the mass anterior to these limbs is not approximately equal to the mass posterior thereto, but very much less, and there must therefore be specialized apparatus for equilibration, which exists in both the hind feet and the tail.

But there are different methods of swimming by the anterior limbs. Thus the platypus apparently pushes the water directly backward with the palm and partly clenches the manus during recovery. The marine turtle holds the axis of the manus almost parallel with the body axis (at least in the more specialized sorts) and pushes the water to the rear by a vertical motion of the arm. The sea-lion also employs an oblique movement of the flipper, but by pushing the water from the *ulnar* border of the manus by adduction of this member in the transverse plane.

There are two features of interest illustrated by mammals which swim by kicking the hind feet in alternation. In one, encountered in almost all aquatic rodents and insectivores, as the hind feet are kicked the tail is involuntarily moved from side to side by the wriggling motion of the hinder end of the body proper. If the tail be of considerable length this member will at first effect all the equilibration that is necessary, and later, as the tail becomes more specialized, it will take over a substantial part, if not the whole, of the function of propulsion. If the hind limbs be relatively small their action in swimming may be assisted by the fore limbs, or if the tail be too short to act efficiently as an equilibrator the pectoral limbs may be kicked in alternation chiefly as balancing agents, just as a man swings his arms when walking.

It is easily seen that if an animal swim by alternate strokes of the hind limbs alone there will be no serious disturbance of equilibrium providing the legs be close together. If the animal be of a tubby shape, however, with hind legs far apart, each kick will throw the body to the side unless there be some separate provision for maintaining equilibration. This may be nicely illustrated by the common mud turtle. If one bind down the fore limbs of one of these animals by means of adhesive tape and then place it in a pool of water it is found that the turtle is very greatly handicapped thereby and will progress by a series of erratic jerks, first to one side and then the other (fig. 1). The reason for this is that in vertical dimension the body is very thin, and in transverse very broad, so that the four legs appear almost as though attached to the corners of a rectangle. If one foot only is kicked it will turn the rectangle, and to prevent such turning there would either have to be a heavy tail, which the animal lacks, or a compensating kick by the opposite limb of the other pair. Thus the mud turtle in reality trots through the water. As specialization advances and the feet become paddles in the marine turtles the tendency is to use the hind limbs as equilibrators and the anterior pair as propellers, these being operated in unison with a sort of flying motion.

The same principle as forces the mud turtle to trot through the water operates to oblige all slightly specialized aquatic mammals with insignificant tails to employ all four feet in swimming, especially if the body be given to corpulency. So apparently swims the hippopotamus, whose legs are far apart, and the capybara, which has no tail; also the polar bear. But I am inclined to think that the culminating development in such mammals would not be the acquisition of four paddles of almost equal efficacy, as marine reptiles so often show, but rather the ascendency of one pair of limbs over the other, such as is now encountered in the Otariidae or Phocidae.

In this connection it may be of interest to mention that aquatic reptiles seem to have had modifications which it is probable that no mammal could develop. Thus reptiles can increase the number of cervical vertebrae to a phenomenal degree—a procedure which a mammal cannot effect. The tendency in marine reptiles always was to develop four paddles, while in mammals but one pair usually becomes very highly modified. Similarly there must be many other reptilian body details which respond to definite stimuli in an entirely different manner than do the same details in mammals. I therefore regard it as very unsafe to draw conclusions anent the Mammalia from data provided by the Reptilia.

The extinct plesiosaurs followed the turtle path for a considerable distance. Their bodies were heavy and broad, as well as somewhat flattened, and it is not improbable that this shape was accentuated in their ancestors. They were equipped with four long paddles, and as they employed a four-cornered method of propulsion they steered with the feet, kicking according as they wished to turn to one side or the other, and neither the tail nor the neck needed to function as a rudder. These two extremities could therefore develop in response to other stimuli without much regard to any duty in swimming. The result was that some sorts were short tailed and short necked, while others had a long tail and a neck that was relatively much longer than any other vertebrate has ever had, so far as we know. They were rather sluggish beasts, evidently, and the long neck was doubtless developed so that the head might dart here and there in pursuit of active prey in spite of the clumsy body. No other instance is known either in reptiles (apparently) or mammals in which a highly specialized aquatic form had

[20]

both a long tail and a long neck, and presumably this development could only take place in connection with a body in which each motion during swimming was counterpoised by a corresponding motion on the opposite side.

The Crocodilia, mesosaurs, mosasaurs and longer-bodied ichthyosaurs really represent different steps toward the same goal in the development of their limbs, and in all these there is very little difference in size between the fore and hind feet. The longer sorts were partly anguilliform in movement and the shorter, fusiform. The Crocodilia use the tail exclusively for speedy swimming, and the hind limbs to a considerable extent during sluggish progression. So probably did the mesosaurs, and the mosasaurs largely so. In the latter the limbs had become true paddles, the fore limbs differing but slightly in size from the hinder ones, but the powerful, laterally flattened tail is clearly indicated as the more efficient propulsive organ and the flippers had probably begun to be used merely as equilibrators and accessory organs, as they finally became in the later, shorter-bodied ichthyosaurs. In the latter, with their very powerful tails, the hind limbs were considerably smaller than the fore limbs, and hence it seems evident that they had long since fallen into relative disuse, but not at a sufficiently remote period to have disappeared altogether.

From a consideration of aquatic mammals it seems that there are certain mechanical conditions which the more specialized sorts are constrained to endeavor to fulfill during the course of their development. There seems to be a decided tendency for aquatic mammals to develop as the primary means for locomotion a single organ or pair of organs, and if the original conformation of the animal allow and nothing divert it from its goal, the situation of this propelling apparatus will most likely be at the hinder end and medially situated. In other words, the stimulus is for the acquisition of the fish-tail type of propulsion, although if we leave out of consideration the question of comparison of the muscles involved, it makes no difference whether this equipment be in the form of a true tail expanded vertically or horizontally, or of the hind feet held with soles adpressed (Phocidae). Pending the development of a tail fitted for propulsion the chief means by which swimming is accomplished will probably, in the majority of cases, be the hind feet. Unless the animal can use the hind feet in swimming after the manner of the Phocidae (and Enhydrinae?), however, these members will probably never develop into the final, primary propulsive organ, one reason possibly being that the mammalian foot is likely incapable of duplicating the mechanical conditions embodied in the foot of a grebe or duck, which is not the most efficient method of swimming anyway. Swimming by oscillations of the hinder end of the body (by feet or tail) is the most economical method partly for the reason that all motions are propulsive ones, without obligation of any braking action by recovery motions, and because each stroke of the rhythmic cycle is of approximately equal power.

If the tail be too small to be adapted for a propulsive organ it is then likely that the fore feet will finally be modified for taking over this function exclusively (as in the Otariidae). Theoretically it is not quite so efficient a method of swimming, for it can never be brought to as high a state of perfection as can that involving the transformation of the tail into flukes, but when a stage will have been reached comparable to that at which the sea-lions now are, the fore limbs will constitute very effective swimming organs indeed, any braking action by recovery motions being almost entirely overcome, as discussed in a later chapter. Whether a seal can actually swim faster than a sea-lion is unknown, for there are no trustworthy figures on the subject. Both are capable of high speed; not to the extent of 60 miles an hour, of course, as I have heard claimed by some illiterate fisherman, nor is it likely that they could outstrip some of the speedier porpoises. On theoretical grounds the seal should prove the faster. If it is not it may be because it does not need to be.

Before closing the present consideration of aquatic progression it will be well to emphasize the fact that there will usually be a regular sequence of swimming methods employed during the evolution of any aquatic vertebrate. This is conveniently illustrated by the alligator. This reptile may "walk" slowly through the water with all four feet, or may progress by means of the pedes alone, while during more rapid progression the appendages are pressed to the body while the tail is lashed from side to side. This is entirely illustrative of a sequence in methods of progression according as one part after another of a vertebrate becomes differentiated for special use. It is also illustrative of another and fundamentally important consideration with which every aquatic vertebrate is obliged to contend at some point in its evolution, and which the reader will readily appreciate after having read this chapter. No method of swimming involving forthright pushing backward of the water by any pair of appendages is ever truly efficient, for the speed thereby attained is limited by the inertia of the limbs and their inability to repeat the rhythmic motions sufficiently fast. For really high speed an efficient pair of anterior (sea-lion) or posterior (seal) flippers, or a properly formed tail (whale) must act obliquely against the water, for the same reason that an ice boat can travel much faster quartering the wind than running directly before it. This principle has governed the whole course of development of all the more highly adapted aquatic vertebrates.

Chapter Two

Aquatic Mammals

THE FOLLOWING members of the class Mammalia are definitely, although not necessarily exclusively, aquatic in their predilections:

MONOTREMATA
 Ornithorhynchus

MARSUPIALIA
 Chironectes

INSECTIVORA
 Talpidae
  Desmana
  Galemys
 Soricidae
  Neosorex
  Atophyrax
  Neomys
  Chimarrogale
  Crossogale
  Nectogale
 Tenrecidae
  Limnogale
  Potomogale

CARNIVORA
 Ursidae
  Thalarctos
 Mustelidae
  Mustela (the mink and sump-
   fotter only)
 Lutrinae (all genera)
 Enhydrinae (the single genus)

PINNIPEDIA (all)
 Otariidae
 Odobenidae
 Phocidae

RODENTIA
 Octodontidae
  Myocastor
 Hydrochoeridae
  Hydrochoerus
 Muridae
  Crossomys
  Hydromys
  Parahydromys
  Dasymys
  Nilopegamys
  Arvicola
  Microtus (partly)
  Ondatra
  Neofiber
 Castoridae (all)
  Castor
 Cricetidae
  Ichthyomys
  Rheomys
  Anotomys

LAGOMORPHA
 Leporidae
  Sylvilagus palustris
  Sylvilagus aquaticus

ARTIODACTYLA
 Hippopotamidae (all)
  Hippopotamus
  Choeropsis

PERISSODACTYLA
 Tapiridae (all)

SIRENIA (all)
   Trichechidae
   Halicoridae
   Hydrodamalidae (extinct)

CETACEA (all)
   Archaeoceti (extinct)
   Odontoceti
   Mysticeti

Perhaps the tapir should not be included in the above list, while there are doubtless those who will consider that in it should be placed other mammals, such as other swamp-loving ungulates.

No attempt will be made to present a complete diagnosis of the families and genera included, but only a brief characterization and the salient points which are likely of significance from the viewpoint of their aquatic specialization. Preceding these, in order to give the general reader a better understanding of the chapters to follow, will be a general consideration of the mammals and their habits. The precise derivation of different aquatic mammals is not here of particular concern, but it seems advisable to touch briefly, without any great array of supporting evidence, upon the probable stock from which the more specialized of aquatic mammals were derived, in order that we may more intelligently follow the probable course of their development.

It is virtually certain that even in those mammals only slightly aquatic there are present valvular mechanisms for the closure of the nostrils and external ears, and these two points will receive no further mention in the present chapter.

MONOTREMATA

Ornithorhynchus—platypus or duck-bill: an inhabitant of quiet streams and rivers of Australia and Tasmania. It feeds upon worms and similar food upon the stream bottoms. It constructs burrows for resting and for raising its young, and is capable of traveling over land somewhat clumsily but at considerable speed if its stream should go dry. The premaxillaries and maxillaries are expanded anteriorly and support a naked beak which superficially resembles that of a duck, and which is covered with a soft and very sensitive membrane. The body is flattened horizontally and covered with a fine, dense underfur, beyond which projects a coat of coarser hairs. The tail is rather short, very broad and compressed. The eyes are very small and there is no external pinna to the ear. The legs are short and well modified for swimming, the toes being webbed and with large claws. The webbing of the forefoot extends well beyond the claws, and this anterior part is folded back when the animal progresses upon land, is digging, or combing its fur. In swimming the forelimbs are used almost exclusively, apparently alternately, while the hind feet are extended laterally and used, with the tail, as equilibrating apparatus.

We cannot rely too greatly upon the internal anatomy of the platypus in judging its changes for a life in the water; and we cannot even be sure that the flatness of the body was not inherited from terrestrial ancestors. The form of its feet, tail, "bill" and external ear are the result of this life, indubitably, but we cannot be sure of its muscles, bones, and certain details of its internal organs. It and the echidna are the sole representatives of an 'exceedingly primitive group of egg-laying mammals, and we have no suitable material, either living or fossil, with which to compare it. As these are the most primitive of mammals the stock must be of tremendous age, and it is extremely likely that the platypus has been aquatic for a great many millions of years. It would doubtless be far more specialized in this direction but for the fact that it is most at home along streams of but moderate size, has few if any aquatic competitiors, and feeds chiefly upon food that is relatively inactive. It is not impossible that its aquatic preferences began even before the placental mammals became differentiated. For the above reasons the internal anatomy of the platypus will either be entirely omitted or discussed with great circumspection.

MARSUPIALIA

Chironectes—water-opossum or yapok: of Central America and northern South America. It subsists chiefly upon small fish and crustaceans, but is attracted by almost anything edible. Its pelage is short and dense and the external ears are well developed. The tail is round but the hind feet are large and very broadly webbed, in some toes to the tip and in others to the last phalanx.

But few details regarding the habits of the yapock—the only aquatic representative of the Marsupialia—are known, but its dependence upon the land appears to be great. The development of the hind feet seems to be its only change for swimming, if we except the character of the pelage, and it is remarkable that the feet should be so highly specialized while the tail is perfectly terete. The base of the latter tapers gradually from the body, as in many other marsupials. Although it is known that small marsupial pouch-young can suspend breathing for many minutes without harm and Carl Hartman has told me of a swimming opossum *(Didelphis)* with live pouch-young having been captured, it would seem that in *Chironectes* the female *must* curtail aquatic activity while she raises a family. Without doubt this factor has been of exceeding importance in limiting the degree of aquatic specialization of this animal, and it has probably contributed very largely to the fact that the Australian marsupials, so able in their plasticity to fill all vacant ecologic niches, have never developed an aquatic form, for such bold individuals as dared venture freely into the water would be likely to drown their young and die without issue. At the same time it is possible that during submergence the pouch might retain sufficient air for the needs of the young.

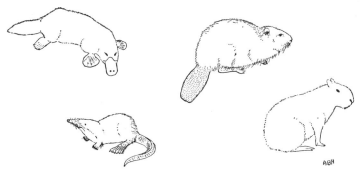

FIGURE 3. Some aquatic mammals. Platypus (*Ornithorhynchus*) (redrawn from Lewin), beaver (*Castor*), desman (*Desmana*) (redrawn from Flower and Lydekker), and capybara (*Hydrochoerus*) (redrawn from Scott).

INSECTIVORA

Aquatic insectivores are all small and few of them can venture into deep water for fear of large fish. Hence they must retain in large degree their dependence upon the land, for throughout their lives along the rills which they inhabit escape by a scamper through pebbly shallows may be just as necessary as by diving. Aquatic forms have the fur even denser than usual: the tendency is for the hind feet to become enlarged and fringed with bristles as well as to be slightly pronated and for the tail to become keeled with hairs, or (and) flattened horizontally. Rarely have the external ears disappeared. Swimming, except in the aquatic Tenrecidae, is evidently always by alternate strokes of the hind feet assisted to some extent by accompanying motions of the tail.

Talpidae

Desmana—the desman, occurs in southeastern Russia. The fur is especially dense and fine, the snout is long and flattened vertically, and there are no external ears. Although partly webbed the forefeet are small, while the hind feet are enlarged, webbed to the base of the nails, and heavily fringed by bristly hairs along the outer border. The tail is robust and laterally compressed. This genus is more modified for an aquatic life than any of the Soricidae.

Galemys occurs in parts of Spain and adjoining France. It is similar to *Desmana* but smaller, snout somewhat longer, tail longer and laterally compressed only near the tip, but its hairs form a faintly defined ventral keel. It is not quite so highly modified aquatically as the desman.

Soricidae

Neosorex and *Atophyrax* are two groups of water shrews inhabiting parts of the United States and Canada. Jackson (1928) has recently placed them with the genus *Sorex* but they have usually been accorded generic (or subgeneric) rank and may be so treated here for convenience. Their habitat is along the borders of little streams and in bogs and marshes, where they secure water insects and small fish. Appreciable modifications for an aquatic life consist only of short fringes of stiffened hairs upon the sides of the hind feet and toes, and slight pronation of these members.

Neomys—the European water shrew, is found in Europe and European Russia. Its aquatic modifications also consist of fringes to the hind feet and toes, but in addition one species *(N. foidens)* has a median keel of stiffened hairs extending the entire length of the under surface of the tail.

Chimarrogale of the Himalayas and parts of China and Japan has the hind feet and toes fringed and the hands slightly so, while the tail has a well marked inferior fringe.

Crossogale of the Malay Archipelago is said to be essentially like *Chimarrogale.*

Nectogale of Tibet and adjacent China is more specialized. In general it too resembles *Chimarrogale* but there are said to be no external ears, the hind feet are webbed in addition to having slight fringes, and the tail is fringed below.

Tenrecidae

Limnogale of Madagascar has a decidedly flattened muzzle and webbed feet. The whole tail is very powerful but only the distal part is laterally compressed. Presumably both hind feet and tail are used in swimming but probably the latter is the chief organ of propulsion.

Potomogale is an inhabitant of the streams of equitorial Africa. External ears are present. The muzzle is flattened and the body is cylindrical, continuing uninterruptedly into the thick, powerful tail, which is excessively compressed laterally. The legs are short, rather weak, and the toes are not webbed, but the lateral border of the hind foot is broadened to a thin edge so that it may be folded back more smoothly against the tail. The latter is evidently the exclusive means of aquatic propulsion, the feet being then pressed against the body and base of the tail.

The aquatic members of the Talpidae and Soricidae may be considered as a unit. The insectivores are among the most primitive of placental mammals and reasonably close to the original prototype. They are notoriously conservative, some of them, we know, having remained virtually unchanged since Cretaceous times. Remains of desmans which

FIGURE 4. The African insectivore otter *Potomogale,* from a mounted specimen in the National Museum.

are considered to be congeneric with the living animal are known from the Middle and Lower Miocene of Europe. So we can rest assured that the aquatic insectivores are no very recent development but that their modifications have been brought about throughout very long geologic periods. That such specialization is not now more marked than merely in the form of the hind feet and at times, slight changes in the tail and ear, may be attributed to the conservativeness of the insectivore phylum and the fact that they are so largely dependent upon the land.

Potomogale is a paradox, for it is extremely unlikely that any land mammal would take to the water and from the very start use its tail as the sole means of propulsion. In other cases a flattening of the tail follows acquisition of webbing by, and at times an increase in the size of, the hind feet. Yet the feet of this genus are unwebbed and remarkably weak. It too belongs to a primitive family of insectivores and its modifications probably began at a very remote time.

CARNIVORA

Ursidae

Thalarctos—the polar bear, is an inhabitant of the Arctic regions. It spends so much of its time in the water that it must needs be included in any consideration of aquatic mammals, yet the only way in which this bear differs from others is in the increased hariness of the soles of the feet, which was probably brought about by the coldness of the ice on which it walks rather than any influence of the water. It also has a build somewhat more rangy and slender—in other words more streamline—than is usual in this family. All four feet are employed in swimming.

Mustelidae

Mustela. In this genus only the American mink and European sumpfotter may be said to be aquatic, and yet there are really no discernible external indications of the fact. The toes are partially webbed but so are those of many strictly terrestrial species of this genus.

Lutrinae. All river otters are very largely aquatic. Of this subfamily the following genera are now recognized by those who have recently worked with them: *Microaonyx, Paraonyx, Aonyx, Amblonyx, Hydrictis, Pteronura, Lutra,* and *Lutrogale.* They occur almost throughout the world on the larger and many of the smaller land masses. The legs are rather short, the body sinuous and cylindrical, and the tail is very stout, especially at base, tapering gradually from the body to the tip. In *Pteronura* there is a lateral flange or keel upon each side of the tail. The webbing of the feet differs somewhat in the different genera, and this is discussed fully in the case of the Old World genera by Pocock (1921). Thus in *Aonyx* and *Paraonyx* of Africa the forefeet are unwebbed, and even in the common otter they are no more webbed than in many of the strictly terrestrial mustelids, although in at least one species of this genus *(Lutra maculicollis)* they are. In *Paraonyx, Microaonynx* and *Aonyx,* the so-called clawless otters of Africa, the hind feet are practically unwebbed, but in most genera the webbing extends almost or quite to the tips of the toes, although the feet are little if any enlarged. The base of the tail is more robust than in terrestrial mustelids.

[30]

The remains of forms supposed to be intermediate between otters and true mustelids are known from the Tertiary, but these are fragmentary, and nothing, of course, can be told regarding any aquatic adaptations which they may have shown.

Enhydrinae—the sea otter, of a single genus and species, with distribution in the north Pacific and south along the American coast into the waters of Lower California. Apparently this animal spends its entire time in the sea save occasionally when it may haul out on the rocks to rest. It is said to give birth to the young upon floating patches of kelp. External modifications for an aquatic life consist chiefly of a reduction in the size of the external ear and of the fore feet, increase in the size of the hind feet, which are fully webbed, with hairy sole and with the fifth toe the longest, in the slight vertical flattening of the moderate sized tail, and the excessive softness and denseness of the pelage. Significant points in its osteology consist of reported increase in the flexibility of the vertebral column, slight shortening of the neck, flaring of the anterior element of the pelvis, slight shortening of the fore limbs and of the thigh and shank as well, possible slight increase in stoutness of the hind limb and enlargement of the pes.

In spite of my best efforts and numerous letters to those who should know, I have been entirely unsuccessful in ascertaining the precise movements of the hind feet whereby this animal accomplishes swimming. Possibilities will, however, be discussed in chapter eleven.

PINNIPEDIA

All the pinnipeds are highly aquatic, to only a lesser degree than the whales and sirenians, and are chiefly pelagic. As in the case of the Cetacea there are many sorts of pinnipeds, especially phocids or true seals, which differ from one another in respects which for our present purposes are of a rather minor character. The order is separable into three well defined families, as discussed below. In addition there are several curious fossil pinnipeds whose allocation may prove puzzling. Because this order is of the utmost importance in a consideration of aquatic modifications its characteristics will be discussed in considerable detail.

As an order the Pinnipedia are mammals of markedly stream-line shape and with very short tails. The elbow and knee are always situated well within the body contour and the crotch is located at or slightly above the level of the heel. The feet are webbed and in some are paddle-like. The fifth toe of the hind foot is approximately as long as the first and both are longer than the three middle toes. The pre-acetabular length of the pelvic bones is always less than the post-acetabular. The humerus is relatively massive and the bones of the forearm very broad at one end, while the femur is much reduced in length and is flattened. There is no clavicle. In adults there is probably always a large hepatic sinus of the vena cava. Nictitating membranes are said to be present, and retia mirabilia occur to an unknown extent.

Otariidae

These are the sea-lions and fur seals or sea bears. The pelage is short, either entirely hairy or with a dense and fine underfur, and the young do not have a coat that is exclusively woolly. The external ear is small and narrow. Both fore and hind feet are used in limited terrestrial locomotion but the former exclusively in swimming. The area of the fore foot is great and the axilla is situated at about the middle of the forearm. The hind foot assumes a plantigrade posture during terrestrial locomotion and the astragalus is without a posterior extension. Each digit of both fore and hind foot has a cartilaginous extension and the palms and soles are naked. The nails of the fore foot and those of the first and fifth digits of the hind foot are vestigeal. The testes are scrotal. The canines are not unusually developed. The vertebral spines of the anterior thorax are well developed. The greater tuberosity of the humerus is higher than the lesser. The ilium is slightly curved—not markedly and abruptly bent laterad—and the femur has a lesser trochanter.

Odobenidae

This is the walrus, of the north Atlantic and north Pacific. This family, containing a single genus, is in most respects nothing but a specialized otariid, although in some points it is fairly intermediate between that group and the phocids. The body is almost hairless. There is no external ear. Both fore and hind feet are used in terrestrial locomotion, which is more limited than in otariids, but the hind feet are used in swimming by being moved from side to side, while the fore feet are also used, but alternately. In major details the feet are most like those of an otariid but the cartilaginous extensions are not as long, the fore foot is relatively smaller and the hind foot broader and more comparable in shape to that of a phocid. The hind foot assumes a plantigrade position during terrestrial locomotion and there is a slight posterior extension to the astragalus. The testes are scrotal. The canines are enormously and phenomenally enlarged, especially in males. The greater tuberosity of the humerus is higher than the lesser. The ilium is moderately curved outward, to a greater extent than in the otariids and less than in phocids, and the femur is without a lesser trochanter.

Phocidae

These are the true seals or earless seals, of which there are numerous genera, are found locally almost throughout the world, even in some large lakes. The pelage is hairy and never with a furry undercoat, but the pelage of the young, at least in a number of genera, is long and woolly, this being shed either shortly after birth or *in utero*. The area of the fore foot is reduced, the axilla falls opposite the wrist, and in most sorts these members are of use only for equilibration, on land being employed only for surmounting obstacles; although in the elephant seal (*Mirounga*) at least, they are usually placed fairly flat upon the ground. The area of the hind foot is somewhat increased, relative to length of limb, these appendages being plainly indicated as the means for aquatic propulsion, which is accomplished by rhythmic lateral movements as by the tail of most fishes. In spite of the apparent inefficiency of terrestrial locomotion, which is accomplished by a caterpillar-like wriggling of the body, with the hind feet elevated, seals have been known to travel overland for many miles, and even through mountainous country, when the freezing of coastal waters has made it advisable for them to seek new aquatic territory.

The astragalus has a posterior projection as long as that of the calcaneum and the foot is prevented from assuming a plantigrade posture by the unusual tension of the flexor hallucis longus tendon. There are no cartilaginous prolongations of the digits and the palm and sole are as well haired as the dorsum of the feet. The nails are all well formed, except that they are absent from the pedes of *Mirounga*. The testes are abdominal. The canines are but moderately developed. The vertebral spines of the anterior thorax are poorly developed. The lesser tuberosity of the humerus is usually higher than the greater, the ilium is markedly and abruptly bent outward, and there is no femoral lesser trochanter.

I have already discussed the pinnipeds at considerable length (Howell, 1929). Their derivation is a matter of controversy, for although both otariids and phocids are known from the Tertiary, they were even then true pinnipeds. It is a commonly accepted belief that they are biserial, and that the walrus is but an aberrant and modified otariid, but as in the case of the whales much of the discussion has been of a profitless nature. The meaning of the term biserial or diphyletic is loose and varies with the investigator. It is presumed that all placentals were derived from a common ancestor, and therefore all phyla eventually go back to this point. It is commonly believed that the pinnipeds were derived from the adaptive creodonts, which in turn were probably derived from a single ancestor. It may be pointed out, however, that because the pinnipeds have so far been unable to sever their connection with the land at the time of the birth of the young

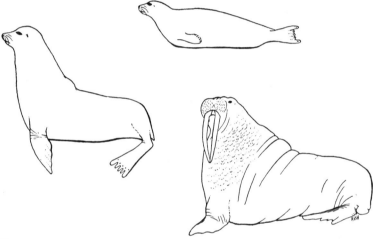

FIGURE 5. Pinniped postures. Seal (*Phocidae*) (above), sea-lion (*Otariidae*), and walrus (*Odobenidae*).

their latter day (geologically speaking) evolutional velocity may have been at a slow rate, and they may well be nearly as old, phylogenetically, as the Cetacea.

Whether or not the otariids and phocids became individually differentiated before or after their ancestors first took to the water is, with the present paucity of fossil material, a matter that cannot be settled and over which one may waste much valuable time and ink. It has been frequently claimed—first by Mivart, I believe—that certain anatomical resemblances which it is not necessary here to repeat at length indicate that the otariids have descended from the bears and the phocids from the otters. In the first place the otariid stock is almost certainly older than the ursine line, and the phocid older than that of the otters. The resemblances are incontrovertible and undoubtedly of real phylogenetic significance, but probably have no greater value than indicating that the primitive ancestors of both otariids and bears, although entirely distinct, had certain characters in common, as now have any number of very diverse carnivores; and similarly with the prototypes of the phocids and otters.

Be that as it may these two groups of pinnipeds are now fundamentally very different and have been distinct for a very long geologic time. Their true phylogenetic dissimilarities are further accentuated by the differences developed by their diametrically opposed methods of aquatic progression. The walrus seems certainly a branch of the otariid stock, modified by its enormous tusks, its feeding upon inactive prey, and its bulk. The way in which it swims both with its fore feet like the otariid and its hind feet like the phocid is extremely interesting and significant.

RODENTIA

No rodent is very greatly modified for an aquatic life, probably because they are either dependent to great degree upon the banks of the streams which they prefer, or feed upon plants which are terrestrial or of the shallow water. In addition many of them are small enough to be eaten by large fish if they venture into deep water. Besides webbing or bristly fringing of the hind feet in the more specialized sorts, these members may be enlarged, the tail may be flattened, the fur dense and soft, and the dorsal outline of the skull in some sorts is less convex than usual, especially in the rostral region.

Octodontidae

Myocastor—the coypu of eastern South America. This large rodent has beneath its outer coat a suit of fine fur that is not so dense as that of the muskrat but still heavier than one would expect to find in any terrestrial rodent of the tropics. The hind feet are quite large but the webbing of the toes does not extend to the tips. The tail is perfectly round and the external ears are rather small. The mammae are located dorsad.

Hydrochoeridae

Hydrochoerus—the capybara of eastern South America is the largest living rodent, attaining a body length of about four feet. The nostrils are situated rather high, the toes are webbed only at their bases, and the pelage is coarse. Like many of its relatives it is without a tail, differing in this respect from all other aquatic rodents, and it is by no means unlikely that this lack accounts for the fact that (from all accounts) it uses all four feet in swimming. It is partial to beds of reeds but seeks the water when disturbed, swimming and diving with facility.

Muridae

Crossomys—one of the Australian water rats, in all three genera of which the fur is dense and soft, and in all the external ears are small but well formed. In this genus the toes are said to be webbed and the tail, which tapers gradually from the body, has a strongly marked swimming fringe below. In these water rats swimming is undoubtedly by alternate movements of the hind feet, assisted by the tail.

Hydromys—a second genus of Australian water rat and the only one which I have examined. The feet are partially webbed, and the dorsal outline of the skull straight (as may be the case in the other two).

Parahydromys—a third genus of Australian water rat, in which the aquatic specialization is said to be somewhat less than in *Hydromys*. The first and last hind toes are practically unwebbed.

Dasmyms—a genus of African rodent which is said to be somewhat aquatic, but the only indication of this which is shown by specimens consists of the pelage, which is short, thick and soft.

Nilopegamys is another genus of aquatic rodent from Africa, known from a single specimen and recently named by Osgood. Its pelage is also soft and dense, the hind feet are slightly enlarged, and the dorsal outline of the skull has a concavity near the interorbital region.

Arvicola and *Microtus*. Several members of the subfamily Microtinae (meadow mice or voles) live in bogs or along the banks of streams and swim freely. Perhaps the most aquatic of these is the European water rat *(Arvicola)*, but it has no discernible modifications that are considered really aquatic.

Ondatra—the muskrat, of the United States and Canada, is the most aquatically specialized of all the Muridae. Its underfur is extremely soft and dense, the hind feet are enlarged and although not completely webbed, the sides of the feet and all the toes are bordered by very heavy fringes of stiff hairs, and the tail is laterally much flattened. Swimming is accomplished by means of alternate thrusts of the hind feet and accompanying horizontal movements of the tail; and I have been led to believe that occasionally the tail is thus used alone, while the feet are trailed.

Neofiber—the Florida water rat, is a smaller edition of the muskrat, but the tail is round, the feet are but slightly webbed, and hairy fringes are lacking.

Castoridae

Castor—the beaver, of North America and locally in northern Europe. The under fur is famous for its softness and density. The nostrils are situated slightly higher than in most rodents, the external ear is reduced in size, the hind feet are large and very fully webbed, and the tail is naked, flat, and relatively broader than in any other partially terrestrial mammal. Swimming is usually accomplished by alternate movements of the hind feet, although occasionally I have seen the tail alone used as a sculling oar. The tail is also used to facilitate quick submergence, being then slapped loudly on the water as a warning signal: but it is not used as a trowel, as has frequently been claimed in the past.

Cricetidae

There are three closely related genera of this family that are partially aquatic. They occur in central and South America and have short, dense fur.

Ichthyomys. The muzzle is slightly flattened and the eyes and ears unusually small. The upper incisors are prolonged into sharp lateral points, which should facilitate the holding of slippery prey, and the dorsal outline of the skull is definitely concave. The hind feet are broad and fringed, but very incompletely webbed. The hairs on the under side of the long tail are slightly lengthened and are dense, forming a slight keel.

Rheomys is very similar to the last genus but hairy fringes are limited to the sides of the feet and outer toes only, and the tail is not noticeably fringed. It is said to feed upon aquatic snails.

Anotomys is said to be similar to *Ichthyomys* also, but the ear opening is reported to be a mere slit.

As with the insectivores the line of aquatic modification which rodents usually follow is in the webbing and fringing of the hind feet, followed by enlargement of these members and a lateral flattening or ventral fringing of the tail. External ears are said to be absent in *Anotomys* but are present, although usually reduced, in all others.

As previously mentioned, it is significant that the capybara is the only aquatic rodent without a tail and the only one in which the hind feet are not indicated as the primary means of propulsion.

LAGOMORPHA
Leporidae

Sylvilagus palustris and *S. aquaticus.* The marsh and swamp rabbits of the southern United States are quite aquatic, readily taking to the water when disturbed and swimming freely. They have no discernible aquatic modifications, however, and the hind feet are even smaller and narrower than in the majority of rabbits. It is not meant to imply that their swimming propensities have resulted in a reduction of foot size, but it is not unlikely that they are descended from some stock of small-footed running rabbits (as is *Romerolagus*) rather than of hopping rabbits, like most other sorts.

UNGULATA
Artiodactyla: Hippopotamidae

This family is the most aquatic of any of the existing ungulates. The animals usually spend the day either in the water or basking upon some segregated sandbar, repairing to the land at night to feed along the banks, and sometimes foraging to considerable distances or traveling overland from one river to another.

Hippopotamus—the common hippopotamus of Africa is highly aquatic in habits. It evidently swims with all four feet but these are not especially modified for the purpose. External specialization for an aquatic life consists of the elevation of the nostrils and the peculiar orbits, which allow the eyes as well as the nares to be kept above water while the remainder of the animal is submerged. The shortness of the legs may, to some extent, also be due to this influence. The short tail is exceedingly flattened laterally, but it is too small to assist in swimming and the reason for its shape is not clear.

Choeropsis—the pigmy hippopotamus of parts of west Africa is much smaller than its cousin and its head is much less modified in the direction which the other genus has taken, the orbits not being excessively elevated.

The hippopotamus is related to the pigs and remains that are hardly distinguishable from the existing species have been found in the European Pliocene. The wonder is that this animal is not considerably more modified aquatically than we find it. However, its critical activities are more terrestrial than aquatic, for it must seek the land for its food, of which it requires a prodigious amount, and this condition is unique among aquatic mammals, for no other spends the whole day in the water and feeds exclusively on land. This, then, is its primary requirement—that it may walk with facility upon the land. At the present time it can have no dangerous terrestrial enemies save the lion and man, and there is really little reason why it should spend the most of its time in the water save that it is probably more comfortable there. And

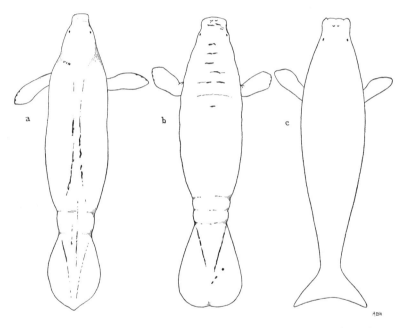

FIGURE 6. Dorsal representation of Sirenia: (*a*) manati (*Trichechus*) from British Guiana (redrawn from Murie, 1880) ; (*b*) *Trichechus americanus* (from Murie, 1872) ; (*c*) dugong (*Halicore*) (from a mounted specimen in the National Museum).

as it should be perfectly safe when in the water from everything except a large crocodile, there seems little reason for the high specialization of its nostrils and especially its eyes, which enable it to breathe and see when every other vestige of its body is submerged.

SIRENIA

The sirenians or sea cows are exclusively aquatic and inhabit rivers and coastal waters where aquatic plants and algae abound. The fore limbs are paddle shaped, there are no external hind limbs, and the tail is excessively flattened vertically and correspondingly expanded horizontally. The nostrils are considerably elevated, the eyes are small and are furnished with nictitating membranes, and there are no external ears. The skin is rugose and almost hairless. The mammae are lateral and postaxillary and the testes are abdominal. All the bones are exceedingly dense and heavy, the external nares are high upon the rostrum and the nasal bones are usually absent (present in the manati). The neck is shortened, there is no clavicle or sacrum and the pelvis is extremely vestigeal. The phalanges are moderately flattened. The lungs are long and shallow (vertically) and the diaphragm is almost horizontal and extends far back. There are very extensive retia mirabilia. Swimming is accomplished exclusively by the tail and individuals (at least of the manati) habitually rest with the back excessively curved and the tail tip touching the bottom (see figure 7). The pectoral limbs are used in feeding and are more mobile than in the Cetacea.

Trichechidae

Trichechus—the manati inhabits rivers in parts of eastern tropical America and locally in tropical Africa. Nasal bones are present and there are but six cervical vertebrae. The nostrils are not quite so elevated as in the Halicoridae, the tail is shovel-shaped, and vestigeal nails upon the flippers are usually present (absent in the form *inunguis*).

Halicoridae

Halicore—the dugong is found in shallow bays and rivers locally from the Red Sea to Australia, and feeds chiefly upon marine algae. The nares, both externally and in the skull, are more elevated than in the manati, and there are no nasal bones. There are seven cervical vertebrae. The tail is notched centrally and the lateral tips are quite pointed rather than rounded. Nails are absent.

Hydrodamalidae

Hydrodamalis—Steller's sea cow inhabited the vicinity of Bering and Copper islands in the North Pacific, where they fed upon marine algae, but they were exterminated more than one hundred and fifty years ago and very little is known about them. The tail took the form of two lateral pointed lobes and the flippers were relatively small and said to have been covered with short, coarse hairs, although the body was naked. The cervical vertebrae numbered seven, and they probably lacked nasal bones. There were no teeth in the adult, their places being taken by horny oral plates.

The sirenians are commonly believed to have been derived from the progenitors of the proboscidian stock. Known remains date back to the Eocene, and even then they were highly modified for an aquatic life. They are not as progressively specialized as the whales but their sedentary existence with no need for seeking active food is not as stimulating for rapid evolutionary change and their aquatic habits are undoubtedly of great antiquity.

CETACEA

Archaecoceti or zeuglodonts were essentially whale-like in body form and general details, but were less specialized in some respects. Thus the anterior limb was less modified and the skull had few of the characteristics which we associate with modern whales. Externally the hind limbs had disappeared, however. According to Kellogg (1928) zeuglodonts are first known from the Eocene, but were evidently a rather unsuccessful experiment for they suddenly disappeared in the Upper Oligocene. Although the narial aperture of the skull shows considerable retrogression, the bones exhibit no indication of the telescoping that is so characteristic of the modern whales. The dentition of the more primitive sorts resembles to a considerable extent that of certain creodonts, but this is believed to be only superficial, and the skull is remarkably suggestive in general conformation of some Cretaceous insectivores, and indeed modern ones as well. The majority of paleontologists, I believe, consider the zeuglodonts and modern whales to have had a common ancestry, but that it is impossible that the latter could have descended from any sort of zeuglodont now known. Most likely this ancestor was very different from either, the modern whales diverging in one direction and the zeuglodonts branching at another tangent in a direction that proved unsuccessful. It is interesting to note, as Kellogg says, that "periotic bones of both the whalebone and toothed whale type have been found attached to skulls of zeuglodonts," and, I may add, no other mammal known, either living or extinct.

Of those zeuglodonts so far known there are two extremes of body type. One of these is represented by *Basilosaurus,* with very long tail which unquestionably was utilized for anguilliform propulsion, and the other by *Zeuglodon osiris,* with a rather short tail suitable for fusiform propulsion. Just as these two swimming types grade one into another in the case of fish, according to body length and slenderness, so may we presume that there were zeuglodonts at some period that were intermediate between the two types mentioned. *Basilosaurus,* however, certainly could not have had flukes like modern cetaceans, for it had not the musculature to control such equipment, as indicated by the low spinous processes. But the tail was very mobile, because the zygopophyses did not articulate, and it must have had some sort of caudal expansion, but of great lineal rather than lateral extent. The same applies in some measure to *Zeuglodon osiris,* except that having a short tail its caudal expansions must have been lineally shorter and possibly somewhat suggestive of the manati.

Modern whales are exclusively aquatic. The body is fusiform, with no real constriction at the greatly shortened neck, and tapers gradually from the thorax to the base of the tail. The latter terminates in a pair of flukes, vertically flattened and laterally expanded into two pointed lobes with a notch between them. There is no vestige of external hind limbs and the anterior limbs are paddle-shaped, the integument being continuous over the digits, which have no sign of nails, and there is no external division of the limb into segments. The skin is smooth and hairless except that there are occasionally a very few bristles about the head. There are no sebaceous glands and the suderiferous ones are absent or much reduced. There is a specialized blubber layer beneath the skin. The eyes, which are specialized to function in salt water, have no nictitating membrane nor lachrymal duct. There is no pinna to the ear and the auditory meatus is a minute aperture. The nostrils open not near the snout (save in sperm whales) but near the vertex of the head, with corresponding alteration in the bones of the skull. The latter is very remarkable in that some of the bones have slid partly over others, resulting in the condition referred to as telescoping. The bones are spongy, the cavities being filled with oil. The cervical vertebrae are often phenomenally shortened and in some species most or all of them are fused. The articular connections of the body vertebrae are much reduced and are separated by elastic intervertebral discs affording great freedom of movement. The spinous processes are always well developed. There is no sacrum and the body and caudal vertebrae are differentiated only by the presence upon the latter of chevrons. The rudimentary pelvis has no contact with the vertebral column. There are

AQUATIC MAMMALS

no clavicles, and all articulations of the forelimb except that of the shoulder are fibrous and virtually immovable. The radius, ulna and digital elements are greatly flattened, and the phalanges of the second and third digit always exceed the normal number (hyperphalangism). The phalanges have epiphyses at both ends. The salivary glands are rudimentary or absent. The diaphragm slopes greatly and extends far backward. The vascular system is characterized by extensive retia mirabilia. The testes are abdominal. The internal ears are highly specialized.

Odontoceti—the modern toothed whales, including the cachalot or sperm whale. The integument of the throat never has numerous grooves. The nostrils unite to form a single external orifice, within which are a number of diverticula. The epiglottis is elongated and tubular. Teeth are present and there is no baleen. The olfactory apparatus is absent or extremely rudimentary. The skull is asymmetrical, the right side being more developed than the left, the nasal bones are rudimentary and the telescoping of the skull is mainly from before posteriorward. Mediad of the angle of the mandible there opens a large space into which fits a specialization consisting of fatty tissue and trabeculated air passages opening from near the Eustachean tube (this may not be present in some sorts). More than one pair of ribs articulate with the sternum, which is composed of more than a single bony element (although these may fuse in adults). The muscles of the lower arm are very rudimentary and the digits always number five. There is no vestige of the femur present except in the sperm whale. There is no caecum except in the Platanistidae.

Mysticeti—the whalebone or baleen whales. The integument of the throat has deep longitudinal grooves in some genera, but not in others. The nostrils have two distinct openings. The epiglottis is not especially elongated. Teeth are present only in the fetal state, adults having the roof of the mouth furnished with plates of baleen or "whalebone." The olfactory organ is much reduced but is present. The skull is symmetrical, its chief peculiarities for the present purpose being the development of the rostrum for the accommodation of the baleen, the elevated position of the nares, and the marked anterior inclination of the occipital shield, the telescoping being mainly from behind anteriorward. The nasals are much reduced but are not strictly rudimentary. But one pair of ribs articulates with the sternum, which is composed of only the manubrium. The muscles of the lower arm are not as much reduced as in the Odontoceti and the digits number either four or five.

There is a vestige of the femur present and, at times, of the proximal tibia. There is a small caecum.

Modern whales are the most highly modified for an exclusively aquatic existence of any mammals that have ever lived. Externally some of the porpoises are, save in minor details, to all intents replicas of ichthyosaurs and some sharks, furnishing perhaps the most spectacular instance of convergence in the entire class of mammals. This fact is proof positive that the whales of the more typical shape are close to the ideal in body form for their habitat. They are excessively specialized —in fact more so than any mammal, living or fossil, of which we know —and their survival bespeaks eloquently of the lack of ecologic competition which they have encountered.

The published discussions regarding the derivation of the Cetacea would fill many volumes, but most of this material is open to the same objection that has been presented in the case of the pinnipeds. Perhaps the majority still believes that they are descended from a hyaenodont stock, more particularly from the adaptive creodonts (a contention refuted by Matthew, Gregory, Kellogg) ; but there exist those who argue from numerous other angles, including their descent directly from the Pro-Mammalia (Albrecht), the ungulates through the sirenians, that their closest affinities are with the pinnipeds, or the edentates, and even that their ancestors never were land mammals but aquatic vertebrates. Some of the most interesting evidence in argument against the carnivorous ancestry of the whales has been presented by Anthony (1926), comprising anatomical details that are notoriously conservative. He stated that the following points are common to some of the whales and certain ungulates: Lateral nasal diverticula; diverticula of the Eustachean tubes; communication of the pleura through the anterior mediastinum; tapetum choroideum of the eye composed of fibrous lamellae; absence of os penis (some whales are said to have this) ; ovaries resembling *Hyrax;* frequent persistence of Muller's canals in males; placenta adeciduate and diffused; hippomanes body present in fetal membrane; usually but one young at birth; long period of gestation; and young large and well developed.

Flower and Kellogg at least have gone on record in their belief that the terrestrial ancestors of the whales must be sought among Cretaceous remains, and this is a most plausible contention. It seems not improbable that the cetacean stock is older than that of the true carnivores or of the ungulates, and therefore this ancestor was presumably a member of the carnivore-insectivore stem that is believed to have been ances-

FIGURE 7. Postures, drawn from life, of the manati (*Trichechus*) (after Murie, 1880).

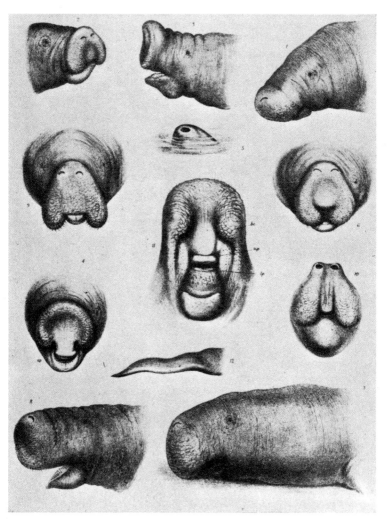

FIGURE 8.   Some of the facial and nasal postures assumed by the manati, redrawn from life, after Murie.

tral to a large proportion of present-day mammals. If this were the case, then naturally there would be significant resemblances of the whales to the ungulates as well as to the carnivores. But far more and far older fossil material must be found before anything at all in this direction can be proved.

Modern whales are biserial as are the pinnipeds, and as in the case of the latter it seems equally futile to argue as to whether the two groups separated before or after they took to the water. Those who are but casually acquainted with whales seldom realize the fundamental distinctiveness of the baleen whales or Mystaceti and the toothed whales or Odontoceti. The former, usually, but perhaps erroneously, believed to be the most recently developed, are known no farther back than the Oligocene, while a few remains of the latter have been detected in the Upper Eocene. The differences that occur in the two groups may, however, mean nothing more than that their habits or environment differed considerably at an early stage of their aquatic career, thus affecting their rates of differentiation. For instance the ancestor of the baleen whales may have lingered in rivers and shallow lakes, retaining considerable dependence upon the land, while the toothed whale ancestor may at the same time have taken to the sea and severed connection with the land at an earlier time; or vice versa.

Gigantism in whales is an indication of overspecialization and could be developed to the degree in which it now at times occurs only in an aquatic mammal completely emancipated from the land. The elephant seal, which may exceed 20 feet in length, seems now at the critical point in this regard. Some increase in size or a slight increase in its aquatic specialization would likely render it unable to leave the water, and we do not know whether it would perish or survive as a result.

Some whales descend to great depths, in the case of the cachalot presumably as far as a mile, where the pressure would be more than a ton to the square inch, and it is purported to be able to submerge for considerably more than an hour. These facts bring up a host of physiologic questions which are extremely difficult of solution.

Chapter Three

External Features

FOR GREATEST efficiency, which simply means the attainment of considerable speed with the least expenditure of effort, certain requisites of a self-evident nature are necessary to an aquatic mammal of high specialization. The body must be of streamline form without excrescences that would offer resistance to the water. The propulsive force *should* be from the rear, as has been learned from a study of shipbuilding, and there must be some sort of apparatus for steering and equilibration. The limitations of mechanical adaptation in a vertebrate necessitates that separate means for steering be situated where this will receive no great disturbance from motions of the body concomitant to propulsion. Hence the rudder must be elsewhere than near the activators for progression. But there may be accessory equilibrators, as a dorsal fin. An additional necessity is that in an active mammal that is thoroughly aquatic there must be an alteration by means of which respiration may be carried on while the animal is traveling at full speed. The above qualifications are mechanically essential to ideal efficiency, a thesis that may be accepted without argument, and is a goal toward which every mammal heads as soon as it takes to the water. Whether every such mammal will attain this goal eventually is a different matter, and is dependent upon its inherent capabilities and inhibitions for such evolution, and environmental factors, which at times deflect it at a tangent, so that later it may be incapable of regaining the straight course leading to ideal aquatic specialization.

A streamline body form is not a fixed shape that must fit into a particular mold, but it can be long or short, narrow or relatively broad, shallow or deep. It must taper at both ends, after a fashion, however. Some terrestrial mammals already have a streamline form, while to attain it others must pass through successive steps. As a standard of perfection in this regard it is safe to take the swiftest of the cetaceans, and we can rest assured that this form is close to the ideal, for it is approached to almost identical degree, with but minor variations, by three diverse phyla of vertebrates—the porpoise, ichthyosaur, and certain sharks—as already mentioned. Other aquatic mammals that differ from this body form either have had insufficient time to attain it, although headed in this direction, or else by inherent limitations or influence of environment, have been or are being prevented from doing so. But all existing aquatic mammals, excepting, if one prefer, the hippopotamus, are of streamline form to a greater or lesser degree dependent upon the stage of aquatic perfection which they approach in other respects. The corpulent bulls of the Steller sea-lion and the walrus are more ponderous than we should judge best for their aquatic efficiency, but the bulk of the bulls of these two animals is doubtless a secondary sexual character the development of which evidently proved to be no critical handicap. Similarly with the female walrus, only slightly less corpulent. The latter pinniped feeds upon inactive prey such as clams. Therefore speed, and the more slender proportions which are requisite to its attainment, is not a necessity in the securing of food, and it is evident that there was not a strong stimulus for it to develop great speed in order to escape from its enemies, else it would have become extinct or more speedy long since. This shows that the adoption of an aquatic habitat will not develop a slender body form, but that the stimulus for this is the necessity to travel through the water speedily either for the purpose of securing fast-moving food or to escape from rapid enemies.

Although it has evidently been aquatic for a very long geologic time the form of the hippopotamus is far from suited to traveling with rapidity through the water, although the celerity of its movements is said to be quite surprising. It has, however, been under the necessity of little besides sinking sluggishly beneath the surface of its pool or moving languidly about. A hippopotamus that could dash about a small river at the rate of thirty miles an hour would in all probability promptly flatten itself against a rock. In other words, there has been no stimulus for this mammal to develop a more perfectly streamline form, or if some slight influence of this sort has been experienced, it would doubtless overborne by the greater importance of its ability to seek enormous quantities of green fodder on shore.

The long-tailed zeuglodonts did not conform to the aquatic ideal of bodily shape as they were anguilliform, and this relative inefficiency may have contributed to their disappearance; but the short-tailed genera also became extinct and the critical factor doubtless lay elsewhere. The shape of modern whales is quite variable. Some of the small porpoises are very speedy and with ease encircle a fast boat, the inference being that they can travel considerably in excess of thirty miles per hour. The larger, pelagic sorts of toothed whales are slower, however. The sperm whale is decelerated by its enormous head and the mass of its sperma-

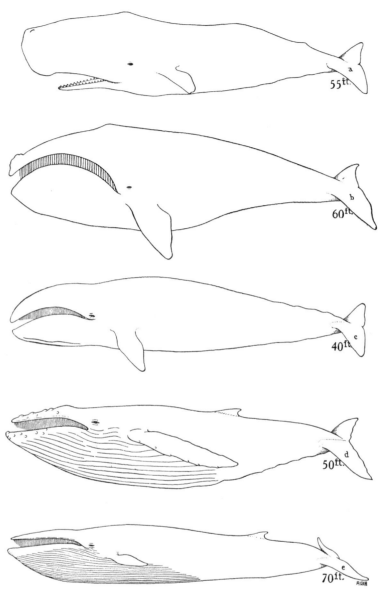

FIGURE 9. Some of the larger whales, to illustrate type (compiled from various sources): (*a*) *Physeter*, the sperm whale or cachalot (an odontocete); (*b*) *Eubalaena glacialis*, the right whale; (*c*) *Rhachianectes*, the gray whale; (*d*) *Megaptera*, the humpback; and (*e*) *Balenoptera physalus*, the finback. The last four are mysticetes or baleen whales.

ceti organ, but its abundance until harried by man proves that speed was not requisite to the filling of its stomach and its lack of capacity for swiftness, as well, indeed, as in the case of other large, slow whales, is doubtless a secondary acquisition. At least most of the larger, pelagic Odontoceti feed largely on squids and other cephalopods, which are among the most agile of sea forms, and how some of the slowest swimming of whales manage to secure the vast quantities which they require is a puzzle.

Among the baleen whales or Mysticeti, the speed of the more slender sorts, the rorquals, is astonishing for a body that sometimes approaches a length of 100 feet. By a possible speed estimated to be above thirty miles per hour they can often escape the killer whale, which is one of the most relentless enemies of the slower genera, and in addition, such speed probably secures for them a relative freedom from ectoparasites. At least it is a fact that these pests are rarely found upon the fast rorquals, while they usually abound upon the slower mysticetes. The baleen whales do not secure their food by direct pursuit but by swimming leisurely with mouth partly open through hordes of crestaceans and small fish. Hence the speed of the rorquals has been developed either for the purpose of escaping from thir enemies, or very likely partly to enable them to travel rapidly from one feeding ground to another; for the half-inch crustaceans which constitute their favorite food are local and may not be easily found in the quantities necessary for such vast appetites, so that frequently it should be necessary for the whales to cover great distances between meals.

The very appearance of these vast cetaceans bespeaks of speed, with their relatively slender bodies and great bulge of muscles and tendons above and below the peduncle. And equally eloquent are the shorter, more ponderous bulks of the gray, humpback and balaenid whales. They are much slower and the grays, at least, often succumb to onslaughts of the killer whale; but in spite of this they occurred within their rather restricted range in prodigious numbers until the depredations of the whalers decimated their ranks so sadly.

When a mammal reaches the stage when it spends most of its time in the water it must either have developed a quality of pelage that will retain the air that fills the interspaces or have modified its skin to withstand any detrimental action of the water. Especially if it be of boreal or even temperate habitat will the tendency rather be for the underfur to take on a fine and particularly dense texture, and accompanying this will usually if not always be fine adjustments in the functioning of the glands of the skin whereby there will be secreted just the correct amount of suitable substances to hinder the pelage in becoming water-logged but yet insufficient for the matting of the coat. All aquatic insectivores have this type of pelage, most rodents have it or seem to be in process of acquiring it, and it is a character of some of the aquatic carnivores. It is not to keep it warm that the *Potomogale* of tropical Africa has an undercoat almost as fine and heavy as the Siberian desman, but for the purpose of protecting its skin from detrimental action of the water during prolonged submersion, or to accomplish flotation with the aid of the air imprisoned in the fur. Among aquatic rodents the most notable, and indeed only, example of the lack of an undercoat of fine fur is the capybara, whose pelage is markedly coarse. Like all rodents of Cavidae affinity its coat was probably coarse to begin with and the easier line to follow was to modify its skin rather than its pelage.

But there are other considerations concerned with the pelage of an aquatic mammal. In the case of a mammal of small or even moderate size the presence of a coat of fine fur whose surface is plastered smooth by the action of the water is doubtless not an appreciable reducer of speed, but although perhaps not impossible it is at least unlikely that a large whale could ever have had the velvety covering of a sea otter, and were a hairy covering present it would doubtless be of wiry texture, which would act as a definite retardant of progression comparable to a weedy growth upon the hull of a vessel.

Yet another possibility must receive consideration. The Cetacea and Sirenia are the most completely hairless of marine mammals and they are the only ones which can not come ashore to bask and dry their hides. If they had coats of coarse hair it seems almost a certainty that during continuous immersion they would than accumulate such a crop of parasites and sessile marine growth that life would be unbearable for the wretched creatures; and resulting scaly and scabby condition of the hide would then inevitably eliminate the hair. The pinnipeds can occasionally haul out upon a rock or the ice, thoroughly dry the hide and remove at least such unwelcome attendants as need continued immersion. The northern fur seal, however, spends many months on its annual migration far from land, and the sea otter very seldom leaves the water, so that the beautiful, soft pelage of these two mammals refutes any implication that frequent drying is requisite to the retention of hair in *all* aquatic mammals.

Among the pinnipeds there are three types of body covering, represented by the walrus, which is almost hairless, the majority of earless seals or phocids and most of the eared seals or otariids, in which the pelage is of course hair only, and the fur seals, famous for the softness and denseness of their undercoat. Nothing, of course, is known regarding the pelage of the pinniped ancestors and it is useless to speculate. Undoubtedly the progenitors of the walrus had a coat of some sort, which was largely lost for reasons unknown. That the earless seals are derived from an ancestor with a more luxuriant and softer pelage is indicated by the fact that at birth some phocids are covered with an excessively long, woolly coat which in some is shed at the age of about one month (and until this time they cannot swim), while in other genera this coat is shed *in utero*. And this is an important reason for considering the phocids to be more primitive phylogenetically than the otariids. Probably contributory to these three conditions of pelage in the Pinnipedia are the facts that the walrus is reputed normally to have the thickest coat of fat and so has less need for hair; but I cannot see that the need of the fur seal for a warm pelage could be greater than that of the Steller sea-lion, and it is probable that there is an additional, obscure reason for the presence in the former animal of this type of coat.

The only aquatic mammals with practically naked and *rugose* hides are the walrus and the sirenians. The former was probably derived from a hairy ancestor and the reason for its present naked condition is obscure. Embryos and young of the latter are said to be more hairy than the adult, which would indicate that the ancestral form was also more hairy. But the sirenians may never have had a thick coat, for this order is commonly believed to have been of the same stock as the proboscidians, and elephants have been prone to a hairless condition. And the latter theory may well account for the rugosity of the hide of these animals. Feeding upon inactive prey and having no great need of speed, a rugose hide and its consequent retardation of speedy passage through the water would doubtless prove of slight detriment. But theoretically no aquatic mammal should attain to high speed and at the same time retain a rugose hide.

The hippopotamus might well be expected to have a rugose hide but on the contrary it is remarkably smooth for an animal of this size, and its present hairlessness, in a slow mammal that spends much of its time on shore, would indicate a hairless ancestor. Its hide appears to be rather tender in spite of its great thickness, and it is not surprising that some integumental provision has been made for the alternation of a whole day spent in the water followed by long terrestrial excursions at night in a hot and at times a dry climate. The glands of the skin excrete a thick, apparently sticky, pinkish substance which formerly gave rise to the belief in the blood-sweating proclivities of this behemoth of holy writ. This is presumably accomplished for the purpose of furnishing a protective covering to guard against dessication of the hide, but for all we know to the contrary it may also function in fitting the skin for lengthy submergence.

It was at one time argued by several investigators (as Abel) that a number of bony plates discovered associated with the remains of a zeuglodont constituted a part of a dermal armor, but it was later shown that these plates belonged to a turtle. Several other investigators, chiefly Kükenthal, have vigorously championed the theory that the series of dermal dots upon the middorsum of *Neomeris,* and to a lesser extent one or two other genera of porpoises, is a remnant of a dermal armament. Kükenthal also found what he interpreted as dermal ossicles upon the anterior margin of the flipper, and scattered over the body of the common porpoise. His most significant evidence was that the dorsal rows of dots were much better defined in an embryo of *Neomeris* than in adults; but this character is individually variable, for I have found the dots exceptionally sharp in a young specimen and not discernible in another of equally tender age (Howell, 1927), nor in an adult female. In none of the animals which I have examined did these dots assume a squarish shape as claimed by Kükenthal. Furthermore, in preserved specimens of porpoises there is often a roughness of the skin upon the anterior border of the flipper caused by the shrinking and cracking of the epidermis, and it is by no means unlikely that Kükenthal was misled by his enthusiasm for his theory of dermal armature into mistaking such roughness for definite ossicles. A histological study showed me that the dots upon the dorsum of *Neomeris* are formed merely by a slight thickening and local cornification of the epidermis, and that it is more logical to consider them as the beginning of some integumental specialization rather than as the remnant of dermal plates —an opinion shared by Winge (1921).

One would naturally presume that to a mammal highly modified for best efficiency in traveling through the water the shape of the head, as the part of the body which chiefly has the function of cleaving this element, would be of paramount importance. In our inquiry regarding what shape of head is best fitted for this use one naturally turns to the Cetacea as being the mammals most highly modified for an aquatic life, but from them we can learn but little. It is, of course, a self-evident fact that an animal with a pointed snout which tapers gradually to a smooth head can cleave the water with least resistance, and conversely that no whale with a broad, blunt forehead which piles up the water before it can attain to highest speed without expending a disproportionate amount of energy. But whales' heads are of almost every conveivable shape, varying all the way from those with an excessively long, tweezer-like beak and small head, to the cachalot with its amazing bulk of rostral tissue. In consequence we are forced to believe that in the Cetacea, no matter how strong the stimulus for a stream-line snout and head, the opposing stimulus for the development of an unwieldly frontal

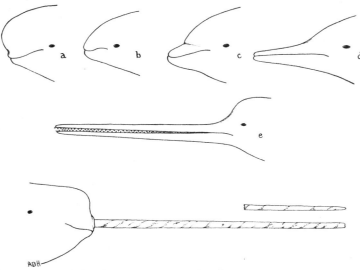

FIGURE 10. Heads of porpoises, illustrating frontal prominences and length of rostra: (*a*) *Globiocephala;* (*b*) *Phocaena;* (*c*) *Tursiops;* (*d*) *Delphinus;* (*e*) the extinct *Zarhachis* (restoration after Kellogg); and (*f*) *Monodon.*

fat organ, or else for a huge and rather blunt snout (as in the Balaenidae) has at times proven the more powerful. The inference is therefore drawn that moderately rapid propulsion through the water for many millions of years need not necessarily bring about a good stream-line form to the head. But it is logical to accept the thesis that the head is just as amenable to stream-line influences as is the remainder of the body, so it may therefore be accepted as a fact that the stimulus that has at times resulted in a cetacean head that offers more resistance to the water than all but a very few of the terrestrial Mammalia, is extremely strong.

In other words, such broad frontal prominences as occur have developed not haphazardly but for a purpose that is of the utmost importance to those species having them.

Broadly speaking the shape of the cetacean head is of two sorts. In the Odontoceti the throat contour passes almost straight back, while that portion of the head above the rostrum offers by far the greater part of the cephalic resistance to the water. In the Mysticeti the rule is for the dorsal outline of the head to extend straight back from the rostrum virtually parallel with the back, so that this part of the head does not offer aquatic resistance, while such resistance as is offered is experienced by the throat or the lateral dilation of the head. An exception to this rule is found in the Balaenidae, with their rostra excessively curved—a phenomenon which is a more recent specialization.

The rostrum of the toothed whales is exceedingly variable. It may be so short that the fatty prominence of the forehead projects beyond it (Globiocephala) or it may extend slenderly for almost four feet (Zarhachis). Presumably those with no beak to speak of should feed on prey that is not particularly agile, while a moderately projecting beak is of advantage in snatching fish that move at considerable speed. But theoretically an excessively long beak such as that of the Miocene Zarhachis would be more of a disadvantage than otherwise. With a mobile neck a beak of this sort could be thrust quickly in all directions largely independently of the position of the body, but in the Cetacea, with their much shortened neck, the beak would have but slight directional mobility, so that to effect a decided alteration in the direction of the beak thrust the whole body would have to be moved. Evidently the ancestor of these long-beaked porpoises responded first moderately to some stimulus for a lengthening of the rostrum but it seems very likely that this modification attained undue evolutionary velocity, developed first beyond the needs of the animals and then became of positive disadvantage, so that it seems likely that the handicap of such an excessively long rostrum may have contributed materially to their final extinction.

The external form of the rostrum in the Mysticeti—at least as we now find it—is probably largely attributable to mechanical needs. We can tell little about it save that presumably the rostrum has developed in a way most suitable to act as a support for the baleen and in response to the need for a large mouth. The external grooving upon the throat and chest of some of the whalebone whales has evidently been brought about in the same way, and will be discussed later.

Presumably the ancestors of the Cetacea were hairy, as seems clearly indicated by the fact that in the fetal state bristles are more numerous about the head than in adults, but the hair of the body has long since disappeared. Adult mysticetes may have a few bristles about the head as well as scattered cutaneous pits, evidently constituting the relics of hair follicles. These are most noticeable, at least in *Balaenoptera*, beneath the chin tip, where they are gathered in a sharply defined cluster, which often shows to good advantage in a photograph. In fetuses there may be a row of widely spaced bristles along the rostrum, and scattered at other points upon the head. Among the odontocetes, *Stenodelphis* is the only one that retains a few bristles in adult life, while *Monodon* and *Delphinapterus* (the white whale or beluga) never have them at any stage in their development. In *Tursiops*, at least, among the porpoises there are scattered pits about the lips which may have follicular affinity.

The specialization in this direction has progressed farther than in the loss of hair, however. The Cetacea have lost sudoriferous glands and possibly the sebaceous as well, although authors seem to be at variance on the latter point. For instance Beddard (1900) stated that the sebaceous glands have begun to vanish. Certainly they are absent from most sections of the skin that have been examined, but perhaps they still persist in certain circumscribed areas.

In a section of the skin of *Tursiops* before me, prepared by G. B. Wislocki, the corneum is tissue thin, and associated with it seem to be even thinner elements of a stratum lucidum. Next comes an excessively thickened stratum germinativum of very homogeneous character. The papillae of the corium are unusually slender and long (for their thickness), while the coreum insensibly merges with the tela subcutanae—the true blubber layer. In reality the blubber layer should probably be considered as a definite component of the coreum, for it is very tough and fibrous, whether relatively collapsed (in an emaciated animal), or gorged with fat cells, and totally different from the simple layer of soft, subcutaneous fat of the seals.

In some ways the whole layer of skin and blubber is remarkably tender. In a freshly caught *Balaenoptera* the corneum is relatively even thinner and more tender than in a porpoise, and may be rubbed off in great patches by the palm of one's hand, the resulting sheet resembling tissue paper. I have easily scored the surface with a finger nail, and after inserting a pocket knife for the full length of the blade, one may cut as easily as through so much cheese. One would suppose that the first

[57]

scrape of so ponderous a bulk against a sharp rock would lay the animal completely open. And yet many hundreds of pounds may be applied to a cable passed through an incision in this whole layer without it tearing away.

The constitution of the blubber layer in the Cetacea certainly appears unusual, but whether its precise histological structure is or is not unique among the Mammalia has not been determined. At any rate the mere presence of a considerable layer of fatty tissue beneath the integument of aquatic mammals is not surprising. Many land mammals store up just such a supply of extra fuel both for warmth and to tide them over periods of food shortage. Any carnivore must be prepared to live through times of hunger and this should be true in the case of large whales which require such a prodigeous quantity of food, often of small size, and which travel so widely. Any sort of mammal which experiences times of plenty alternating with periods of hunger will quickly acquire the ability to accumulate a reserve of fat. Some sort of insulation of the body is probably of critical necessity to an arctic marine mammal, for although it never experiences an aquatic temperature lower than a few degrees below freezing, the conductivity of the water is 27 times that of air. But there are probably other advantages derivable from the presence of a fat layer in such mammals. The tropical sirenians surely have little need for insulation and yet they are not only abundantly supplied with fat, but this seems singularly inefficient in insulating the body, for the manatis of Florida are known quickly to succumb to any unusual lowering of the air temperature. Furthermore, we cannot know just how advantageous it is to an aquatic mammal to have the added buoyancy afforded by an extensive deposit of fat. The blubber layer itself may be significant of nothing more than stated above, but the excessive oiliness of other parts of the body in marine mammals and of the proneness of the Odontoceti to develop fat organs of some sort seems to point to the probability of all this fat and oil serving some other and very important physiological purpose of which we do not know. This question will receive further consideration elsewhere.

It is perhaps proper here to discuss the question of a dorsal fin, although little can be stated in this connection. Some whales are without it (Balaenidae, *Rhachianectes,* Physeteridae, *Delphinapterus, Monodon, Phocaena*) ; in some it is barely indicated and too small for function, while in most odontocetes it is quite large, culminating in the killer whales, in which it may attain a length of several feet. All that it is now safe to say is that all aquatic vertebrates are prone to develop such a fin, almost certainly as an aid to equilibration. It should be mentioned, however, that a dorsal fin can hardly act very efficiently as an equilibrator in an animal swimming by a vertical motion of the tail (as whales) as it can in one employing a horizontal motion (as most fish), for equilibrators should be situated at a right angle to the direction of the swimming force. In the Cetacea, therefore, the situation of the flippers

FIGURE 11. Some of the extinct aquatic reptiles: (*a*) *Mesosaurus;* (*b*) *Elasmosaurus* (plesiosaur); (*c*) *Clidastes* (mosasaur); (*d*) *Ichthyosaurus;* (*e*) *Trimacromerum* (plesiosaur): (redrawn from Williston).

renders these appendages by all odds the most efficient organs of equilibration and it seems that a dorsal fin would not be of sufficient practical importance for all sorts of whales invariably to have developed it. It may be mentioned that theoretically a dorsal fin would be of greater use to the Phocidae in pure equilibration than the fore limbs, but these mammals are not yet sufficiently specialized to have developed the former.

[59]

Other external features, such as position of the narial openings, length of neck, tail, and form of appendages will be discussed in detail in other chapters. Suffice it to say here that in the highly modified aquatic mammal the tendency always has been for the elimination of any unevenness of body contour that could offer resistance during progression through the water. The pinna of the ear disappears, the scrotum is eliminated, and the appendages that are not used definitely either for propulsion or in equilibration will first atrophy and then sink beneath the surface of the body.

Chapter Four

The Senses

VISUAL SENSE

CERTAIN adjustments in the optic equipment of mammals that habitually seek their food beneath the surface of the water is to be expected. The platypus, most of the insectivores and some of the smaller aquatic rodents secure their food largely through the sense of touch. It seems that they must actually keep their eyes closed for a large part of the time which they spend submerged, and the result is the same as is found in fossorial mammals, namely, a tendency toward reduction in the size of the eye. Among Cetacea it is interesting to note that *Platanista* at least has followed this same course, for it has become virtually blind after having lived for a great length of time in the waters of muddy rivers.

Other sorts of mammals may have experienced a pressing need to watch for enemies above the water while keeping a maximum amount of head and body hidden from view. The result of this is best exemplified among mammals by the common hippopotamus, with its dorsally protruding orbits. The pigmy hippopotamus *(Choeropsis)*, however, does not have markedly protruding orbits but appears more like a young individual of its larger cousin, indicating that the latter is probably derived from a more generalized ancestry—not vice versa. The hair seals (Phocidae) show a tendency to acquire dorsal direction of vision to a considerable extent and where this is pronounced there is osteological indication of it in extreme interorbital constriction. The upward pointing of the axis of the eye is to be noted in the sea-lion (Otariidae) and walrus (Odobenidae) also but to a lesser degree. The eye in at least the majority of the Phocidae is relatively larger than in other pinnipeds, possibly because these seals are more in the habit of seeking smaller prey in the subdued light of deep waters, although it must be admitted that the dorsal direction of the eyes would hinder such action. Incidentally I have found scores of half inch shrimps in the stomach of a *Phoca hispida,* which fare is probably too insignificant to interest an otariid.

[61]

In the Sirenia the dorsally directed axis of the eye is also quite apparent, but the size of the eyeball is very much reduced and in consequence it is inferred that visual efficiency has become much impaired.

The beaver probably has a greater dorsal inclination of the eye than any other rodent, as might be expected, and the muskrat, capybara and other aquatic sorts exhibit this character to some degree, but in rodents this may be without much significance from the present standpoint, for some species that are strictly terrestrial have this character considerably developed. Among these are the Microtinae or meadow mice, most of which follow runways through grass and must watch above for approaching danger.

There may be marked visual differences, however, in eyes that have moved dorsad. One extreme is represented by *Hippopotamus* in which the protruding bony orbits are directed chiefly laterally with a slight forward trend, thus indicating vision that is largely monocular. The well developed supraorbital prominences of the platypus also obliges vision in this animal to be completely monocular. The tendency of some rodents and the hair seals, however, seems to be for the acquirement of dorsal binocular vision, although it is probable that none of these mammals has the power of true stereoscopic vision, in which the image of one eye is exactly superimposed upon that of the other and the two function together perfectly as a unit.

Such binocular visual powers as the seal may have are doubtless of a character somewhat intermediate between monocular and stereoscopic vision. In order that the visual sense shall function perfectly in a reflex manner the physiology must be quite complicated, and our knowledge of the mechanics of the sort of monocular vision which must be employed by an animal having the eyes upon opposite sides of the head is still incomplete. Thus it is not certain whether the animal must first give its attention to an object upon the right and then to another upon the left, or whether both eyes can function as separate units with equal efficiency simultaneously, in which case the optic colliculus of the brain would have to function in a manner differing in some unknown degree from that of man. Also it is unknown whether the apparatus for accommodation of each eye can operate entirely independently of the other, so that the animal can simultaneously focus one eye on a near, and the other on a distant, object with an equal degree of perfection.

It may be presumed that a tendency for the elevation of the optic axis of the character encountered in the seal is useful for the purpose of detecting enemies that are prone to pounce down from above, while

[62]

the hippopotamus has had need to watch for foes approaching horizontally along the river banks.   At any rate it seems likely that only those aquatic mammals which have experienced a definite need for eyes directed more dorsally will have acquired this characteristic, and one would not expect it to develop to as high a degree in the great majority of cases as it now occurs in the anomalous hippopotamus. For one thing, unless the need for it was extraordinarily strong, by the time that dorsal vision had become moderately efficient it is probable that aquatic specialization in other respects would usually be so far advanced that the animal would have but little reason to fear enemies approaching above the surface of the water.

As soon as the last mentioned stage in aquatic specialization has been reached the occular stimulus would automatically change, for the chief need, and probably the only one, would then be for the discernment of submarine food and enemies.   In the case of a mammal whose eyes had already turned upward to a greater or lesser amount, there would then be a secondary migration downward of the direction of vision, so as better to detect food which it was approaching.   For all we know the progenitors of the whales may have passed through just this visual cycle, in which case there would have been left some complication in skull development which one cannot hope to decipher by means of the fossils now available.

Horizontal direction of vision may be either forward or lateral, or even backward.   If the former then it must presumably be binocular, and if lateral, monocular.   We have no means of knowing which of these sorts of sight might prove most useful to an aquatic mammal but it seems likely that binocular vision would never be developed by any reasonably active mammal of high aquatic specialization.   This would entail a forward direction of the orbits where the eyes would receive the full force of water pressure as the animal progressed—a position which obviously would prove of considerable, and perhaps critical, detriment.   As a matter of fact the eyes of the Cetacea are directed at practically a perfect right angle to the body axis, where they receive the minimum of friction and irritative interference by the water.

Eyelids that are at least partly functional are retained by aquatic mammals.   In pinnipeds, serenians (apparently) and mysticetes these are largely comparable to what are found in terrestrial mammals, with lids moderately wrinkled, indicating that they may easily be closed and held in that position without strain as long as the animal wishes. This statement should perhaps be qualified as regards the whalebone whales. From examinations of many freshly killed specimens of the latter I am convinced that it is so, for the lids may be moved with the fingers very easily; but Putter has stated that they are immovable, as is also the case in odontocetes. It is true that in at least the majority of toothed whales the lids are unwrinkled when the eye is open. It therefore seems likely that their mobility is very much reduced. Complete closure of the eye is perhaps impossible, but I would be loath to believe that the lids are without power of movement entirely. The narwhal *(Monodon),* and perhaps some others, has a very peculiar modification of the eyelids, which will be described in detail by Ernst Huber.

An optic tendency resulting from a life in the water is for the elimination of eyelashes, but their disappearance is slow. Lashes are entirely lacking in the Cetacea and I understand that they are practically so in the Sirenia. All other aquatic mammals have them in various degrees. One might expect that there would be a great advantage in the acquirement of a nictitating membrane, but the only sorts which have them are the Sirenia and Pinnipedia, so far as I have been able to learn. There should hardly be any use for a functional lachrymal duct in aquatic mammals and this shows a tendency to disappear (Cetacea, Sirenia and Pinnipedia).

Frequent or constant submergence will probably bring about some sort of alteration in composition of the lachrymal fluid in order that this will not so readily be washed from the eye, and very possibly in order to counteract any irritative influence of the sea water; and Putter has determined that this is actually the case. It should be noted in this connection that cetaceans are perhaps the only mammals which cannot rub the eye against some part of the body in order to free it of foreign particles, including parasites.

Very probably there need be no optic stimulus in addition to the above for aquatic mammals inhabiting fresh water. The refractive index of the latter is not so different from that of air but that the normal eye can see well beneath the surface and there is no need for it to withstand any considerable pressure. In a marine mammal such as the whale, however, matters are entirely different. The refractive index of sea water differs so much from that of fresh or of air that when the normal eye is submerged in it very little can be done but to distinguish objects by their high lights and shadows, because of the fact that the point of focus for the image then lies far behind the retina. Hence, for a whale to see properly when submerged means that there will have been quite profound changes in its visual apparatus; and in addition there must be alterations to enable the eye to withstand the enormous pressure to which it is at times subjected.

Putter (1902) has investigated this subject more thoroughly than anyone else and it is upon him that we must rely for our facts. He found that pinnipeds and sirenians also have the eyes importantly modified, but his material representing the latter group was not altogether satisfactory. The changes which he found to have taken place in the eyes of pinnipeds, sirenians, mysticetes and odontocetes he has summarized as follows:

Optic adaptations
1. Lens almost spherical.
2. Refractive index higher than in any terrestrial mammal; almost as great as in fish.
3. Relationship of the neural elements of the retina; unusual number of rods connected with a single ganglion cell.
4. Superfluous ganglion cells in outer granular layer of retina.
5. Extensive tapetum lucidum.
6. Enlargement of fundus at the expense of the pre-equitorial zone.

Thermal adaptations
1. Diminution of the cornea in proportion to size of the bulbus.
2. Form and number of lymphatics in cornea propria; large and relatively few in number.
3. Unusual development of choroid and of the perichoroidal lymph spaces.
4. Form of optic orifice: is so much reduced that only the cornea is visible.
5. Tremendous development of musculature, with immovable bulbus.

Hydrostatic adaptations
1. Curvature of the cornea so as to receive support laterally.
2. Lateral thickening of the cornea.
3. Epitheleal cornification of cornea; cornified substance unites directly with elastica anterior.
4. Thickening of sclera; tremendous at equator and above fundus, slight at corneal sulcus.
5. Thick optic sheath; supporting bulbus like a column.
6. Arterial and venous network of the ciliary blood vessels (retia mirabilia).

7. Location of the bulbus away from the bony walls and embedded in muscular, fatty and glandular tissue.
8. Acquisition of a special hydrostatic sensory organ in the Odontoceti.

Chemical adaptations
1. Development of glands to produce a fatty, oily secretion.
2. Increase in size of Harter's and lachrymal glands, and development of a subconjunctival stratum of glands.

Putter suggested that as the eyes are immovable, the very large retractor muscles may have developed a specialized function and that it is possible that they may now act to develop heat to warm the eye. But there is apparently not the slightest basis for this theory. Even if these muscles were continually contracted and relaxed the heat furnished could hardly be a fraction of that supplied to the eye by the blood through the retia mirabilia. Similarly with the structure in the eyes of odontocetes which he interpreted as a hydrostatic organ. This was a bit of neural epithelium in the connective tissue of the sclera at the angle of the iris and isolated from the retina so that it could receive no stimulus by means of light. I can see no reason for considering this to have any hydrostatic function.

In summarizing the aquatic specialization which the whale's eye has undergone Kellogg (1928) has stated that "in its gross features the whale eye differs from that of a land mammal in having an eyeball immovable, eyelids without eyelashes, no tarsus or supporting cartilage in the eyelid, no Meibomian glands, and a downward direction of the eye axis. As the result of an aquatic mode of life whales have acquired a more spherical lens and a greatly thickened sclera. The ciliary processes and their muscles are reduced in size and have lost their original function of controlling the shape of the lens. The tension of the suspensory membrane (the *zonula Zinii*) is not great enough to flatten the anterior surface of the lens, and as a general rule the latter retains a more or less spherical shape. Whales thus lack the power of accommodation."

The more spherical lens projects the image upon a retina that is relatively much nearer the lens than in the normal eye. This is accompanied by an enormously thickened sclera, and it is this which stands the optic stress experienced during deep diving.

The above lack of accommodation is not unique, for it is a character shared by a number of land mammals.

The manner in which Putter found the eyes of pinnipeds to differ from conditions in the Cetacea may be summarized as follows: The pre-equitorial segment is thick and the equitorial segment thin, while the latter is thick in whales; the fundus segment is thick in all. The choroid is thin in pinnipeds and mysticetes, but thick in odontocetes. The tapetum lucidum is poorly developed. The ciliary muscle is feeble in pinnipeds, absent in mysticetes, and represented by a few fibers in odontocetes. The ciliary processes are moderately long in pinnipeds but short in cetaceans. The lens is larger than in whales and the rods of the retina very long. Although the sheath of the optic nerve is not so robust as in whales, it is much thicker than normal. The tarsus of the lid, while poorly developed, is more so than in whales. Meibomian glands are absent in all three. Nictitating membranes are present in pinnipeds (and sirenians) but absent in whales. The eyeball in pinnipeds is slightly movable, which is not the case in cetaceans. In the former the axis of the eye is horizontal or directed somewhat dorsad, while in the latter it is horizontal or directed somewhat ventrad. It is further mentioned that of all mammals sirenians may be said to have the cornea least developed.

As Kellogg has suggested, adaptive visual changes, when the need for them has arisen, are of critical importance to a marine mammal. He pointed out that the atrophy of the optic chiasma in the zeuglodont brain indicates a failure in their visual apparatus very possibly because of an inability to adjust their vision to salt water requirements, and this may very well have been the deciding factor in their eventual elimination.

Kellogg's investigations of cetacean eyes were based on mysticete material and he informs me that there is not the slightest doubt but that the eyes of this group are well nigh useless for seeing above the surface of the water. Putter stated that the eyes of odontocetes and mysticetes differ in many important respects and to a degree that indicates that in each the eye has experienced its own particular type of specialization from a more generalized optic equipment. Having never worked on this question myself I am in no position to point out any erroneous conclusions which this author may have reached, but it seems that he was mistaken in ascribing to mysticetes, odontocetes and pinnipeds a common inability to see effectively through an atmospheric medium, very likely for the reason that apparently he did not determine mathematically the refractive powers of their visual equipment. At any rate it is well known that when a watchful seal is basking it is quick to discover the approach of a polar bear or other enemy, and that not only the killer whale but the cachalot is in the habit of thrusting the head above the water for the purpose of ascertaining what may be of interest above the surface. Hence it is evident that there yet remain for settlement several fundamental questions regarding the optics of aquatic mammals.

ACOUSTIC SENSE

There is a tendency for the elimination of the external ear in aquatic mammals, partly following the law that ultimately the aquatic life will eliminate superfluous prominences upon the body, and partly because of the ultimate disuse of the pinna as an accumulator of atmospheric sound waves. In the case of those sorts which inhabit streams this reduction of the pinna may be barely or not at all appreciable, partly because they are not very highly modified aquatically if they are still stream dwellers, and partly for the reason that although hearing has ceased to be an aid in capturing aquatic prey, they must still be on the alert for terrestrial foes. Also it must not be forgotten that mammals of this category are almost always densely furred, and that in these the ears are usually hidden in the pelage, constituting elimination of pinna as far as concerns the external contour. But this need not be of aquatic significance since many terrestrial representatives of these mammals (among Insectivora and Rodentia) exhibit the same condition.

The platypus, anomalous in so many ways, has rather peculiar ears. Although it lacks the pinna, the musculature enables it to "cock" the orifice forward, as is shown to excellent advantage in plate 6 of Burrell's (1927) book. This authority also states that the auditory "orifice lies at the posterior end of a facial furrow, the eye lying at the anterior end, while the furrow is incompletely divided into two by an oblique fold of skin. The edges of this furrow act as a long pair of lids, by means of which both eye and ear may be tightly closed at the will of the animal. The aural aperture may also be dilated and contracted while the eyes are open."

The aquatic Talpidae—*Desmana* and *Galemys*—also lack the pinna, which has no aquatic import, for they belong to the earless family of moles. *Nectogale* is the only other insectivore which is said to lack a pinna. And *Anotomys* is the only such rodent, but I question this without verification (I have not seen this genus). Among pelagic mammals the Cetacea, Sirenia, Phocidae, and Odobenidae have no external ears.

[68]

It is thus seen that mammals which inhabit fresh water usually retain the pinna, while pelagic mammals have mostly lost it, both because they are more highly aquatic and because they have less need for it. It is retained by the sea otter and Otariidae, but in the latter the pinna is but a remnant and too slender to act as an acoustic aid.

If an aquatic mammal show by the position of its eyes that it has experienced a need for peering above the surface of the water with the minimum of its body exposed to view—in other words, if its eyes have assumed a somewhat dorsal direction, then has it similarly experienced a need for hearing in this position, and the external orifices of its ears will also be situated more dorsal than usual. This is not apparent in any insectivore and is at all marked among rodents only in those aquatic forms having a rather flat skull, in which the external auditory meatus is relatively close to the dorsal profile (as *Castor*), and in such this development may have nothing to do with an aquatic modification, as many terrestrial rodents exhibit the same character. The well formed ears of the hippopotamus are markedly dorsal in position. In all Pinnipedia the auditory lumen does not extend directly laterad from the meatus but turns quite sharply upward, reaching the body surface considerably dorsad of the meatus. This is most marked in the Phocidae and least so in the Otariidae, and is an expected condition. I strongly suspect that this is also the case in the Sirenia. In the Mysticeti (at least in *Balaenoptera*) the auditory tube extends practically in a direct lateral line to the surface of the body. At least in *Neomeris,* among the Odontoceti, there is a sharp bend of the lumen upward for about an inch. If this should prove to be a uniform character among the toothed whales it should be of considerable phylogenetic importance.

As already mentioned the ability to close the ear so as to exclude water is one of the first acquirements of an aquatic mammal and it is confidently believed that all mammals that may be so classed have it. It is one which is easy of accomplishment. Most terrestrial mammals which burrow also have it, for the purpose of keeping loose soil out of the ear. Perhaps most often this closure is accomplished by the pulling into the orifice of a valvular plug, probably homologous with the antitragus, specialized for this purpose. This may be seen to excellent advantage in the case of the seal *(Phoca)* but the precise functioning of the mechanism is not uniform and the result may be attained in different genera, and even in well differentiated species, by complex variations in the actions of the small auricular muscles, which are relatively very well developed. A somewhat different method for closure is employed by the hippopotamus, for it may be seen that directly preceding submergence this animal contracts the entire base of the ear, which evidently compresses and shortens the lumen.   The extremely vigorous and repeated twitching of the ears upon its reappearance is perhaps indicative that the closure is not directly at the orifice, but slightly deeper, so that some effort is necessary to dispose of lodged water.

Still another way of closure is employed by the sea-lions.   The cartilage is very slender and somewhat furled, and muscular action results in further and tighter furling, and perhaps some longitudinal contraction of the lumen.   For this method of excluding the water at least some remnant of the external pinna would seem to be of distinct advantage, and it is not unlikely that this is the sole reason for the retention of a part of the external ear by otariids, while the lack of such need has hastened its disappearance in phocids and the walrus.

Among Sirenia I have had an opportunity for examining only a hardened dugong.   The auditory aperture in this was relatively smaller than in a seal, but larger than in the Cetacea, and appeared to be fairly intermediate in other respects.

At first, in the early history of a hypothetical aquatic phylum, the closed position of the ear would be retained with some muscular exertion, and later with ease.   A point would inevitably be reached, at a time when the mammal had ceased to leave the water for more than short periods, spending the greater portion of its time with the ears beneath the surface, when the action of the auricular musculature had so changed that the closed position would be the one maintained involuntarily, while muscular effort would be required to open the ear.   Whether or not any pinniped has yet attained this stage is unknown.

Thence it would be but a relatively short time until, through disuse of the musculature involved, the ear could not be opened at all, the auditory lumen would first remain permanently water-tight, and because its apparatus for closure had throughout a long period been operating with but little or no opposition from the opening musculature, its caliber would be reduced.   This in effect, is just what is now to be found in the Cetacea.   I have introduced a match stick into the auditory tube of a *Balaenoptera* 75 feet long, but it would not accommodate anything larger; and in some odontocetes it is even smaller.   In the latter, as would be expected because of the continued action here in the past of the musculature for closure, the lumen is smallest at the distal end, expanding to a slightly greater diameter proximad.   In the Mysticeti the external orifice is also small, the lumen thence expanding for a short distance until it is sealed entirely, once more becoming open and again expanding farther proximad.

It is usually stated in the literature that auricular musculature has disappeared entirely in the Odontoceti, and is vestigeally represented by one or two remnants in the Mysticeti, but few investigators have specialized sufficiently in the facial musculature to do a thoroughly satisfactory dissection of this portion of the Cetacea. A slip, believed to be a vestige of the ear muscles, was found by me in *Neomeris* but because of the bad condition of the specimen I could not be positive. Ernst Huber has found remnants of several auricular muscles in the narwhal, and I confidently expect that there are actually more than the two usually stated to occur in mysticetes. A vestige of the auricular cartilage has also been found in some whales.

It is extremely doubtful if any aquatic mammal swims with the auditory tubes open. Hence, as the tube is closed when beneath the water a mammal can use its ear in normal manner only for air-borne sounds. Thus the ears of a mink or beaver are probably largely inoperative under water save as sound waves may be transmitted by resonance through parts of the head. When, in an aquatic mammal, the opening mechanism of the ear has ceased to function and the lumen remains closed, normal use of the ears will cease forever, for then neither air nor water can transmit extraneous sounds directly to the ear drum. The latter situation now obtains in the Cetacea. But there must be a gradual accommodation to this change—a gradual increase in ability to receive water borne vibrations and a gradual decrease in the power to receive those transmitted by air. And there must be a change in the quality of reception also, for it is unthinkable that during the thousands of years since the abandonment of atmospheric hearing in the Cetacea they receive sounds under water only after the same fashion as do we when the head is submerged. It is not unlikely that the Pinnipedia are now undergoing this auditory alteration, and that the change is more marked in the Sirenia.

There is abundant evidence that whales are sensitive to certain water borne vibrations which cannot possibly be transmitted through their auditory or Eustachian tubes. Hence these waves must be transmitted through some solid part of the head, but we have no means of ascertaining which part is most resonant. There is further proof of cetacean keenness of hearing in the high development of the internal ear, and in the size and character of the acoustic colliculus, which in a porpoise G. L. Streeter (in Kellogg, 1928) found to be four times the size of the optic colliculus.

The middle and inner ear of the Cetacea are sufficiently modified for us to be sure that the function is altered. Presumably the ear of the Sirenia has also undergone modification, but little or nothing is known about it. It is also permissible to surmise that in the Pinnipedia some change is going on, perhaps of two sorts, for the auditory bulla of most genera of seals is considerably inflated, while that of the sea-lion is not, appearing shrunken and rugose. Gray (1905) has reported on the former. He stated that the inner ear of this mammal is larger than in any other except the walrus. He found that two otoliths of remarkable size were present in the vestibule and as these were unlike those of any mammal yet recorded he surmised that they must have some particular physiological function. It is unknown whether otoliths are present in the Cetacea. That the large vestibule of the Phocidae is not a simple corollary of the aquatic life is indicated by the fact that in the Cetacea this is particularly small.

The ear bones of the Cetacea are characterized by extreme hardness. Kellogg (1928) said "the tympanic bulla is the relatively dense and heavy sounding box fastened to the periotic by two thin pedicles, which can be set in vibration. Vibrations set up in these pedicles produce a corresponding amount of motion in the malleolus, whose anterior process is likewise fused with the bulla between these pedicles, and it in turn transmits these vibrations to the incus and stapes." The malleus is rigid and the stapes is immovable in the vestibule. Kellogg further said that "water-borne vibrations transmitted to the air contained in the tympanic bulla cause it to function as a sounding box, and its vibrations reach the cochlea by way of the ossicular chain and vestibule." Kernan and Schulte (1918) stated that Denker (1902) thoroughly demonstrated vibration of the ossicular chain to be impossible, but the latter merely claimed that vibration cannot be activated by the tympanic membrane, for this is too lax for any such function. Kernan and Schulte also mentioned what is well known clinically, that an increase in the conduction of sound through the bones of the head accompanies diminished function of the middle ear (this is the condition in the author).

The ear bones of the Mysticeti and Odontoceti are of two distinct shapes and this is accompanied by other differences. In the former the ear drum looks like the finger of a gray leather glove, moderately pliable when fresh but inelastic, the part representing the finger tip extending distad into the auditory tube. In the Odontoceti the tympanic membrane is not at all finger-shaped but is gently bowed and often partly calcified. In the whalebone whales (at least in *Balaenoptera*), the anterior part of the bare bulla, without membranous covering, projects into a fossa the size of one's two fists, and in freshly killed specimens this is entirely filled with a coarse foam of albuminous, rather than greasy, texture. Whether this is so in living specimens cannot be demonstrated, but presumably it is, and the foam may have some function in determining the quality of sound reception. There is free communication between this fossa and the choanae. In the odontocetes there is a different but analogous system of air sinuses adjoining the inner ear and connecting with an intricate labyrinth of ducts. Authors have been very vague and cautious about describing these ducts, and with good reason, for without the injection of a suitable colored mass into this part of a freshly killed specimen their proper definition is uttterly impossible, as their finer ramifications are otherwise not to be distinguished from adjoining blood vessels and oil ducts. It must therefore suffice to say that this system of air sinuses communicate with the choanae and apparently send trabeculated branches ramifying through the peculiar fatty tissue that occurs in the odontocetes within the angle of the lower jaw.

As already stated it is well known that cetaceans are sensitive to certain sorts of water borne vibrations. It has been reported that porpoises are peculiarly sensitive to the waves that are transmitted by the sonic depth finder and will disappear in great haste and apparent discomfort form the vicinity of a vessel when one of these contrivances is put in operation. This fact suggests that the Cetacea may be sensitive to water-borne sound waves of a character and after a fashion that we do not yet understand. The transmission of the sounds that reach them and the ears themselves are so different from anything connected with our own acoustic apparatus that my personal opinion is to the effect that we know nothing whatever about the matter.

The function of a hydrostatic organ or depth gauge has been assigned to the air sinuses and passages about the cetacean ear, but it has become the practice to assign this function to any part of cetacean anatomy which seems in any way unique or peculiar. I cannot see that any sort of a hydrostatic organ would be necessary in this order, for they can always tell when water pressure upon the body becomes too great for comfort or safety.

Numerous writers have concerned themselves with the question of how the air within the inner ear is equalized to correspond with the great external pressure experienced during deep diving. I cannot see that any equalization is necessary. There is no tension of the cetacean tympanic membrane and hence no need for nice adjustments of pressure within the ear to that of the auditory tube, such as is often so annoying to us during speedy ascent of a mountain. No water, and hence no pressure save that applied to the body surface can reach the ear through the auditory tube. Judging by the powerful rush of the air which leaves the cetacean blow hole as soon as the nasal orifice is opened, the air is evidently retained between breaths at considerable pressure—greater than the external pressure when the body is at the surface, and it seems that this air pressure must reach the inner ear through the Eustachian tube, at least during the beginning of expiration. It is therefore logical to assume that the inner ear has been simply adjusted to withstand any pressure experienced without the necessity for nice muscular adjustments for the equalization of pressure.

OLFACTORY SENSE

It is strikingly apparent that a mammal which seeks its food exclusively beneath the surface of the water and comes to land only for brief periods of basking or to have its young on some safe island retreat, from whence it can dive into the water without an instant's delay, has virtually no need for a sense of smell. Hence it is rather remarkable that the Pinnipedia are so well equipped with olfactory apparatus. The olfactory bulbs are not as well developed as in most terrestrial carnivores, according to O. R. Langworthy (MS), but the sense of smell is by no means vestigeal in this order and turbinal bones of considerable complexity are retained. It accordingly seems likely that a fairly well developed olfactory apparatus is a character which is not readily relinquished, at least by mammals of this sort, and that it will be retained considerably after any great need for it has passed.

Probably no mammal less specialized for an aquatic life than the pinniped has had the olfactory apparatus appreciably reduced. In the Sirenia it is considerably more reduced than in the Pinnipedia, and Owen (1868) mentioned that the olfactory nerves are fewer and the cribriform plates smaller in the dugong than the manati, as might be expected from the greater aquatic specialization of the former. But Murie (1872) stated that the size of the olfactory bulbs of the manati indicates that its sense of smell is "fairly well developed." It is in the Cetacea, however, than one finds the greatest olfactory alteration occurring among aquatic mammals. The whalebone whales retain an olfactory apparatus, although it is vestigeal, but for what purpose it is difficult to understand, for ages ago they must have ceased to have the slightest need for it. The olfactory nerves are simple and after piercing the cribriform plates, are distributed to the mucous membrane of the narial passages. Needless to say, water cannot come into contact with these nerve endings, so they could not have taken over any aberrant taste function, and it is well nigh inconceivable that any message which they might receive by this means from the atmosphere would have any significance for them. Kellogg (1928), however, thinks that "the retention of the sense of smell (in the mysticetes) may be due in a larger measure to the actual mechanical construction of the skull than to the need of such organs."

Olfactory apparatus has ben found in fetal Odontoceti but the adults have lost it and the cribriform plates are imperforate. Again Kellogg says that in modern toothed whales "mechanical changes in the relations of the component parts of the odontocete type of skull appear to have restricted at first and finally prevented the physiological functioning of the olfactory apparatus."

Chapter Five

Mouth and Nose

MOUTH

Except for the teeth where already noted the only aquatic mammals whose mouths are of noteworthy interest are *Ornithorhynchus,* the Cetacea, *Hippopotamus,* the walrus, and the Sirenia.

Ornithorhynchus, or the platypus, has a mouth that is unique among Mammalia. Except for its extreme broadness it superficially resembles the bill of a duck, being used for the same purpose, but whereas the latter is horny, the bill of the former, although firm, is covered with soft, moist, and extraordinarily sensitive skin. As the bones of both the mandible and rostrum are bifurcated, some play is permitted and the animal, according to Burrell (1927), can pucker the mouth sufficiently to form a central suction tube whereby small life of the stream bottoms may be acquired with the minimum of mud. Besides the serrated, horny part of the lower lips which develops after the early loss of the teeth, there is a pair of longitudinal structures in either jaw which Burrell designates secateuring ridges. Cheek pouches are also found, which the same authority considers are used mainly for holding fine gravel as an aid in masticating chitinous or other hard food.

Never having seen a living sirenian my conclusions regarding their feeding actions have been derived from the literature. They are in the habit of using the fore limbs for pulling aquatic growth toward them, the action then being, according to reports and illustrations, for the herbage to be grasped and thrust into the mouth by a unique interaction of the lips and a pair of fleshy folds, one upon either side, which may be designated as the side lips. There is surely considerable difference in this labial equipment between the manati and dugong, but the various postures which may be assumed (see figure 8) so utterly and ridiculously change the expression of the creatures that we can judge little from the illustrations, for the posture shown by one animal may or may not be one that the other can adopt. Certainly the labial musculature of the sirenians is very remarkable, and of it Murie (1872) wrote: "Were the trunk of an elephant cut short at the root, or better still, left entire, but contracted to a minimum of its long diameter, and with the

[76]

terminal tactile appendage aborted, structurally the manatee's naso-labial organ would assimilate with it." Being derived from proboscidean stock (according to belief) it is perhaps not astonishing that sirenians should have such a highly developed and unique labial equipment. Neither is it surprising to find that in consequence (evidently) of this great mobility of the lips this order is incapable of protruding the tongue.

In sirenians the ingestion of food is aided by bristles upon the inner lips, hairs and bristles upon the oral cheek surfaces, and fibrous papillae upon both the floor and roof of the oral cavity. These papillae are better developed in the dugong than the manati, and are said to have been best defined in the extinct Steller sea cow. In the dugong at least, a skinned head of which I examined in the National Museum, the papillae constituted rasp-like surfaces that must be extraordinarily efficient in tearing up vegetable food. This individual also had a most peculiar oral development consisting of a subglobular prominence the size of a small hen's egg upon the ventral part of the rostral tip. This was smooth, and hard and tough in the preserved specimen, but in life it was evidently soft and mobile, functioning as a sort of soft, accessory tongue tip. From Murie's figure the tongue proper of the manati is relatively smooth.

The adult male walrus is so constructed that it cannot use the fore limbs as an aid to ingestion of food, and unless a morsel is small enough to slip readily between the bases of the huge tusks, difficulty is encountered. The midsection of the upper lip is relatively immobile, and the lower lip seems but slightly muscular and incapable of protrusion. The difficulty is therefore partly overcome by the mystaceal pads. These are greatly developed, project considerably, are mobile, and fitted with short vibrissae of such large diameter that they are veritable spikes. By a medial contraction of these, food may be forced into the mouth. There is record of a captive walrus having killed and partly eaten a seal, using the bristly pads to tear off flesh—a feat which it would have been entirely incapable of accomplishing without their aid.

The apex of the walrus tongue is rounded and entire, but cleft in all other pinnipeds, according to Sonntag (1923). In comparing pinniped with fissiped tongues he stated that in the former the tongue is "shorter, wider, thicker. Apex cleft in all except *Trichechus (Odobenus)*. Edges lobulated or have laterally-projecting papillae. Mucosa of pharyngeal part folded. Many glandular orifices present. Vallate papillae frequently absent. Lytta absent. Frenum slight. No trace of a spinous patch or papillae clavatae. Lateral organs variable." He also stated that *Phoca* has a fuller complement of gustatory organs than any other pinniped which he examined.

The mouth of the hippopotamus is remarkable for its size. It has evidently been developed thus partly by the stimulus of excessive use, for it feeds upon succulent vegetation of very low nutritive value, and hence must consume an enormous quantity. Indicative of this is the fact that the stomach is upward of 15 feet in length, and even more complicated than is usual in Artiodactyla. But the huge mouth is not altogether an efficient apparatus, for mastication is a slow and laborious process, and the animal has much difficulty in properly handling the pendulous angles of the mouth.

In the Cetacea the lips are always immobile, smooth and rubbery. In Odontoceti the width and size of the mouth is never excessive, but is quite variable, largely according to the size of the snout. Thus in *Kogia* the very short mandible makes the mouth relatively very broad, while in the long-snouted porpoises it is more beak-like. In the cachalot *(Physeter)* the enormous mass of the head dwarfs the mouth by comparison, and yet the animal feeds upon the most bulky fare of any cetacean. The tongue of the toothed whales is variable, but may be said to be normal in function. It often has a fringe of papillae. Sonntag (1922) said that "the tongues of the Cetacea have their glandular organs better developed, but their gustatory and mobile functions are less, than in other Mammalia." The tongues of Mysticeti agree with those of Odontoceti in having filliform papillae scanty or absent, the mucosa more or less corrugated, no trace of foramen caecum, lytta, frenal lamella, lateral organs, and apical gland of Nuhn. They differ as follows:

Mysticeti	Odontoceti
tongue soft	tongue firm and hard
intermolar elevation present	absent
much oil in tongue	absent
apex massive	not so
marginal lobes absent	present
lateral borders ill-defined	well-marked
glands less numerous	very numerous
muscles slight or large	well-developed
mobility slight	variable

Salivary papillae and plicae fimbriatae are absent in both. In odontocetes the surface of the tongue is rather parchment-like. It usually fills

[78]

the intermandibular space, but not always *(Monodon)*. The esophagus is not noticeably reduced in diameter in this group, and the muscles for deglutition, although strong and compact, are not remarkable.

In the baleen whales the oral condition is very different. In this group the mouth is of enormous size to allow for the unique feeding habits, but the details differ considerably. In all forms the lower jaw is very much broadened and bowed out, but the actual size of the mouth varies so as properly to accommodate the baleen equipment. The mouth is apparently larger, relatively speaking, in the humpback *(Megaptera)* than in any of the true balaenopterid whales, and of fairly prodigeous size in balaenids, for in the latter group it may be more capacious than the thoracic and abdominal cavities combined (Flower and Lydekker, 1891). The reason for such oral disparity is evidently correlated with food habits, but it is difficult to observe precisely the finer points of such actions below the surface of the water, even upon the rare occasions when one can approach sufficiently close to a feeding whale. It is usually claimed that the mysticetes feed by swimming slowly along, the rostrum level with the surface and the mandible dependent, the mouth being wide open. It is not logical that the mouth should be held in this position for long or at speed. Except possibly in the balaenids, if these animals maintained the mouth *wide* open while feeding upon small food, the mandible would fall below the ends of the baleen plates, so that all food would pass out of the mouth with the free flow of water. Even such an experienced observer as Andrews (1909) is noncommital in regard to the precise action of the mouth during feeding. He has stated, however, that when feeding on shrimp both *Megaptera* and *Balaenoptera* turn upon the side, a posture which he has illustrated by photographs, and that when the jaws are closed the water rushes out in streams. Regardless of the position of the body the procedure of a balaenopterid when shrimp are being consumed is presumably for the animal to swim slowly through a swarm of the crustaceans with the mouth opened for but a foot or two. Then only will the baleen operate efficiently as a strainer, for the oral cavity will then act as a closed filter, the water flowing away laterally through the baleen fibers and leaving the shrimp in the mouth. The form of the mouth and length of baleen plates in the balaenids (see figure 9) allows the mouth to be opened much farther for best efficiency, though no one can say why this is the case.

If balaenopterids cannot find shrimp, they will feed on anchovies, or if these are unavailable, then on smelts, or small mackerel. In other words, they prefer the smallest food they can get. I am unaware of the method they employ in securing fish as large as one-foot mackerel, but have been told of how the humback *(Megaptera)* obtains them. An experienced and trustworthy whaler has assured me that off the coast of Lower California he has often watched these whales feeding on fish of this size. According to him, they "stand on their tails" in the water with snout protruding, presumably in a dense school of fish. With mouth opened at a right angle the water is fanned with the greatly elongated flippers and the fish (according to his theory) mistaking the dark cavern of the great mouth for a safe haven, rush thither. During this process he assured me that the mandible was twisted from side to side in so outlandish a fashion that it was a wonder the animal could ever get it back into place. Then with a heave, the whole head was lifted above the surface, and as the jaws were closed, fish and water cascaded from the mouth. I believe this statement, but give it for what it is worth. Certainly fish of this size must be secured after some other manner than that employed in catching shrimp. The bow-head whale is said to feed largely on jelly-fish of several sorts. This larger fare, being less crowded in the water, may have had something to do with the development of the enormous head, which may reach one-third of the total length, or it may be because of some other stimulus.

In the Mysticeti the tongue is of two distinct sorts. In the balaenid whales and the gray whale it is of large size, apparently even higher than broad, firm and highly muscular. It is not a fortuitous accompaniment that in these animals the external throat is either entirely without furrows or practically so (*Rhachianectes* has two or three short, shallow grooves, and it is of interest to note that among the toothed whales this is also the case in the Ziphiidae). This indicates that the gular region is not extensible, and it does not need to be. When, while feeding with the mouth partially opened, it is wished to swallow the food that has been captured, all water is forced from the mouth by simply elevating the tongue, just as we ourselves would do.

This procedure in whales with grooved throats (at least in *Balaenoptera,* and probably others) is entirely different. It has been frequently stated in the literature that the tongue of these whales is enormous. Of course it is, in an animal of this size, but not relatively so. On the contrary it is extremely small and so weak that it may really be considered as nonfunctional. In an eleven-foot fetus of *Balaenoptera* the tongue was slightly muscular (considerably less so than in the normal mammal)—much more so than in the adult. It almost filled the mouth and was at least a quarter the size of the adult tongue, although the whole head was perhaps but one-hundredth of the bulk. Transection of the adult tongue shows but few muscle fibers and these are partly disassociated. Connective tissue is soft and flabby with much soft fat of oozy consistency, so that when a large piece of the tongue is turned over on the ground by means of an iron hook it is strongly reminiscent of a large bladder partly filled with mercury. In a freshly killed specimen the tongue lays flabbily at the base of the oral cavity, appears shrunken and almost shapeless and, as said, is clearly almost nonfunctional. So there is ontogenetic reduction of the tongue in this group, indicating definite lessening of function. Incidentally it is hardly likely that just this change in the character of the tongue could take place without a corresponding reduction in the size of the esophagus, and hence, in the function of deglutition.

Now the integument of the throat in these whales has an intricate system of longitudinal plicae, grooves, or folds, extending from the chin to the mid-ventral line for half its length or more, and from the angle of the mouth to the base of the fore limb (see figure 9). Their presence has given rise to fanciful theories, as that the grooves are highly vascular and so in this manner oxygen is secured from the sea water, but anatomical facts clearly illustrate the reason for their being. After death the gular musculature is often relaxed, allowing the throat to bulge, but occasionally it is more tensed, as in life. Unfortunately I did not dissect the fetus which I examined so my acquaintance with the gular musculature of this group is limited to what I could learn by hasty observation while numerous finbacks were being cut up. It is evident however, that neither Schulte (1916) nor Carte and MacAlister (1869) properly interpreted all of the conditions which they encountered in this group of whales. That Schulte's designation of his mylohyoid as a part of the three-sheeted gular musculature is erroneous is shown by its innervation as well as by its position in relation to adjacent muscles. Furthermore it seems that the muscle which he called a geniohyoglossus was in reality the mylohyoid, his hyoglossus the geniohyoid, and his sternomandibularis probably the specialized digastric—at least this seems, from its position, the most logical interpretation; but lack of precise statements as to the innervation of all muscles prevents incontrovertible interpretations. At any rate it seems fairly certain that the three main sheets of the gular musculature comprise a specialization of at least two sheets of a primitive sphincter colli, now occurring as a sphincter colli superficialis, a sphincter colli profundus, and a remarkable third sheet—Schulte's longitudinal muscle of the ventral pouch—which may be either a superficial and extraordinary division of the mylohyoid, or an equally astonishing platysma. The action of these three is undoubtedly assisted by what seems to be the digastric, specialized for this purpose and continuing caudad onto the sternum. It is interesting to note that the caudal limit of the long gular muscle (either mylohyoid or platysma) coincides exactly with the extent of the external plicae.

The feeding actions, then, of those whales which have gular furrows is apparently as follows. When the mouth is opened for the purpose of securing food while the animal is slowly swimming, the inrush of water distends the gular region (as noted by Lillie, 1910), which is abundantly allowed for by the furrows and elasticity of the musculature involved. Then, when the whale wishes to swallow what it has captured, the weak tongue lies quiescent but the specialized gular musculature, including the digastric, contracts in the powerful way for which it is fitted, expelling the water from the mouth and restoring the throat and its plicae to their normal, static state of tone. The conditions occurring clearly indicate that this is fact—not theory.

Of passing interest here is the presence upon the under side of the rostrum, near its tip, in the mysticetes of a pair of shallow pits which, in a specimen of *Balaenoptera physalus,* Lillie (1910) found led into two narrow tubes which ended blindly about two inches from their mouths. These have been termed naso-vomerine organs and are popularly believed to be the vestiges of Jacobson's organs, now separated from the nostrils by a distance as much as ten feet.

The esophagus of the Odontoceti may be said to be of normal size, but that of the Mysticeti is remarkable for its small diameter. That of a finback of 70 feet was not larger than five inches at the most, and down its length was a close-ranked procession, in single file, of foot long mackeral, all headed to join the host of their fellows in the huge stomach. The latter is complex, and even more so in odontocetes.

TEETH

The teeth in aquatic mammals as in other sorts have developed to conform to the character of the food consumed and the manner of obtaining it, and as most aquatic mammals feed chiefly upon fish, the usual tendency is for the teeth to become simplified into sharp points for the grasping of slippery prey.

Among less modified sorts the rodent *Ichthyomys* has the lateral edge of the upper incisors prolonged into a sharp point. This undoubtedly facilitates the capture of small fish, but the development may have been fortuitous, for the members of the terrestrial subgenus *Mictomys* exhibit the same character. It is caused by the very slender lower incisors working against the broader upper ones, leaving the lateral edges largely unworn. The upper incisors of the beaver are especially modified to act as chisels for cutting down trees—a character that really has no bearing on aquatics.

Two aquatic mammals other than whales lack teeth in the adult state. Both the platypus and Steller sea cow *(Hydromadalis)* shed the molars at a very tender age, their places being taken by leathery or horny plates—a development accompanying the food habits of the former, feeding upon worms and similar soft foods, and of the latter in consuming soft marine algae. In the young *Halicore* there are four pairs of teeth beneath the horny plate of the mandible but these are absorbed before maturity. In this genus a pair of the upper incisors have developed into small tusks in the male, but these are diminutive and non-protruding in the female. In the manati *(Trichechus)* incisors are said to be present only during the fetal stage. In this genus but in no other sirenian, either living or extinct, there is a progressive succession of the molariform teeth comparable to conditions in the elephant. This takes the form of a constant forward displacement and loss of the tooth from the front of the series as a new back one develops. After studying a considerable series of skulls of different ages Thomas and Lydekker (1897) came to the conclusion that during the span of a manati's lifetime there must be not less than 20 teeth in each series, which is the same as stating that this process of succession will continue ceaselessly until death, or at least until senescence. If this be the case the teeth can hardly be homologized with those of the normal mammalian complement, the whole process of succession being different.

Another interesting alteration of the molariform teeth is that encountered in the walrus, which feeds largely upon mollusks, and the back teeth have become very broad and flat for crushing these. Somewhat similar in general respects are the teeth of the sea otter, which favors echinoderms or sea urchins as an article of diet. The molars of seals and sea-lions are simpler than those of the average terrestrial carnivore, although whether they have always been so or have become secondarily simplified from a more complex pattern has not been established, but is a question of some controversy. Their precise pattern need not concern us here, but it may be mentioned that the dental armature of the sea-lion appears to be extraordinarily powerful for the use to which it is put in coping with fish, large ones being torn asunder by a powerful twist of the head rather than by shearing with the teeth proper, and small ones swallowed entire. The strong canines may be chiefly for the purpose of fighting among themselves.

Among aquatic mammals, there are several exceedingly interesting modifications of the front teeth. In the fetal narwhal there is found within a socket in either premaxillary a small, nonprojecting tusk. In males the right tusk does not ordinarily develop further, but the left one grows, projecting straight forward from and through the upper lip, and attains a length of eight feet or even more. It is twisted, with spiral grooves running in a sinistral direction. Occasionally the right tusk also develops and this too shows a sinistral twist. Females do not normally have a projecting tusk, though very rarely one develops to some extent. No definite use is known for the narwhal tusk. It has been claimed that the males sometimes fence with them in play or in battle, and the theory has been advanced that they are used for digging in the sea bottom, but they grow too long to be handily used for such purpose. The proper explanation probably is that the tusk at first developed to a moderate length of say, one foot, for the purpose of fulfilling a definite need, such as rooting in the mud or as an ice pick. But tusks, antlers, and appendages of this sort are exceedingly prone to acquire undue evolutional velocity and to develop beyond the point of real use, ultimately becoming a hindrance, and hastening extinction through overspecialization (as some extinct elephants, the saber-toothed tiger, Irish stag, etc.). It therefore seems probable that the tusk of the narwhal is too long for any useful purpose and that it is more of a handicap than a help. The enormous tusks of the male Pacific walrus belong in the same category. Originally the canines were undoubtedly used for prying up clams and developed in response to this stimulus until they were ideally fitted for this purpose, with a length of several inches. But the development was not checked and ran wild, resulting in a length and bulk of tusk in the male that must surely prove most unwieldly, and if continued, threatens the extinction of this pinniped through constriction of the mouth to a degree which will hinder the ingress of food. The female walrus is provided with tusks also, but these are of but moderate length.

It will be noted that as with terrestrial mammals with tusks, as well
as in the case of the majority of mammals bearing antlers 'or horns,
these develop to a considerably greater size in the male. The claim is
usually made that this is for the reason that the males may more effectively do battle with one another for the females, but this natural
selection resulting is likely only secondary. Rather does it seem that
the male sex hormone almost invariably stimulates excessive growth of
tusks, horns, or other excrescences of a secondary sexual character to a
far greater degree than does that of the female, just as it frequently
stimulates growth of body in the male. But this is an uncultivated
field and I believe there are those who claim that there is no such
stimulus in the male, but rather an inhibitional factor in the female.

Of a somewhat different sort is the development of the large teeth
of the hippopotamus, for these are equally developed in both sexes,
but overspecialization is indicated here as well. The size of the teeth
has evidently kept pace with size of jaw and the result is that the
front teeth are so large and unhandily placed that they can serve but
little useful purpose and seem detrimental.

There is much dental variation among the toothed whales. The
teeth are usually simply conical and attenuate, but the crowns may be
chisel-shaped (*Neomaris, Phocaena*) and the tusks of the ziphioid
whales are at times curiously flattened and twisted. There may be well
over 200 teeth present (*Eurhinodelphis, Stenodelphis*), but two, situated in the mandible (some ziphioids), or a single large tusk (*Monodon*). So far as known, however, nonfunctional back teeth are always
present in the fetal state at least. The teeth of living odontocetes are
always of a single or homodont pattern—never heterodont—and this,
as well as the enormous increase in the number of the teeth of some forms
over the normal mammalian complement has given rise to endless discussion. To account for the latter condition there was first advanced
the theory of the intercalation of milk teeth into the series, but this
was abandoned when Kükenthal discovered indications of tooth succession in embryos; and at any rate this theory could not account for
the great number of teeth occurring in some sorts. Abel has argued
strongly that the original teeth were split up into numerous simpler
units; but this theory is unsupported by any good evidence and is viewed
none too favorably by many. In fine, we know absolutely nothing about
the matter. There is not always perfect alternation of upper and lower
teeth in the closed jaws, and the smaller teeth near the jaw tips are often
crowded and insecurely attached, indicating that if the jaws become

further elongated, more teeth may be acquired at the end of the series to fill the space provided. But we do not know how or why.

Tooth buds are present in fetuses of the Mysticeti or whalebone whales, but these are absorbed when the baleen starts to develop, or even before. Flower (1893) considered that the baleen plates developed gradually over their entire present area from the oral ridges of the roof of the mouth, which are present to some extent in most mammals and are highly developed in ungulates. The baleen is analogous to the ridges but Flower was mistaken in his premise and they are not really homologous. I have examined a fresh eleven foot fetus of *Balaenoptera* in which the start of the baleen growth was to be seen to good advantage. Slightly sunken within a groove which corresponded in position exactly with a maxillary dental arch was a soft, whitish tissue abruptly differentiated from the normal oral epithelium. This was about 3 centimeters in width and continuous save for a brief interruption anteriorly at the midline of the rostrum. It was plain that as this strip grew to form the young baleen there would be an accompanying widening in a medial direction so as to cover most of the roof of the mouth, which at the stage examined was formed of normal mucous membrane. So Tullberg (1883) was correct in principle in stating that baleen develops from a growth of papillae along the outer margin of the upper jaw.

Of extreme interest in the above connection is a condition in *Phocoenoides dalli* to which Miller (1929) has recently called attention. In a preserved section of the upper jaw of the specimen which he had the tooth tips appeared as being sunk in small pits and were below the gum surface so as to be nonfunctional. In compensation the gum along the dental arch and immediately adjoining on either side had developed rows of cornified papillae which had obviously taken over the grasping function of the decadent teeth. Histologically the structure of these papillae certainly seems to be homologous with baleen. So we appear to have an illustration in an odontocete of just how and why the baleen first began to appear in the mysticete ancestry.

Whalebone whales may have nearly 400 plates of baleen on either side. Flower and Lydekker (1891) stated that whalebone consists of modified papillae of the mucous membrane, with an excessive cornified epithelial development, there being at the base and between the blades an intermediate substance consisting of a softer epithelium. The latter is white and of a cheesey consistency, and projects into the base of each blade. When an animal is examined in the flesh the baleen appears as a series of rather triangular blades, set close together, the outer or labial borders being almost vertical, smooth and firm. The formation is of long fibers cemented together. From the ventral tip of each blade obliquely upward and inward to near the median line of the roof of the mouth the softer cementing substance wears or dissolves away, leaving the fibers to form a brush-like inner border to each plate. These diffuse, intermingle with those of adjoining plates, and form an effective apparatus for straining from the water and retaining within the mouth the half inch crustaceans which form the favorite item of food. The baleen of some whales, as the California gray *(Rhachianectes)* is short, coarse, and commercially valueless. In this whale, as well as probably all baleanopterids, the baleen equipment fits into the closed mouth without bending. In the balaenid (as the right and bowhead) whales the plates are of fine texture and may reach a length of more than 12 feet (bowhead), the ends folding upon themselves at the bottom of the oral cavity, but because of their excessive elasticity, they at once spring straight when the mouth is opened and pressure upon the blade tips is released.

We may, because of the present condition in *Phocoenoides,* follow the probable course of the development of the baleen with some little feeling of assurance. First starting as short epithelial papillae along the maxillary dental arch, they were used in place of the disappearing teeth to hold and retain small active prey. As they increased in length, water could be squirted out between the papillae so as to separate small fish and similar food held within the mouth, and the further transition of the baleen was but a matter of the ability properly to respond to the stimulus provided, and of sufficient time. The degree to which the baleen has developed in the large balaenid whales surely constitutes an overspecialization at the present time.

NOSE

Probably the first modification which an aquatic mammal undergoes is the acquisition of the ability to close the nostrils, for any mammal would be decidedly handicapped during under-water activity by having to guard continually against the sudden entry of water into the open nose. As in the case of the ear, there will tend to be a progressive adaptation in this respect, consisting first of closure with difficulty and then with ease, followed by the time when the closed position is the involuntary one and opening is the voluntary. Cetaceans and probably the Sirenians belong in the last category, while pinnipeds, the otters,

Potomogale and the Hippopotamidae have either attained it in some degree or are approaching it.

The procedure of closure is accomplished in a variety of ways, and probably no two sorts of different mammals have precisely the same mechanism for attaining this end. Unfortunately, observation of the live animal is not always illuminating in this respect, and the interaction of the small nasal muscles is so nice that dissection does not always show to our entire satisfaction the exact method followed. Broadly it may be stated that closure of the mammalian nose is effected by actions of the Mm. naso-labialis and maxillo-naso-labialis (of Huber), complicated by sundry specializations of these and the help of pads, flaps or valves. In the pinnipeds, for instance, the naso-labialis arises from near the middorsal line above the orbit, and, diverging slightly fanwise, inserts into the mystaceal pad. Its contraction lifts the pad dorso-caudad and crowds it mediad. The maxillo-naso-labialis arises from the zygomatic process of the maxilla below the infraorbital foramen, and inserts into the mystaceal pad deep to the naso-labialis. Its contraction pulls the pad latero-caudad and opens the nostril. Hence, mass of pad and contraction of the naso-labialis (or its normal tone when the specialization is greater) closes, and contraction of the maxillo-naso-labialis opens—really a very simple arrangement.

In most insectivores and rodents practically nothing can be told about the mechanism for nasal closure—at least without a long period of very painstaking work. They are too small to watch properly in life and similar difficulties are met in their dissection; and many others are not available, either dead or alive. Closure mechanism may, however, be divided into three classes. In one it is effected by a flap or valve; in the second, by the crowding of a fibro-muscular pad; and in the third, by fibrous or muscular tension from two or more sides. But action may partly combine two of these modes.

I am unacquainted with the precise method of narial closure employed by the platypus, but the external apertures remain open, so there must be a deeper valvular arrangement.

There is a tendency in many aquatic mammals (*Potomogale,* fissipeds and pinnipeds) for the broadening of the muzzle. This is popularly supposed to be a specialization in the direction of perfection of aquatic bodily form, but it is doubtful whether there is any real logic in this. The reason for its acquirement may be partly tactile, because of an increase in the sensitivity of the vibrissae, and hence in the branches of the infraorbital nerve that extend to their bases. But the increase in size of these pads is chiefly for the purpose of maintaining the nares in a closed position. The large pads of *Potomogale*, otters and phocids, by their bulk and elasticity crowd the narial apertures so that in their fullest development little or no muscular effort is necessary for closure. It effort be necessary, then it can be supplied by the radiating fibers of the pad itself and by tension of the naso-labialis muscle. In the fissipeds and pinnipeds the pad pressure is largely in a medial direc-

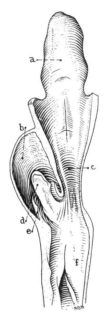

FIGURE 12.   Pharyngeal region of the pigmy sperm whale *Kogia*, from above (redrawn from Kernan and Schulte); (*a*) tongue; (*b*) nasopharynx; (*c*) oropharynx; (*d*) larynx; (*e*) arco palato-pharyngeus; and (*f*) esophagus.

tion, while in *Potomogale* it would seem to be chiefly anteriorly. The reason for the latter condition is that in this insectivore otter there is a bilobed rhinarial shield, smooth and evidently firm in life. The nasal passages are situated posterior to this while the openings are to the sides of the shield.   Hence, pulling the very broad mystaceal pads to the rear opens the nostrils, and relaxation presses them firmly against the shield.

I have for hours watched the breathing actions of various pinnipeds. By far the greater part of the time phocids inhale rapidly and then close the nostrils, although occasionally they are kept open for the duration of several breaths, especially when the animal comes to the surface after a period of active swimming. And a definite impression is given that the closed position is the relaxed one. In the sea-lions the mechanism for involuntary closure seems to be less perfected, but the mystaceal pads are not so broad, the animal is more prone to maintain the nostrils open between breaths (and, incidentally, to breathe through its mouth), and one is unable to decide from observation whether the opened or closed position is the involuntary one. Similarly with the otter. Although the muzzle of the latter appears very broad this is partly attributable to the rostral breadth of the skull and the pads are really not as large as they seem.

I have been unable to tell by observation of both American and Asiatic tapirs just how the nasal passages are closed. Apparently there is a contraction of the proboscis and consequent crowding of parts of the passages deep to the apertures. In this group of mammals there are diverticula of the nasal passages comparable to those existing in the Equidae, but no use is known for them.

In the Hippopotamidae the nostrils take the form of a pair of slits, closure seems to be partly voluntary, and opening accomplished by muscular pull along both borders of the slits.

I have been unable to acquaint myself with the exact processes of narial action in the Sirenia. The apertures are each somewhat crescentric and closure is evidently chiefly involuntary and largely of the valvular type, like a modification of the condition in the Odontoceti. The nostrils are located at the angle of the muzzle and are evidently farther back in the dugong than the manati. Their present position may be perfectly ideal for the needs of these sluggish creatures and it is by no means certain that there is any stimulus whatever for further migration of the nostrils toward the rear. The extensive rostral basining in this order indicates a high development of the naso-labial musculature, but from Murie's (1872) descriptions I judge that this concerns chiefly the lip movements. The muscular complexity which exists in this region may well be an inheritance from proboscidean ancestry more properly than a relatively recent specialization in response to generic needs.

For a proper understanding of the narial equipment in the Cetacea it will be necessary to consider the lungs and progress therefrom anteriorward. Wislocki (1929) has announced that in the porpoise *Tursiops* the bronchioles with a diameter of less than 0.5 mm. are provided with numerous muscular sphincter valves and that the terminal air sacs are guarded by the same means, as discussed more fully in the last chapter. Heretofore investigators have been obliged to consider the possible means whereby the nasal, or laryngeal, equipment of the Cetacea was enabled to prevent the escape of air from the lungs under the enormous pressure to which the animal is subjected during deep dives. Wislocki's discovery puts an entirely different aspect on the matter, however. Although the individual sphincters of the bronchioles may be assumed to be relatively feeble in action, the amount of air which each imprisons is so minute, and the valves occur in such prodigious numbers, that this equipment alone seems entirely adequate to prevent the escape of air even into the esophagus against the wishes of the animal. Presumably these valves close at the end of inspiration and open at the initiation of expiration, and it is reasonable to assume that their presence is characteristic of all Cetacea. The simultaneous relaxation of these small sphincter muscles at the same time that the nostrils are opened would account for the way in which the air, imprisoned in the lungs under the pressure applied by the relaxed thorax plus the pressure of the surrounding water, rushes forth with a veritable pop. But there is no reason for believing that bronchial sphincter valves *must* remain closed between breaths, and the animals may at times inflate a part of the nasopharynx before the nose is actually opened, as I have been led to believe by watching the escape just preceding expiration of small bubbles of air around the external slit of the blowhole.

In the Odontoceti the larynx is prolonged in a tube-like manner, projecting the epiglottis into the nasopharynx, and this may be closely clasped by the soft palatopharyngeus muscle. This is quite remarkably developed in the Physeteridae and Kogidae especially, in which the nasopharynx branches from the oropharynx upon the side and the larynx is correspondingly situated. Lillie (1910) stated that the latter was situated upon the left side in two specimens of cachalot. Raven (MS) makes the same statement, as well as Kernan and Schulte (1918) in the case of *Kogia;* and the dissimilarity in size of the choanae is undoubtedly the precise reason for the laryngeal asymmetry.

Kernan and Schulte have shown the pharyngeal region of *Kogia* (fig. 12) to good advantage. They said that the conditions are such as to prevent the entrance of water from the mouth into the nasal passage, and this may be a superficial function also, but the construction of the arc of the palatopharyngeus is clearly such that the greater the air pressure from the lungs the more securely it should clasp the larynx. Thus in these sperm whales at least, the air under pressure may be forced from the lungs through the larynx and into the nasopharynx, thus pressing *backward* against the palatopharyngeus, closing it the more tightly the greater the pressure, its suggested primary function then being to prevent the escape of air from the nasopharynx into the mouth. Some such procedure may be necessary in these animals in connection with the remarkable specialization of the right nasal passage.

H. C. Raven's preparations of *Monodon,* however, suggest that in this animal conditions are different, for the palatopharyngeus is very heavy and different from what Kernan and Schulte showed for *Kogia,* which may be due to the fundamentally different narial equipment of these two genera. It has usually been assumed that odontocetes at least may breath while at the surface at the same time that they swallow, and that when submerged they could not expel air through the mouth even if they so wished. But in the *Monodon* and *Physeter* which H. C. Raven (MS) dissected the larynx had been withdrawn and the epiglottis was in the oropharynx. Accordingly this authority is of the opinion that the larynx is thrust upward only temporarily for breathing or other purpose. If this actually be the case then it is certain that the small sphincter valves of the lungs act alone in preventing the escape of air, for the formation of the epiglottis is such that it could offer but little hindrance to the involuntary escape of air from the trachea into the mouth.

It may be mentioned that an elongated larynx for thrusting the epiglottis into the nasopharynx is not in itself an aquatic adaptation, for it is a character that is developed to a considerable extent in some marsupials and ungulates. The odontocete ancestry may have had this provision in moderate form to begin with, or else have evolved it completely for their particular needs.

The Mysticeti do not have this tubular type of larynx. Benham (1901) and others have described it but from these and existing sketches I am unable to envision the conditions with satisfactory clarity and I neglected to investigate the matter myself when I had the opportunity. Apparently the arytenoid body closes the epiglottis from above, the latter is triangular rather than tubular, and of very moderate length.

In a consideration of the precerebral portion of the nasal apparatus of the Cetacea it must be remembered that in most mammals the muscles of the face proper have a variety of functions to perform, such as involved twitching of the nose, snarling of the lips, and a host of others

[92]

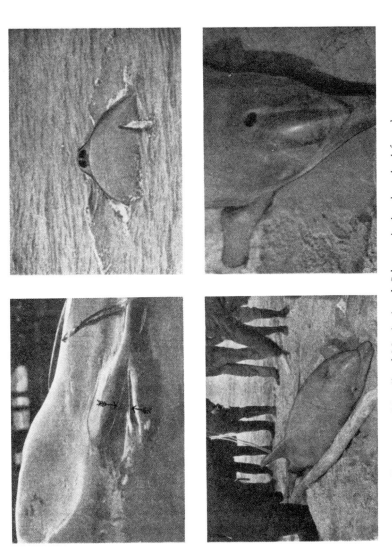

FIGURE 13. Nostrils of Mysticeti and Odontoceti, closed to the left and open to the right: *Balaenoptera* above, after Andrews; *Tursiops* below, from photographs by G. B. Wislocki.

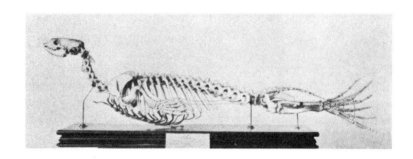

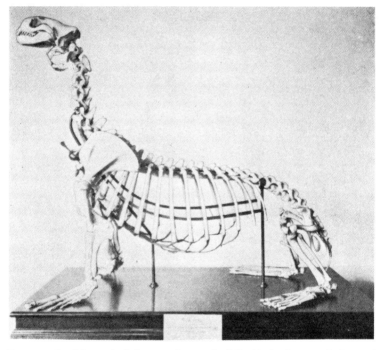

FIGURE 14. Mounted skeletons of a seal (*Phocidae*) above and fur seal (*Otariidae*) below, in the U. S. National Museum.

[94]

having to do with facial expression, while in the whale all these muscles are subservient to the single purpose of opening the blowhole. Such as cannot readily lend themselves to this use have become virtually nonfunctional. It is therefore of small wonder that we find the nasal mechanism as perfected as it now is in the odontocetes. Furthermore, in view of the bronchial conditions as reported by Wislocki it is necessary that we free our minds of the assumption that a special provision for the purpose of retaining air under great pressure is situated in the blowhole. Nor is a complicated contrivance for excluding water essential. The external area of the closed blowhole is small and it has been a simple process for this to acquire such a form that the greater the pressure the tighter is automatic closure.

Precranial narial conditions in the Cetacea are really of three sorts, the most simple being in the Mysticeti, a more involved situation in most Odontoceti, and a remarkably complicated one in the Physeteridae and Kogidae. In mysticetes the nostrils are in the form of two slits running in a sagittal direction (see fig. 13), but converging to form a rather steep-sided V with angle pointing forward. They do not actually join, however. Along the margin of each slit is a raised area, the elevation being slight along the medial margin and considerably higher along the lateral. To the touch this region is, like the rest of the body, tough and elastic. I have run my arm to above the elbow down one nostril and found that there was decided, though not heavy, pressure upon my arm for perhaps one foot below the surface, and as my arm was withdrawn, the elasticity of the tissue closed the passage completely. This involuntary closure may likely be assisted, does the need arise, by voluntary tension at both ends of the nasal slit, and by a downward contraction from the surface. But the conditions are such that the greater the surface pressure the more securely do the nostrils close, and I regard it as unlikely that the need ever arises for voluntary closing action in order to exclude sea water. In mysticetes there are no widely divergent diverticula. There is well known to be a "subcircular diverticulum from the dorsal wall of the respiratory passage," as stated by Schulte (1916), marking the olfactory region, and a "spritzsac" along the anterior wall, but from published descriptions I am unable exactly to envision the conditions. By manual investigation within the passages of a fresh adult I could not discover any true diverticula but only a slight folding and wrinkling of the mucosa rostrad and to a lesser extent laterad. Schulte considered the arrangement of this to be such as to aid closure when pressure is supplied from without.

The interaction of the specialized facial musculature in opening the nostrils of mysticetes has not yet been investigated with sufficient exactitude for us to be sure of all the actions involved. Of course the facial musculature is very highly modified, as it must be to control nostrils situated upon the top of the head, but the intricacy is not nearly so great as in that of the odontocetes. There are sufficient good photographs (see fig. 13) of breathing mysticetes to gain an understanding of the external results when the opening mechanism of the nostrils is in operation. In these there is apparent a marked lateral dilation of the outer lips of the nares, and a surprisingly high elevation of the anterior margins—an arrangement which evidently operates as an efficient barrier to the entry of water within the respiratory tract while the animal is in locomotion. There must be definite mechanical provision for this elevation, either in excessive elasticity of the tissue deep to the anterior margins or in wrinkling, and the latter is exactly what we find in the spritzsack, to which theory Schulte also subscribed.

The latter authority, in his study of *Balaenoptera borealis,* indicated uncertainty regarding just how the elevation of the anterior narial lip was instigated. From what I could learn from adult finbacks there is no great difficulty in this. The closed nares are suggestive of a V with apex directed forward, and the open nares are each broadly oval, with the long axes almost parallel. It is clearly evident that the pull of superficial muscles converging to the apertures from rostrad and laterad open the nares. The elevation could easily be provided by superficial muscles pulling rostrad upon the anterior and lateral margins while at the same time antagonistic action is supplied by deeper muscles pulling largely caudad, and relatively nonyielding tissue anterior to the narial passages. This seems to be the general principle of the mechanism, although in reality the actions are likely very intricate.

The narial conditions in the Odontoceti are very different. In all of them the external aperture is single, in all but the sperm whales medial, and almost always it is crescentric in shape, the concave aspect being directed anteriorly. The only known exception to the last detail is in *Platanista,* in which the orifice is said to be in the form of a longitudinal slit, but in exactly what manner this is opened is unknown. Excepting the latter for the time being, narial development of the toothed whales is of two definite sorts—that represented by the Kogidae and Physeteridae or sperm whales, and all other odontocetes. The latter, as the least intricate, will be discussed first. In perhaps no two distinct sorts is the facial musculature arranged exactly the same, for as the genera diverged from a common ancestry, more pronounced differences developed each in its own particular way as specialization increased. It is thus easily seen that the narial musculature of a beaked porpoise with rather low forehead, such as *Tursiops,* should be considerably different from that of a nonbeaked form with large frontal bulge, such as *Globiocephala* (see fig. 10). Of the first type I have examined only *Tursiops,* and of the second, *Monodon* and *Neomeris,* and the last was too hardened to be satisfactory. In this only did I attempt to dissect the facial musculature, for in the others this was done by Ernst Huber. I will therefore not go into precise details but will discuss only the generalities of this feature.

In the majority of odontocetes, then, the external nares take the form of a single crescentric orifice. Within this, at a distance varying with the genus or species, the passage is divided by a membranous partition into two. In this, as well as in mysticetes, the supracranial part of the nasal passages is practically vertical or slopes gently forward as it approaches the surface. But between the skull and the orifice in odontocetes there is a system of diverging diverticula or membranous folds (see description of *Neomeris,* Howell, 1927). These apparently vary with the species, or even with the individual, but as I understand their underlying principle they had better be described as two different parts of the nostrils. Lying forward of the bony nares and in contact with the premaxillae in a slight hollow of these bones, the extent of which can be determined upon the skull, is a diverticulum of each nostril which may be designated the premaxillary diverticula. These and the supracranial part of the nostrils proper describe an angle of, say, 90 degrees or less. Within this angle the tissue is very soft because of the soft fat of which it is largely composed. At the apex of the angle this soft tissue fits over the bony narial openings, effectively sealing them, especially when there is any pressure from without. When the animal wishes to take a breath the soft tissue, acting as a plug, is drawn forward by muscular action in a way that is greatly facilitated by the premaxillary diverticula. This action is strongly suggested by conditions in an adult narwhal *(Monodon)* the entire head of which H. C. Raven had sawn lengthwise in thin sections. When examining figure 15, however, it should be clearly understood that the sketch is only roughly diagrammatic and the degree of expansion of the deeper part of the passages and diverticula is unknown. Neither is it meant to imply by the above statements that these deep parts are *always* kept closed between breaths. Perhaps the complete sealing of the entire mechanism is indulged in only at some depth.

[97]

The remainder of the nasal diverticula is more puzzling. They occur somewhat nearer the surface and may be either forward, backward or to the side of the main nasal passage. Histologically the mucosa of these diverticula consists of simple, stratified epithelium without speciali-zation. To them has at times been assigned the function of a hydro-static organ, which theory I consider to be untenable. They may orig-inally have been a phylogenetic relic, like the nasal diverticula of the Equidae are supposed to be, since become somewhat more complex but still without useful function; they might serve in some slight measure to divert water that might slop over into the open blowhole, which would hardly be of sufficient importance to account for their existence.

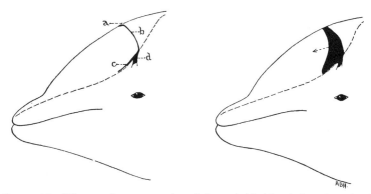

FIGURE 15. Diagramatic representation of the probable blow-hole action, closed and open, of the porpoise *Tursiops*: (*a*) external orifice; (*b*) supracranial part of nostril; (*c*) premaxillary diverticulum; (*d*) bony nares; arrow in-dicates direction of movement in opening, and broken line the outline of the skull.

Winge (1921) considered that "the air, which is exposed to strongly varying pressure and temperature, has a tendency to provide itself with greater space by widening out the nasal passage and Eustachian tube wherever it meets with least resistance." For all we know this may be the case and no one can prove the contrary; but I believe there has been a more-primary stimulus for the initial development, or at least retention from a primitive ancestor, of the system of nasal diverticula of the odontocetes. Let us liken the narial passage to the soft, elastic, rubber tube which it so greatly resembles. The intricate and laminated muscle layers which converge to operate its opening cannot function as dilators with precisely equal force at each and every point, however,

[98]

but their pull must be very unequal. Some must pull more strongly than others, or act obliquely, and there would result a bulging stimulus at some points which, throughout the ages, might well have resulted in the formation of the diverticula as we now encounter them. It hence seems likely that they are now a part of the opening mechanism of the nose.

The external aspect of the blowhole of, say, *Tursiops,* is crescentric and somewhat greater than the half of a circle, the concavity facing forward. Its outer margin is firm, elastic and of rubbery texture. Within this curve is a sort of valve that is much softer to the touch, intrinsically muscular and mobile, which is really a part of the deeper, soft, valvular plug, so that in the living animal while the blowhole is closed one or another part of its surface may move or "work" slightly. The "hinge" of this valve stretches transversely between the points of the arc, and when expiration is desired the posterior part is merely withdrawn within the spiracle and sinks into the highly elastic tissue anterior to the nostril.

As far as concerns surface indications of muscular action during respiration, all we can tell is that the nasal valve is opened by contraction of certain of the rostral muscles, and that in closure there is tension from laterad of the blowhole, but whether the latter is voluntary or involuntary is not clear. Of course there is intricate action and interaction of the deeper nasal musculature, but discussion of this I shall leave in the capable hands of Ernst Huber.

There still remains for discussion the astonishing nasal passages of the Kogidae and Physeteridae or sperm whales, reported on by Pouchet and Beauregard (1885), Le Danois (1910), and Kernan and Schulte (1918). The conditions are so excessively involved, however, that written descriptions can hardly be clear without extensive diagrams, which so far have not been published. For a proper understanding of these I am indebted to H. C. Raven, who has most generously placed at my disposal a description and diagrams of the head of a young cachalot which he recently dissected. He will himself report upon this in detail so it is wished here to describe briefly only such points as are necessary for a general discussion. In the cachalot, then, the left nasal aperture of the skull is huge and the right relatively minute. From the skull the lumen of the left passage extends for many feet along the left side of the spermaceti organ, opening upon the left side of the antero-superior part of the snout in a somewhat S-shaped orifice. The right nostril leaves its small cranial orifice and expands laterally to extraordinary proportions. It is situated fairly between the spermaceti organ and the adipose cushion below. It has a large diverticulum between the organ and the skull and another anterior of the organ, and joins the left nostril just within its orifice. The mucosa of the left nostril is plicated but that of the right is smooth in *Physeter*. In *Kogia*, whose spermaceti organ and nasal passages are relatively very much shorter indeed, the mucosa of the right passage is thrown into complex folds, evidently in compensation for its shortness. These folds, according to Kernan and Schulte, are highly vascular.

It seems almost certain that originally the narial conditions of the sperm whales were very similar to those found in other odontocetes. At least we are justified in assuming so, for the conformation of the skull indicates that there was retrogression of the external nares to the top of the head in the sperms as in other toothed whales. What I believe to have since taken place in the case of the latter is a secondary displacement forward of the blowhole proper and its controlling musculature by crowding of the spermaceti organ, although H. C. Raven is of the opinion that the organ began to develop in front of the nostrils and intruded between them as it increased in size. Ordinarily expiration and inspiration in this group must be accomplished solely by means of the left nostril. At least it seems certain that because of the disparity in size between the right and left narial apertures of the skull an insignificant amount of air could pass through the former during the short time occupied by the act of breathing. But the enormous dilation possible in the case of the right passage can hardly be fortuitous and must have some function. If it be dilated with air, as seems certain, then this should occur chiefly between breaths. It might hold an accessory air supply, but hardly enough to account for the long submergence of these whales, which are purported to remain below in excess of an hour. It is not impossible that the right nostril may be emptied of air to facilitate deep diving, and filled from the lungs so as the more easily to bring the animal straight up from the depths, as suggested by H. C. Raven (MS). And one must not overlook the possibility that when beneath the surface for a considerable period a part of the air in the lungs might be forced into the right passage and again withdrawn for some particular physiological purpose which might or might not have something to do with the spermaceti organ.

No discussion of the position of the external nares in aquatic mammals has yet been made in the present contribution. The position will correspond to the sort of stimulus experienced, the length of time that it has been in operation, and the capabilities of an animal for responding to it. Thus the whales are relatively speedy, compared to swimming land mammals; most of them cover much territory; and we can well see that they would be greatly handicapped if they could not renew their air supply at full speed in a choppy sea, but had to slow down and twist their nostrils above the surface at every breath. Hence the nostrils would be expected eventually to occupy the position upon the top of the head that they now do. Sea-lions and otters might be expected ultimately to follow the same development. On the contrary an animal like a sirenian or hippopotamus which does not seek active prey but only pokes the nose cautiously above the surface, usually while stationary and in a sheltered lagoon or river, might never attain a like nasal development.

Nothing much in this respect can be told about the platypus. The nostrils are dorsally located upon the "bill" but we have no means of judging the condition in its terrestrial ancestor. No migration of the nostrils is apparent in any of the insectivores, and in few rodents. Thus the nostrils of the coypu, capybara, beaver and muskrat are considered to be located slightly more dorsal than usual, but the alteration is really very slight. In the hippopotamus the nostrals are definitely upon the dorsal side of the snout, so that when the remainder of the body is submerged the animal may breathe with only the nostrils, eyes and ears exposed; or when thoroughly frightened, the nostrils alone may be quietly thrust forth. If the latter were situated farther back on the head this could not be accomplished without showing the eyes also, and this may militate against any farther migration caudad of the nares. The same may be said of the Sirenia. The nostrils might conceivably fuse to form a single blowhole and this might protrude somewhat so that breathing could be accomplished with the very minimum of exposure, but unless there was a radical change in habits there could be little stimulus for a migration of the nostrils to the top of the head. At present they are not even as dorsally situated as in the hippopotamus, for the animal evidently does not make much use of its eyes for peering above the surface, and all it need do is extrude the muzzle, with the axis of the head at an angle (not parallel) with the surface.

In a different class are otters, pinnipeds and whales. These either pursue active prey, travel at speed from place to place, or both. The otter is not as yet sufficiently modified for more than a slight change in position of the nostrils to be apparent. And unless there is a great increase in size throughout future geologic time the stimulus for posterior migration of the nares will doubtless be very slight. This animal can hold its nose above the surface with slight effort because of the small size of its head. A slight tilt in the cranial axis is all that is necessary. If it had a blowhole upon the forehead it would be but two or three inches from the present position of the nostrils and I judge that the difficulty it might experience in keeping water from slopping into the nose would be even greater than at present. Therefore small size should be a retardant of nasal migration. On the other hand, a long-snouted, large-headed beast with a body of ten or twelve feet would encounter a different set of conditions. The snout would travel through a larger arc for breathing, nostrils upon the forehead would be of more definite advantage than in the otter for the reason that they could then be thrust higher out of water with but slight exertion. The critical factors here would be size of head and speed; but the forces at play are intricate. Thus, in a mammal propelling itself from the rear the head (if the neck be shortened as in the whales) is a critical part of the swimming apparatus. If the head be large it cannot be swung about to thrust the snout above the water without acting as an undesired rudder. Further, if the neck be shortened to facilitate speed of locomotion the effort to elevate the snout would be disproportionately great. To me these facts point to the probability that when the whales were undergoing the most marked migration of their nostrils they were of a size perhaps comparable to a sea elephant or larger, with relatively large heads and necks that had already become markedly shortened. The difficulty both muscular and mechanical, of raising the snout while swimming at speed obliged them to elevate and thus retract the nostrils to their full ability, and this stimulus was largely critical in the migration of the latter to their present position.

The nostrils of the Phocidae, or those which I have observed, are slightly elevated, and judging from other aquatic specializations of this group they might be expected to show more marked modification. According to my experience the true seals are not very prone to breathe while swimming at speed, but are more often in the habit of doing so at rest, pausing after swimming beneath the surface quietly to renew their air supply before starting off afresh. If this be really characteristic and is persisted in they may never undergo more extensive migration of the nares.

The sea-lions (Otariidae) with which I am familiar habitually breathe while swimming rapidly at the surface, but the alteration in the position of the nares is even less marked than in the seals. For one thing there seems to be a different set of stimuli here. Swimming with the anterior limbs, the head and anterior part of the neck do not constitute such a critical part of the natatory equipment. Hence the head may be thrust moderately in any direction without disturbing equilibrium providing the hinder end acts in compensation. Furthermore the mouth is used to a considerable extent for breathing, and we cannot be sure that there is any really marked stimulus experienced for the retraction of the nares.

ADIPOSE CUSHION

Because of a complex of reasons, chief among which are doubtless the size and breadth of the head, and feeding habits, the speedy balaenopterid whales carry the flattish dorsum of the head parallel with the body axis, so that there is the minimum of cephalic resistance offered to the water. The slower *Rhachianectes* and balaenid whales have a downward-curving rostrum, but the base of the rostrum and posterior part of the head are carried parallel with the body axis also. In these there is no adipose cushion in the frontal region, unless the "bonnet" of *Eubalaena,* consisting of a raised, warty area upon the anterior rostrum, could be considered as the beginning of such a structure.

All odontocetes, without exception I believe, have some sort of recognizable adipose cushion equipment upon the forehead. In these whales the outline of the head from snout to vertex is never a straight line that can be carried parallel to the body axis so that the only resistance to the water is offered by the tip of the snout. The latter is always carried slightly depressed. In the beaked porpoises (as *Tursiops* and *Delphinus,* fig. 10), the dorsal outline is concave where the rostrum meets the cerebral part of the head, while in the short-snouted sorts the forehead is protuberant, but in all, the frontal region is anteriorly presented so that it receives the impact of the water during swimming. This, being the front of the braincase, is a region of rather critical delicacy, and it might be expected to respond to this stimulus by building up some sort of protective buffer or shock-absorber: but the subject cannot be abandoned with any such simple statement.

The adipose cushion in the majority of odontocetes is a thick pad of soft, elastic tissue upon the forehead in front of, and even partly surrounding, the blowhole. It need not be sharply defined, but the fibrous tissue and soft fat of which it is composed is surrounded and to some degree penetrated by the facial muscles which converge to or toward the blowhole. Directly anterior to the blowhole it is composed of pure, soft fat that oozes oil when transected. This fat is not the same as

[103]

that of the blubber layer but, together with the fatty tissue within the angle of the jaws, is of different consistency, and hence specialized, the refined oil from these regions being the finest lubricant for precision instruments so far known, and correspondingly valuable. The development of this frontal fat body is variable in different sorts of toothed whales. In long-snouted porpoises such as *Tursiops* and *Delphinus* its development is but moderate, while in the adult *Monodon* it is large, and is relatively so great in *Globiocephala* (fig. 10) as to cause the forehead to bulge forward beyond the jaws.

In the latter case the frontal bulk must act as a definite retardant to speedy locomotion. If the adipose cushion, or "melon" as it is called in the trade, had the function simply of a shock-absorber it would hardly develop to such proportions, nor would it have its present consistency, but should be more gristley and perhaps, as whales are prone to fatness, have some deposit among the fibers of the same sort of fat as constitutes the blubber layer; but hardly a particular grade of fine fat of the same composition as that found within the angle of the jaws.

In explanation of the above state of affairs it seems to be most likely that originally the water pressure against the anterior braincase of the toothed whales stimulated the formation of a simple fibrous thickening in this area to act as a buffer. This development once having begun, the region should then have been in a state plastic for further adaptations, and I believe that its present condition indicates a specialized physiological function of the adipose cushion to which that of a shock-absorber is now of secondary importance. This is but a personal opinion, however, and entirely unproven.

Sharply marked off from the other toothed whales in details of the facial regions are the sperm whales (Kogidae and Physeteridae). In the cachalot the fatty tissue of the head is of two sharply differentiated kinds. Occupying the entire bottom of the facial basining and projecting beyond the bony snout is a prodigious mass of fibrous tissue so tough that it must be hewn with an axe. This holds fat as a sponge holds water, and from it may be secured about 10 or 12 barrels of oil. It is the adipose cushion, or "junk" of whalers and seems largely homologous with the adipose cushion of other toothed whales. The spermaceti organ proper is a huge ovoid body occupying the upper half of the rostral basin and separated from the junk by the expanded right narial passage. After surrounding tissue has been removed it is said to be composed of an envelope of extremely strong, fibrous tissue of tendinous aspect. Within this is a zone of partly solid oil held by a spongy network, and in the center is liquid spermaceti oil. This spermaceti is very different from the blubber oil, just as is the case with respect to the adipose cushion of other odontocetes. The spermaceti organ is said to be a closed, ductless system, but it must be supplied by blood vessels. In a diagram with which H. C. Raven has kindly furnished me it is shown as resting for its entire length upon the right nasal passage, so phenomenally expanded in lateral direction. Diverticula of this passage are situated directly in front of and behind it, so when the passage is expanded with air, the spermaceti organ would seem to be suspended upon a pneumatic cushion upon three of its sides. This act is not as difficult as might be supposed; for it is not improbable that the organ is lighter than the water which it displaces. Occasionally, because of poor condition of the animal or for some unknown reason, the spermaceti case is empty, but ordinarily it contains up to 15 barrels of oil. Shortly before my visit in 1926 to the whaling station at Trinidad, California, a large sperm whale was captured which yielded 27 barrels of oil, presumably from the case and the adipose cushion combined. The size of the spermaceti organ and junk should be emphasized. In a large male the head constitutes two-fifths of the total length, so the cephalic fatty equipment must be at least one-third the length of the entire animal, or say 20 feet. It may be noted in this connection that in drawings of this whale the end of the snout is usually represented as truncated and ending even with the tip of the lower jaws, whereas all photographs show that the snout is bluntly and evenly rounded, and that in large males it projects for several feet beyond the lower jaws. Conditions are evidently relatively the same in the pygmy sperm whale *(Kogia)* except that the perfectly formed spermaceti organ is very small, and the Ziphiidae—notably *Hyperoodon*—are said to have some sort of fatty frontal organ, but precise descriptions are lacking.

We know nothing about the manner in which this great cephalic oil equipment began to develop except by inference. In other odontocetes the adipose cushion is situated in front of the blowhole, and the junk of the sperm whale is probably homologous with this. The present conformation of the nasal passages in the cachalot is to me strongly suggestive of the theory that the spermaceti organ is not homologous with any part of the adipose cushion of other toothed whales, but is a distinct development, originally having had its inception back of the blowhole, which latter was forced farther and farther forward as the organ increased in size. This belief, however, is not shared by H. C. Raven (MS) who is of the opinion that the spermaceti organ is the part that is homologous with the adipose cushion of other odontocetes, and that as it developed in the more primitive Physeteridae it forced itself rearward *between* the nasal passages.

Be that as it may, the markedly posterior position of the bony nares in the cachalot is suggestive, to me at least, that the blowhole of early physeterids had receded almost as far to the rear as we now find it in other odontocetes before the spermaceti organ had attained even moderate size. In other words, in the sperm whale ancestors there was likely a primary recession of the nares as in other odontocetes, but the spermaceti organ forced a secondary migration of the spiracle in a forward direction. But no fossil sufficently primitive to settle this point is yet known.

I think it is perfectly clear that this organ, developed to the proportions which it now assumes in the sperm whale, cannot be a fortuitous modification. No animal would ever acquire such a cumbersome contrivance which would be so great a handicap in swimming unless there were definite physiological need for it. It may be, and probably is, an overspecialization at the present time, its evolutional velocity out of control, but it undoubtedly began in response to some definite need which it now fulfills. It seems extremely unlikely that it has any primary function for the storage of a surplus of fat against periods of food scarcity, for the animal is abundantly supplied with blubber, or that it is used as an aid to the flotation of the head. But what the true function of this remarkable organ may be is entirely unknown.

Chapter Six

The Skull

SEVERAL volumes might be written descriptive of the skulls of the Cetacea alone, and indeed the literature on this subject is already so voluminous that it would be unjustifiable here to do more than mention the most salient points and discuss those details which are believed to have particular bearing on aquatic modifications.

Skulls of the Monotremata are so unique that there are but a few details in that of the platypus that we can be sure are largely the result of its aquatic habits. These, briefly, are the dorsal position of the orbits, and the form of the broad, bifurcated rostrum and mandible as the framework of the spatulate bill. No insectivore has any well marked cranial modifications for an aquatic life, and almost the same can be said for rodents. In this order it is popularly presumed that there is some tendency for aquatic sorts to exhibit shortening of the nasals, to allow for a more dorsal position of the nostrils, and flattening of the dorsal side of the skull to permit of more dorsal vision; but these characters should be evaluated with caution. Evidently a straight dorsum means little in itself. This part of the skull in such an essentially aquatic genus as the muskrat *(Ondatra)* is indubitably convex, but is straight in *Hydromys*. In the terrestrial wood rat subgenus *Teonoma* the interorbital region is equally straight or gently concave, however. In *Nilopegamys* it is also concave, and still more so in *Ichthyomys*, so it is probably justifiable to accept this tentatively as an aquatic character in rodents. In *Hippopotamus* the nasals are somewhat shortened and there is marked elevation of the bony eye sockets. The peculiarities of the skull of the tapir probably have nothing to do with its slightly aquatic habits.

The skulls of aquatic fissipeds and of pinnipeds show only slight aquatic adaptations in the details where one might expect them to be most marked. They have rostra that are perhaps shorter than in the majority of terrestrial fissipeds; but so have the cats and others. The same may be said about the recession of the anterior margin of the nasals. This is more pronounced in *Phoca* than *Zalophus*, but in relation to total length of skull it is no more marked than in the Canidae, for instance. The lachrymal bone has virtually disappeared in the pinnipeds, although it is occasionally present in Otariidae and possibly more rarely in the Phocidae, and is imperforate. Aquatic modification seems apparent in the interorbital region of otters and pinnipeds. The interorbital septum is thinner in the former than in most fissipeds, is 11 per cent of the skull length in *Zalophus* and may be as little as 3 per cent in *Phoca,* in which it is so reduced that the ethmoturbinals have forced their way to the surface of the bone. This reduction in the interorbital width is ostensibly for the purpose of permitting more dorsal vision, best accomplished by reducing this width so that the eyes may roll farther mediad and hence dorsad. The size of the orbits in the Phocidae are usually larger than in the Otariidae to accommodate the larger eyes of the former, and this has had a tendency to force the postorbital processes of the zygomatic arches farther caudad. The temporal fossae of the Otariidae indicate temporal muscles which are much more powerful than one would expect to find in a fish-eating mammal—an anatomical arrangement which is also the case with the masseter muscles—and these conditions are reflected in the rather robust mandible. The skull of the otter is also well muscled and that of some of the Pocidae, but the tendency in most of the latter is for a reduction in the size of the temporal fossae, which may not meet the sagittal line; and corresponding weakening of the mandible.

No trustworthy conclusions may be drawn anent the palatal and pterygoid regions, nor regarding the bullae, for these may or may not reflect conditions in terrestrial ancestors. Although no more marked than in many terrestrial forms, the occipital condyles in *Zalophus,* for instance, are rather narrow, for this animal has occasion to twist its head in all directions—movements which are facilitated by a circumscribed joint. In *Phoca* the condyles are somewhat wider, for the different method of swimming which it employs militates again great mobility of the head during progression, and broader condyles make a stronger joint.

It is in the occipital region that the pinniped skull shows the greatest results of the aquatic stimulus. In the usual fissiped, as in the otter, the occipital plane is either practically certical or else slopes rearward while in *Zalophus* the slope is definitely forward, and still more so in *Phoca.* The reason for this is discussed under the Cetacea. In *Zalophus* the lambdoidal crest bordering the occipital plane is strongly developed, for strong muscles (cephalohumeral, sternomastoid, trachelomastoid, rhomboideus anticus, splenius and semispinalis capitis) which, as a whole, extend uninterruptedly from the vertex to the mastoid process. In *Phoca* there is no crest formed in the lambdoid region, and the semi-

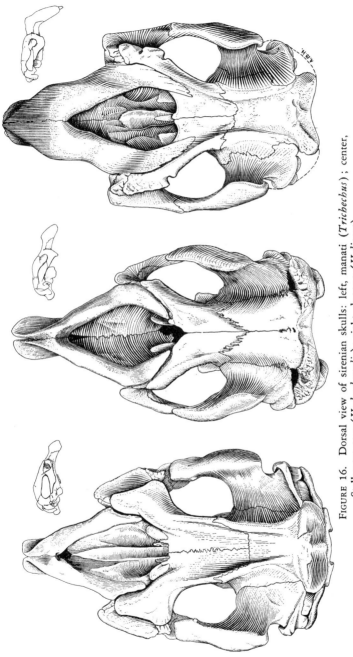

FIGURE 16. Dorsal view of sirenian skulls: left, manati (*Trichechus*); center, Steller sea cow (*Hydrodamalis*); right, dugong (*Halicore*).

spinalis capitis, rhomboideus anticus, humerotrapezius and cephalohumeral are confined to a narrow area near the vertex, while the sternomastoid, trachelomastoid, and splenius are confined narrowly to the mastoid process. Thus, in this animal, the most powerful muscles for motivation of the dorsal region are confined to two circumscribed areas, one for vertical movement of the occiput and the other for lateral movement, which is an excellent indication that *Phoca* has very much less need for such complex cranial actions as are indulged in by the sea-lions. As already indicated, because of the fundamentally different positions of the primary swimming apparatus in these two pinnipeds, the otariid can move its head in all manner of ways without disturbing equilibrium, or else this may be compensated for by movement of the hind feet; and furthermore, the head and neck, by swaying motions, are of decided use in the terrestrial locomotion of this animal. On the other hand, the true seals travel on land by vertical undulations of the body, in which the head and neck do not play an important part, and because the swimming organs are at the rear, the neck, and consequently the head, can be moved only in moderate degree during swimming, as discussed in succeeding chapters.

The only cranial modifications of the order Sirenia which may with certainty be ascribed to aquatic influences are the exceeding density of their bones, the recession of the nares, probably, and the slight forward tilting of the occipital plane, conforming to the usual posture of the head. In addition, adults of the extinct Steller sea cow *(Hydrodamalis)* were toothless, the place of the teeth being taken by horny oral ridges developed for masticating soft marine algae. The phenomenally long, depressed snout and mandible of the dugong is doubtless for the same purpose, as these support horny rugosities of the membrane which should be equally efficient for this purpose. Why bones of the Sirenia should be denser, harder and heavier than in other aquatic mammals is unknown, but presumably they have become so in response to a long continued need.

In Sirenia the recession of the bony nares is marked (fig. 16), and although a lengthy description is hardly necessary it may be mentioned that in this group the posterior narial border is well behind the orbits proper, and that anteriorly there is a great narial basin between the premaxillaries. But the external position of the nostrils is near the anterior angle of the snout, to allow for breathing at the surface with no other part of the head exposed, a position that is definitely forward of any part of the narial basining of the skull. So it is seen that in this group the recession of the bony nares has been far posterior to the position indicated for them by the external nostrils. Conditions are not comparable to those in the seals, for instance, but more like those in such mammals as the tapir and moose *(Alces)*, in which, in spite of the fact that the nostrils are situated upon a prolonged proboscis, the nasal bones have retreated far backward to allow for complicated musculature at the base of the proboscis to give it the mobility required. The narial basining of sirenians—chiefly its broadening—is likely traceable to the same stimulus, so that the muscles embracing the nasal passages may have broader, firmer attachment upon its margins. To me these facts are suggestive of the possibility that originally, in the sirenian ancestors, external evidence of complicated nasal conditions may have been more marked and that possibly they were equipped with a well developed proboscis or comparable narial equipment. Strength is added to this postulation not only by the fact that in the Eocene sirenian *Eosiren,* with very long rostrum, the nasal basining extended to the posterior part of the orbits, but also by the present sirenian equipment consisting of highly specialized and mobile "lateral lips" and by the almost universally accepted belief that they are of prohoscidean ancestry. The conclusion suggested is that mammals of proboscidean derivation are unusually amenable to the specialization of the narial and other facial muscles in the production of probosceal or comparable equipment and that the present result of this in the case of Sirenia is found in peculiar labial specialization, and in the muscular conditions at the base of the narial passages as indicated by the recession of nares and narial basining, this being in part a relic suggestive of former and more specialized (in one sense) nasal musculature.

According as the trend of specialization be in one way or another we might expect that eventually in Sirenia the nasal basining might either be reduced in size by the recession backward of the suture between the two premaxillae, and possibly by a narrowing as well, or else that it might become shallower by a reduction in height of its bony borders, and final broadening, the logical conclusion being a facial condition approximating in some respects that now occurring in the Odontoceti.

Small, thickened, ovoid nasal bones at the side of the middorsal line are present in *Trichechus*; there is no place for any such bones in *Halicore,* and *Hydrodamalis* seems to occupy a middle position in this respect. It is not known for a certainty that they were lacking in the latter genus, for they would naturally have fallen from the weathered skulls that are now available, and this cannot be settled by an examination of the specimens. Some have considered that nasal bones were present but this I doubt. On the whole it is difficult to judge in which of the three sirenian genera the posterior border of the nares lies farthest caudad, as the conformation of the details of this region differs considerably, but it seems to be most pronounced in *Halicore.* We know that in bodily form *Hydrodamalis* and *Halicore* are more perfected for aquatic locomotion than *Trichechus,* and from the skull it seems that *Halicore* is slightly more modified aquatically than *Hydrodamalis.* In all three the orbit proper is comparatively small, following reduction in the size of the eyeball. The temporal fossae indicate that these muscles are quite well developed. Deserving of passing mention is the peculiar hypertrophy in *Trichechus* of the zygomatic processes of the squamosals.

Among the Cetacea the skull of the sperm whale departs most widely from that of the usual terrestrial type than that of any mammal living. But in some respects it occupies an intermediate position between that of the mysticetes and of other toothed whales. A better idea of the cetacean skull can be obtained from the accompanying illustrations than from a description and only some of the points of most importance in the present connection will be discussed.

On the whole the bones of Cetacea are very spongy, and oily, and this is more marked in the Mysticeti. Perhaps the latter circumstances is in compensation for the fact that this group is not so prone to develop particular fatty areas upon the head, nor cavities and ducts for the accommodation of oil between certain muscles and beneath the blubber layer. Winge (1921) has stated that "the effect of water pressure is to develop unusual strength in those rostral bones which project farthest forward", and many others have been of the same opinion. I cannot subscribe to this belief. Of course the bones are heavy, as they must be in large mammals, but I cannot see that they are relatively as heavy, or rather that they form as strong a complex, as in land mammals.

The rostrum of the usual land mammal is affected by several strong, fundamental stimuli; as a support for a nose which must be well developed as an apparatus for taking in air and odors, as a framework for mystaceal pads with vibrissae that may be highly tactile and for the anterior facial musculature, as a housing for intricate turbinal bones, and as a support for dental armature. In addition certain mechanical elements are introduced following the fact that the lower rostrum supports the arch of the nasal bones covering the nasal passages. The Cetacea are the only mammals in which none of these stimuli is present, save very simple and usually weak teeth. Small wonder that in this order the rostrum departs widely in form from the average mammalian type.

[112]

Unless some complication be introduced there is fundamentally little need for a rostrum of more mass than the mandible, and this is the case in *Platanista*. Unless there be some antagonistic influence in operation the tendency should be for the cetacean rostrum to diminish in width and perhaps increase in length to a reasonable extent, thus acting as a cut-water. This it has done in some instances, but whether for this reason or because of food habits is of course unknown. In *Platanista* the two sides of the maxillary dental arch are so close together that they partly blend. The extinct *Zarhachis* had a rostrum five times as long as the cranium proper, and in *Eurhinodelphis bossi* the rostrum constituted nine-elevenths of the total skull length (Kellogg, 1928), although in some specimens of the last the mandible was considerably shorter. Theoretically a long, tweezer-like beak would be best for capturing small active fish which dart about in schools, but it is difficult to see how a beak of such length as that of *Eurhinodelphis* could be effectively used in conjunction with the limitations of a short cetacean neck; and it doubtless constituted an overspecialization. Because there have surely been many complications, however, we find rostra of all shapes, down to exceedingly short, broad ones such as in *Globiocephala* in which the stubby mandible, overhung by a rostral frontal bulge, appears ideal for scooping up relatively inactive, bottom-living food; and yet these whales are said to feed on cephalopods.

The rostrum of a porpoise with moderate beak, such as *Tursiops*, composed of the maxillaries and premaxillaries, is relatively thin in vertical dimension, and certainly not as strong as the average land mammal with equal rostral length would require. The killer whales, with their carnivorous habits, need a stronger rostrum and have it. But the rostrum of *Physeter* is certainly not as massive as one would expect to see as the support of its huge frontal fatty equipment, and the rostrum of the balaenid whales appears very fragile as a scaffold for their remarkable baleen armature. Mechanically the attachment of the Mysticeti rostrum to the rest of the skull is very weak, for the principal sutures are largely in one plane where they could all be acted upon by a single oblique force. But it has been entirely adequate for the needs of the animal and that is all that is requisite. On the other hand, the attachment of the *base* of the odontocete rostrum to the cranium is mechanically of exceeding efficiency. It is broadly distributed in laminated, squamous sutures in a manner that is stronger than could be accomplished by dentate sutures of the usual sort. And this is not surprising, for the stress which the cetacean rostrum must undergo is applied almost exclusively in a posterior direction. Hence the force of the water pushing against the tip is distributed over the anterior part of the skull segment next caudad. Even in the baleen whales is this largely true, for when the mouth is open and the baleen is subjected to water force the animal is presumably always swimming at low speed.

If we examine the stimuli experienced by other parts of the head we find that the eyes are reduced and unusually placed, the function of the ear has altered, the masticatory musculature is reduced in odontocetes and altered in all, the cranium proper must, with the base of the rostrum, act as an anchor for the narial musculature, and the need for supporting the head has been simplified by flotation. Hence almost all of these are different from what the usual land mammal meets.

In zeuglodonts the bones of the skull retain essentially the normal mammalian relationship, although by a recession of the nasals the bony nares have become elevated from one-half to two-thirds the distance to the orbit (Kellogg, 1928). But in some other respects, as reduction of the hind limb, these mammals at the time of their extinction were almost as highly aquatic as most existing whales. Hence it is evident that they were either inherently lacking in the ability to respond to the stimuli that have resulted in the telescoping of the cetacean skull, which, under the circumstances, appears unlikely, or else that they did not experience those stimuli that were most critical in bringing about telescoping. What these may have been will be discussed in the case of existing Cetacea, but it may here be mentioned that the mere fact that the gradual recession in zeuglodonts of the anterior border of the nasals to a position from one-half to two-thirds the distance to the orbits does not necessarily mean that the external nostrils must have shifted to this position. This has been accepted without question, it seems, but the bony nares lie far to the rear of the nostrils in sirenians, and the same may have been true in zeuglodonts. Not only that but the recession of the nasals may, as in the moose, indicate that they had a small proboscis, and so the nostrils may actually have been situated anterior to the tip of the rostrum. If this were really the case, then the zeuglodent skull lacked all those stimuli which I now believe to have been at all important in bringing about the condition of telescoping, except that of backward water pressure against the head.

As Miller (1923) says, in "modern Cetacea the most conspicuous facts are these: (a) That the telescoping of the skull was far advanced in the earliest known extinct genera, and (b) that this process has developed according to two different plans". Briefly, in odontocetes the telescoping, or the sliding of some bones over or under others so as to result in bony laminations and shortening of some of the cranial elements, has clearly been from before in a backward direction against the cranium. The maxillaries and premaxillaries have spread backward and to the side so as to override the frontals, the parietals have been crowded far to the side and the interparietal eliminated. In some forms (*Kogia*) this is so marked that the maxillaries or premaxillaries may actually meet the supraoccipital. There is no tendency for the elimination of the lachrymal as in the pinnipeds, but like the Sirenia this bone becomes thickened and more massive. In practically all living forms the mesethmoids and ectethmoids fuse into a flat bony plate to form the posterior wall of the nares and the nasals are reduced to thickened ossicles above these. The rounded occipital plane has a marked forward inclination, but this is most moderate when compared to conditions in the Mysticeti. Other peculiarities often exist, as the fact that in *Platanista* the palatines meet the maxillaries and the latter have a remarkable development in broad extensions stretching upward, while in adult males of *Hyperoodon* there is a comparable development of the premaxillaries. It is not unlikely that these bony extensions are either cause or effect of unusual specialization of the blowhole musculature attached to them. In life the bony passages are approximately vertical, although the angle, from below upward, which they describe with the long axis of the skull is slightly more than a right angle. Thus the nares have migrated as far to the rear as the braincase will allow, and farther, I may add, than would be possible without the recession of the olfactory lobes of the brain. One bony narial aperture is usually slightly larger than the other and this varies individually. In all odontocetes without exception, however, the right half of the facial region of the skull is larger than the left. In other words, the measurement from between the nares to the lateral border of the frontal is always larger upon the right side than the left. And the Odontoceti are the only mammals either living or extinct in which this cranial asymmetry is the normal condition. In some forms it is much less marked than in others, while in the Physeteridae, because of accompanying asymmetry of the bony nares, it is carried to an extreme. All manner of theories, most of them somewhat fanciful, have been advanced to account for it. Abel claimed, without any convincing argument, that it was due to the atrophy of the nasal bones and shortening of the braincase. Kükenthal believed that a sculling motion of the tail tended to turn the animal to the left, resulting in a thickening of the cranial bones upon that side and a consequent broadening of the right

l

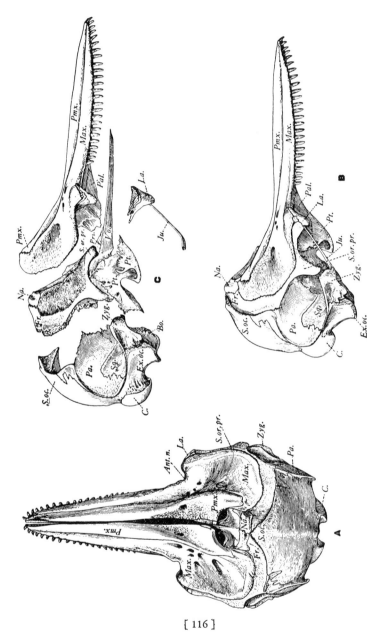

FIGURE 17. Skull of the porpoise *Tursiops truncatus*, after Kellogg.

side. Lillie (1910) inferred that the laryngeal asymmetry in the Physeteridae caused the asymmetry of their skulls, but failed to indicate how.
I attempted to discover whether there was not a twisting action of the
facial muscles in closing the blowhole so that those of the right side developed more strongly than those upon the left, but the results were
entirely negative. Nor has Ernst Huber found any differences in the
facial muscles of the two sides that might contribute to the condition.
In short we know absolutely nothing regarding the cause of this unique
cranial asymmetry and no logical theory for it has yet been advanced.

In regard to the Physeteridae or sperm whales, Kellogg (1928) says
it is confidently believed that "generalized sperm whales had been differentiated from the main odontocete stock subsequent to the elimination
of the postorbital constriction, but at a time long before the beginning
of the Miocene". In the cachalot (*Physeter*) the telescoping has also
been mainly from before backward, as in other odontocetes, but there
has been a basining of the whole facial region, this extending, for the
accommodation of the huge spermaceti organ, well back of the bony
nares to the cranial crest, which is vertical to the condyles. The maxillaries reach this at some points, over-riding the frontals, but while the
left nasal passage is huge, the right is but a fraction of its size. In consequence, extraordinary osteological peculiarities have resulted. While the
right premaxilla almost reaches the cranial crest, the left is much shorter,
stopping at the blowhole (fig. 18). Upon the left, the flattened, expanded nasal takes the place, posterior to the nares, of the position occupied upon the right by the backward extension of the premaxillary,
and the right nasal bone has been eliminated.

In the case of the odontocete mandible the condyle is reduced and the
articulation peculiar in the spreading of the ligaments, but this is more
marked in mysticetes. There is a broad, flaring aperture to the dental
canal, opening to the rear and partly filled with fatty tissue. The anterior mandible is usually relatively weak in accordance with reduction
in tooth size, but in ziphioids having well developed tusks it may be
larger and of peculiar form. There is a true symphysis menti, which in
most forms is small in area and lacks strength, but especially in some
of the long-beaked, extinct forms, was large. Thus in *Argyrocetes*,
Kellogg (1928) said that the mandible had a length of about 35 inches,
three-fifths of which was symphysis. Why the mandible of such forms
as *Eurhinodelphis* stopped far short of the rostral tip is unknown. In
some ziphioids and mysticetes the mandibular tip projects beyond the
rostrum.

[117]

Cranial conditions in the Mysticeti are more difficult to describe and the reader should turn to the illustration of *Balaenoptera* (fig. 19). It is seen in this skull that although the premaxillae and maxillae are long and of very specialized shape, they really do not extend farther backward, using the position of the eye as a criterion, than in the normal mammal. As in odontocetes it is the central elements of the skull—frontals and parietals—that have suffered crowding, but in a different manner. In both does this section appear to have been squeezed between the rostrum and occipital, but whereas in the odontocetes the occipital seems to have remained more or less stationary while the base of the rostrum has been forced caudad, in mysticetes the appearance is that the rostrum maintained its position while the occipital did the pushing, over-riding the parietals and overhanging the inferior part of the frontals, which are expanded to form the floor of the temporal fossae in an extraordinary manner. Beneath this is situated the eye. The relative weakness of the rostrum should again be emphasized. Reference to the illustration (fig. 19) shows that envisioning the mysticete skull as though it were but a few inches in length one would never suspect that it was anything more than a purely edentulous mammal whose rostrum was not subjected to many vicissitudes of strain—certainly not that it acted as a support for an enormous armature of baleen (especially is this the case in the balaenid whales). But like other parts of the animal the weight of the baleen is largely counteracted by the flotation of the water and the lack of much resistance applied to it, such as is experienced by the grinding of an upper dental armature against a lower, save in a fore-and-aft direction by water pressure, allows for the relatively moderate strength of the whole rostrum.

The nasal bones of mysticetes are much reduced but are otherwise approximately normal. As usual in terrestrial mammals they roof over the nasal cavities, which are more gently sloping, and not vertical as in odontocetes. As in odontocetes the interparietal has been eliminated in mysticete adults, although Ridewood (1922) found it in small fetal balaenopterids, but not in those of *Sibbaldus* or *Megaptera*. It was also present in the *Balaenoptera borealis* dissected by Schulte (1916). The palatine is excluded from the wall of the nasal cavity. The skull is symmetrical, although Kükenthal, with characteristic persistence in his theories, attempted to prove asymmetry. He did show fractional differences, naturally, for it is rarely that any mammal skull has the two sides precisely alike, but these were not sufficient to be designated as asymmetry.

The mysticete mandible is profoundly modified by the necessity for its enlargement and bowing to accommodate the baleen equipment. The posterior processes are much reduced and the articulation with the skull altered in that the articular ligaments are greatly spread so as to allow a phenomonal amount of movement. The latter is further facilitated by the lack of a true symphysis menti, the connection of the mandibular tips being highly elastic; so that the mandible is perhaps capable of a greater variety and degree of twisting, distortional movements than that of any other mammal. This is doubtless of economic value in the securing of food, and to facilitate the replacement of the baleen tips in the eventuality that these become extruded from the mouth in some manner.

The development of the muscles of mastication in the Cetacea deserves brief mention. In the extinct squalodonts, with heavy teeth clearly fitted for actively predaceous uses, the temporal muscles were quite well developed, if we compare their fossae with the relative size of the braincase. In most living odontocetes, however, these fossae are much smaller; but it is interesting to note that they still occupy the area from zygomatic process of the squamosal to the frontal border, as is usual in terrestrial carnivores. In other words, the reduction in size of the temporal muscles has kept pace with lateral displacement of the parietals (in *Tursiops* for instance). This may or may not have any real significance. It is well known (see, for instance, Anthony, 1903), however, that mammals of small brain capacity are very apt to have powerful temporals reaching the sagittal crest, which may be very high, and that as the brain becomes larger and more highly specialized, the tendency will be for a definite reduction in size of the temporal fossae. The latter condition is probably not primarily a result of the increase in brain size, but is a reflection of the fact that to all intents a mammal to which great power of the dental armature (to which condition large temporal muscles are necessary) and consequent development of the anterior face, is critically requisite, is hindered in the development at the same time of a large brain. Again, this may or may not be of significance in the case of odontocetes.

Following the reduction in size and masticatory importance of the teeth the masseter muscles of at least some odontocetes are decadent. In fact they may not have any definite attachment to the skull at all (*Neomeris*) and but a very slight one to the mandible. But they were probably much better developed in the extinct porpoises with phenomenally long beaks.

The huge mandibles of mysticetes, being supported to a considerable degree by flotation, do not need a great mass of muscle for simply opening and closing the mouth, but with mandibles of such mass there must be a large supply of reserve muscular power, else a chance encounter with a companion, backed by all the momentum of its huge body, would have disastrous results in the shape of dislocation or fracture. Hence we would expect to find the temporal muscles of considerable size. Of course they are not as powerful relatively as in a terrestrial carnivore which is used to rending sinew and crushing bone, but they are, nevertheless, well developed. At present in balaenopterids they extend far forward of the level of the eye and quite to the base of the rostrum, this being permitted by the position of the eye beneath the intervening supraorbital plate of the frontal, above which the temporal muscle lies. Kellogg's illustrations (1928) indicate that there has been progressive forward extension of the temporal fossae since the Oligocene. In *Patriocetus* of this period the over-riding of the frontals by the temporal muscles is not apparent or has only just begun; in Miocene *Cetotherium* the fossae do not quite include the entire supraorbital part of the frontals; while in living adult *Balaenoptera* they apparently extend beyond and onto the maxillae, although in the fetal state they are naturally more restricted.

There now remains for discussion the question of the telescoping of the cetacean skull. This telescoping may be divided into two categories: the sliding of one bone over another, and the crowding by the two terminal elements of the skull upon intervening elements so that the latter are reduced or eliminated. The latter may well be but a result of the former, for presumably the central elements have been mechanically the weaker and have been largely overcome by a stronger force applied from before or behind.

Miller (1923) is the latest to have considered at length this process of telescoping. He unqualifiedly subscribes to the belief that it has resulted from the backward push of the water against the head as the animal moved forward, combined with the forward push supplied by the moving body. In other words, that the head has been squeezed between the compressed wall of water against the snout at one end and the neck at the other. To account for conditions in odontocetes he considers that these two stresses have been relatively uncomplicated and that the base of the rostrum has overspread the braincase for the reason that originally the sutures in this region were of the squamous type, such as now occur in the fox, and hence would more easily over-ride the frontals than could the occipital, this presumably having had a dentate type of suture.

He considers that the simpler type of water-and-body force acting upon the odontocete skull was complicated in mysticetes by the downward pull exercised by the enlargement of the head for baleen armature, the downward pull offered by the latter when the mouth is wide open for feeding, and the upward pull at the back of the skull needed to counteract these forces. He considers that the telescoping in this group has taken the form largely of an overthrust of the occipital element from behind forward chiefly because, like the present sea-lion, the occipital suture of the mysticete stock was of the squamous type, over-riding the braincase with more ease than could the maxillary-premaxillary element which presumably had a dentate type of suture with the frontal.

In final analysis this theory of Miller's rests solely on the premise that the only stimulus for the telescoping of the skull in whales of all types has been the backward, or at times, oblique, pressure of the water through which the animal moves. He advances as reasons for lack of telescoping in the skulls of other aquatic mammals the facts that their heads are relatively smaller, offering less resistance to the water, they are not held so stiffly or uniformly pressed into the resisting element, and that they all swim at lower speed. In explanation of the importance of the last contention it may be mentioned that the resistance offered by water to a body moving through it increases as the square of the velocity. Hence the pressure against the head of a porpoise swimming at 30 miles per hour would be 36 times as great as against a sea cow (did it have a head of the same shape and size) moving at five miles per hour.

There is certainly a definite stimulus supplied by strong, backward water pressure which we would expect to find reflected in cranial details. It is perhaps certain that no mammal that had not been accustomed to moving through the water with speed for long ages could exhibit telescoping of the skull in a form as perfected as we now find it in odontocetes. We may therefore put this down as an essential stimulus. But I am very loath to believe that this was of sole importance in producing telescoping. We will therefore proceed to hunt for stimuli in this direction which may also have been essential to the attainment of this condition, although perhaps in themselves of insufficient strength to have produced it without the aid of water resistance.

It is clear that there is no need of great strength at the tip of the cetacean rostrum. At least they get along very well without it, and its attenuate form is permitted by the elimination of strong front teeth, the mystacial pad with vibrissae, functional facial musculature in this region, and the recession of the nares and narial musculature. We have no means of knowing just what effect these alterations alone would have upon the skull—what proportion of present conditions in Cetacea should be ascribed to them and what to water resistance—for in no other mammal have they all operated; but it is evident that they would have some definite effect. At least this has been followed by some atrophy of the rostral tip in vertical dimension, and it is not unreasonable to look for a compensational hypertrophy in these elements at the rostral base. In mysticetes this is found to some extent in the broadening of this part of the rostrum. But in *Patriocetus* of the Oligocene both the widening of the rostral base and the attenuation of its tip is more marked than in any living mysticete, showing that both of these details have experienced secondary modification through requirements of the baleen equipment as this developed. In odontocetes this rostral change has taken the form both of the broadening of the rostral base and the spreading of the maxillae over the frontals. This as a lone stimulus would probably not be of very great importance to telescoping, but it must be considered as contributing in some degree to the total of stimuli for skull change.

The odontocete skull is so formed that both the rostral tip and the forehead receive the backward pressure of the water, while in mysticetes the forehead does not receive it. The concomitant condition of the telescoping of the rostral elements in the former group is not believed to be purely fortuitous. At least it is presumed that the water pressure at this critical point would be largely instrumental in strengthening the anterior wall of the braincase. Admittedly this might be accomplished as effectively by a mere thickening of the bones, but the fact remains that the end has been attained by a process of laminations. The theoretical effect in this connection of the presence of a frontal adipose cushion is of course unknown, but it would naturally have some sort of stimulus, probably in broadening the face. These stimuli are absent in mysticetes and there has been no lamination of rostral elements over the frontals.

In skulls of toothed wales (excepting present conditions in the Physeteridae and the Kogidae) the nasal passages have migrated just as far to the rear as the anterior wall of the braincase has allowed. With them has shifted a complicated muscular mechanism for the control of the blowhole. This is no system of small muscular slips, weakly attached to the bone, as in many mammals, but a robust complex, firmly anchored to the skull over the entire breadth of the forehead and converging to the blowhole and a part of the adipose cushion. These have dragged with them the main trunk of the facial nerve, which now turns abruptly backward over a groove in the maxilla and lachrymal at the base of the rostrum and so to the dorsal surface. A like condition obtains in the case of the usual anterior opening of the infraorbital nerve of the maxillary branch of the trigeminal, the infraorbital foramen having been forced first dorsad and then caudad, where the nerve now emerges in separate branches from several foramina, which may be located directly above the eye (*Tursiops*) and even within the bony narial orifice. Whether or not this actually be the case it appears superficially very much as though the posterior extension of the premaxillaries and the expansion caudad and laterad of the maxillae had been stimulated to this action by the migration in these directions of the facial muscles which are so firmly attached to these bones;—that the attachments of these muscles had not shifted from one bone to another but had dragged the bones with them, just as anterior migration of the occipital muscles alters conformation of the occiput. And that the more circumscribed area which these narial muscles needed in attachment to the frontals had permitted all but a narrow border of these to be covered by the maxillae. At any rate this excessive migration of the nostrils has had a profoundly stimulating effect upon osseous conformation. This is beyond question.

The mysticetes indicate an entirely different state of affairs. There is no backward pressure of water against the anterior wall of the brain-case, save as applied through the rostral tip, and there has been no occasion to build up a thickened or laminated bony wall for its proper resistance, nor is there an adipose cushion in this region. Unlike odontocetes the baleen equipment has retained considerable growth-stimulus in the rostrum proper, and the nostrils and their musculature have not been forced as far backward against the braincase as they could go. On the contrary the bony nares are more nearly comparable to the usual mammalian condition. And most important of all, there has not been such a pronounced migration backward of the narial musculature, and the requirements in this respect are not such as to necessitate any great broadening out of bony elements to accommodate a pyramid of muscles as broad as any part of the skull, for in mysticetes the intertemporal region is quite narrow.

Kellogg (1928) considers that the mysticete maxillae cannot overspread the braincase because the posterior part straddles the supraorbital processes of the frontals, and he presumably accepts this as a fundamental reason why these whales do not exhibit more telescoping at the base of the rostrum. I doubt whether this is a definite hindrance. If the stimulus in other directions were of sufficient strength at least the superior part of the maxillary could extend farther to the rear. Nor do I consider that the possible presence of a dentate type of suture between the maxillae and the frontals of the mysticete ancestor would be a permanent inhibitor of marked telescoping in this region, as contended by Miller. Sutural changes cannot be instigated in the adult skull, in which the attachments of all bones are secure, but begin in the younger animal whose sutures are not securely locked; and the individual time of sutural closure is notoriously variable. As far as present knowledge goes the cranial sutures will change according to the needs of the animal and it seems

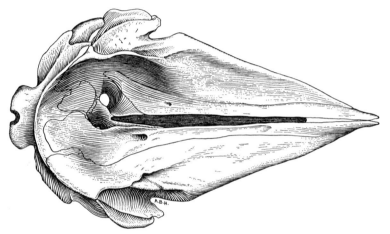

FIGURE 18. Dorsal view of the skull of the cachalot or sperm whale, *Physeter*.

unlikely that they could have been so conservative and tenacious of type in the cetacean ancestors as to predetermine the whole course of cranial development in the two groups.

There is little or no sliding of one bone over the other in the posterior elements (occipital, parietal and squamosal bones) of the odontocete skull, and no details that may not be understood rather readily. In the existing sperm whale the occipital plane is vertical, which gives the appearance that all force has been from before toward the rear, but this is probably a secondary result, for in the Miocene *Diaphorocetus*, and doubtless other early physeterids, the facial basin appears to have been not quite so well defined and there was a marked forward inclination of the occipital. Some of the more peculiar types of ziphioid whales also have an occipital that is almost vertical: but in odontocetes of the porpoise sort there is marked forward inclination of the occipital plane (discussed later). This is definitely rounded and the supero-medial part extends to meet the facial plane. This crowding has forced the parietals far laterad. The restricted temporal fossae project backward to indent the occipital expanse, and the lower part of the exoccipitals and lateral part of the basioccipital are projected into marked processes. The bony border resulting is not continuous but has a deep fissure for the jugular vein and nerves IX, X, and XI (glossopharyngeal, vagus and spinal accessory), with the condyloid foramen (for the hypoglossal nerve) upon its border. Upon this falcate border of the basi- and exoccipitals is inserted the great dorsal scalene muscle, which is probably for the purpose of providing static strength in the same way that the rectus capitis superior is so used upon the opposite side of the neck, but it doubtless also assists in swimming motions. At any rate this falcate border is not present in Mysticeti, the downward motions of whose heads are (presumably) largely governed by the specialized gular musculature, which has leverage clear from the tip of the mandible.

The occipital condyles of the Cetacea are massive but are not any broader in relation to the width of the skull than in many terrestrial carnivores.

On the whole the cetacean brain is of a very high type—quite extraordinarily so when we consider that as far as can be judged members of this order have less need for such equipment than almost any other mammal. It is perfectly apparent upon inspection of skulls that the brain capacity of an ordinary porpoise such as *Tursiops* is relatively considerably greater than in a large baleen whale. But this may be merely in conformity with Haller's law, which is to the effect that in related animals the relative size of the brain is greater in the smaller sorts, and that endocranial capacity decreases as size increases. Dubois (1924) has stated that volume of brain in such comparable material ("species of equal organization") increases as the 5/9 power of the body weight, and that in mysticetes this cephalic index (not actual capacity of the brain case) is only one-third of what it is in porpoises. It is therefore difficult to determine whether there is actually a difference in the endocranial capacity between equal-sized representatives of the two cetacean groups, for the skulls of the larger toothed whales are curiously distorted in some respects and the smallest mysticetes are depauperate and not representative of the group. At any rate, regardless of this point, the brain capacity of a porpoise is relatively much larger than of a finback whale. In the former at least the brain shows evidence of having been compressed in an antero-posterior direction and is broader than long, this having been caused evidently by the recession of the anterior cranial elements against the posterior ones; and one result of this great brain size is the curvature of the occipital plane. In effect, then, the stimulus for a large brain in the porpoise has been greater than any muscular stimulus which might have been productive of a flat, ridged or crested occipital for the accommodation of unusually powerful muscles, and greater than this endocranial stimulus in large baleen whales. In the sperm whale, however, the osteological crowding that has resulted from the frontal basining for the accommodation of the spermaceti organ has forced the cerebral cavity to occupy a restricted space at the apex of the angle formed by the vertical part of the occipital elevation and the horizontal rostral portion. I judge that the size of the brain in an adult *Physeter* is not at the most over four or five times as great as in the porpoise *Tursiops,* and the foramen magnum passes downward (in an anterior direction) at an angle of more than 45 degrees.

Weber (1904) has stated that the cetacean supraoccipital is paired, just as it is in *Tatusia* (Edentata) and *Erinaceus* (Insectivora).

The occipital conditions in the Mysticeti are very different indeed from those in the Odontoceti. Tilting of this element in the former group is excessive and the forward inclination of the occipital plane more accentuated than in any other mammal. In whales of the balaenopterid type this tendency has forced the parietals forward so that considerable portions of them surmount the frontals, and the occipital covers all but the most anterior part of the parietals. As is frequently remarked, in whales of this type (including *Sibbaldus*) parts of the nasals, premaxillae, maxillae, parietals and frontals occur in one transverse plane. Among living mysticetes the large balaenid whales (*Balaena* and *Eubalaena*) usually are said to exhibit the most pronounced tilting of the occipital, but this statement needs qualification, as explained elsewhere.

It is difficult to decide whether the occipital overthrust in mysticetes is now actually on the increase or wane. Kellogg (1928) has remarked that a greater overthrust occurred in one of the Pliocene balaenopterids than in any whale since. He is also of the opinion that in mysticetes the forward overthrust of the occipital elements has not had precedence over the backward interdigitation of median rostral elements, for in Miocene cetotheres there was less occipital tilting and more interdigitation of rostral with cranial elements. But these details hardly should

[126]

be compared in this way for the stimuli involved were doubtless very different. Observation tells us that at least the stimulus for occipital tilting in modernized mysticetes has been one of the strongest experienced by the skull for the reason that it is more pronounced than in any other mammal.

The difficulty experienced in explaining telescoping of the odontocete facial elements because no other mammal exhibits an intermediate condition in this respect is not so applicable in the case of the telescoping shown by the mysticete occipital. In the first place the latter is of a simpler, less astonishing sort, with the lamination of the bones not marked. In fact it is just about what one would expect to see result from an excessive forward tipping of the occipital plane in any type of mammal. Several other mammals show this to a modified degree, and several exhibit incipient telescoping of the posterior cranial elements. Among terrestrial forms occipital tilting is perhaps most marked in the European blind rat (*Spalax*), but in all rodents sufficiently specialized for a fossorial life as to be blind or nearly blind (Spalacidae, Rhizomyidae and Bathyergidae) this character is very pronounced, while in burrowing mammals not quite so strongly modified (fossorial octodonts, murids and geomyds) the occipital change is but moderate or slight. It is also a character of some few ungulates.

That part of telescoping which is exemplified by the broadening of the surface of contact between two bones is shared by *Spalax* also, and by most if not all of the more specialized fossorial rodents, this consisting in a broadening of the squamate sutural contact of occipital with parietal. In varying extent it is found also in the pig tribe, which roots in the ground, in *Arctonyx* the hog badger, which does the same thing, and in the fur seal (*Callorhinus*), but not the hair seal (*Phoca*). In all of these as well as in the Cetacea the head frequently or almost continually experiences a posteriorly directed force applied against the muzzle. Water presses against the head of the aquatic sorts as they swim, and the remainder push against the ground, rooting, tamping the earth of excavated runways with the nose (*Spalax*) or using the incisors as a pick to loosen the earth of their burrows after the manner of many fossorial rodents. Mechanically it would be very easy for the forward tilting of the occipital to bring about a broadening of occipital-parietal suture. Alone it might not have this effect, but the condition may well have been helped by the backward pressure against the rostrum experienced by those mammals which exhibit this tilting to best advantage. That forward tilting of the occipital is not a requisite of marked broadening

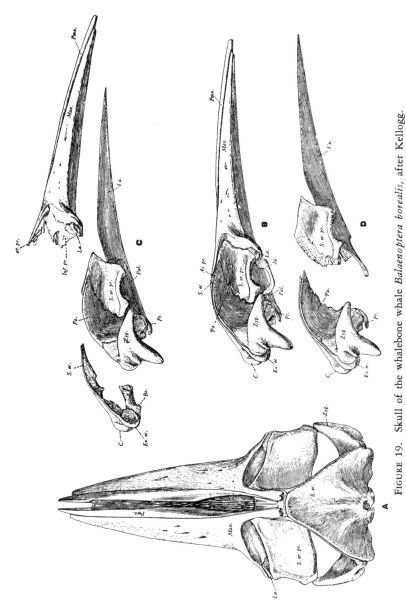

FIGURE 19. Skull of the whalebone whale *Balaenoptera borealis*, after Kellogg.

of the occipital-parietal contact is shown by the fact that the latter situation obtains in pigs, in which the occipital plane is tilted quite far *backward*. And that no stimulus of such sort is absolutely necessary is shown by the fact that broadening of sutural contracts between parietal and squamosal is well developed in at least some of the Kangaroos (*Macropus*) and very likely some other terrestrial mammals. The inference is therefore that such telescoping of the posterior cranial elements as results purely in the extension of one bone over the other so as to broaden the line of contact may be either a character of no consequence (as far as we can see) other than possibly a phylogenetic one: may be largely caused by a forward tilting of the occipital, backward pressure frequently applied to the skull, or both.

Invariably, I believe, it has been stated that the rostral tilting of the occipital plane of the mysticeti, and inferentially of fossorial rodents, is for the purpose of muscular strength, at which point the topic is abandoned without qualification or explanation regarding the quality of strength that is meant. Such strength is of five sorts, or is indicated in five separate ways, which fundamentally affect the conformation of the occipital shield by reason of the fact that the supraoccipital muscles seem utterly incapable of migrating from one bone to another, so that when they shift the supraoccipital must shift with them. In order properly to understand the cetacean conditions in this region of the skull it will be necessary to scrutinize the situation in certain pertinent terrestrial mammals, under five separate headings as follows:

(*A*) There is the strength exemplified in the occipital conditions of the lion or pig. The former must have powerful neck muscles of a sort to enable the head to be strongly twisted in all directions in order that it may, for instance, break the neck of a zebra. The pig roots in ground that is often hard, can do this all day long, and can lift a truly astonishing weight with its snout. This sort of *active muscular strength* is accompanied by high and strongly-defined lambdoidal ridging along the supraoccipital border.

(*B*) A different situation is encountered in mammals which we know must have great strength of occipital musculature for the support of a heavy equipment of horns or tusks, but which have no particular need for cephalic activity, embodying strong twisting motions. Although not occurring in its perfected, simplest form, this sort of *passive muscular strength* is illustrated by the moose and others of the deer tribe with large antlers, large-horned bovines, and old male elephants. The bony indications of this character of muscular strength do not take the form of high ridging, but on the contrary the occipital border is unridged and superficial examination would give the impression of definite muscular weakness in this area. The muscles need not be thick, or "powerful" in the usual sense, but they are either shorter or the muscular fibers less contractible, and the muscle must be tougher. This latter is best accomplished by the presence of numerous tendinous fasciculi or bundles, and in the perfected state of this specialization, which no mammal has yet attained, these occipital muscles would be nothing but tendinous sheets, practically lacking muscle fibers, and resulting in the incapacity for muscular movement of the head in a dorsal direction.

Certain modifications of either *A* or *B,* or a combination of both, may be effected mechanically by body conformation (a longer or shorter neck, or size of head), or by the provision, for instance, that has been developed by some ungulates in quite perfected form. This may be well demonstrated in a small antelope, which has need while grazing to lift the head perhaps every half minute to watch for enemies. The great muscular energy that would otherwise be expended in this action during the course of a day is conserved by the presence of a highly elastic nuchal ligament extending middorsally to the head. If one props up a dead individual in standing attitude, presses the head to the ground and then releases it, the spring of the nuchal ligament will either automatically raise the head or almost do so. This provision thus economically takes the place of much more powerful cephalic muscles that would otherwise be wasteful: but no aquatic mammal has this equipment.

(*C*)   Length of leg in a land mammal largely influences length of neck (partially excepting browsers). This, correlated with posture induced by life habits, to some extent, determines the normal position for carriage of the head. The giraffe, with its long neck, carries its head with axis at a right angle to the neck, while many bovines carry the cephalic axis practically in a line with those of the neck and body. This naturally has an effect upon the occipital musculature. The pig, while rooting, begins to apply the muscular force necessary when the cephalic axis is almost at a right angle to the neck. If the occipital plane were tilted forward or even vertical to the skull axis the occipital muscles would have no effective lever arm on which to work. Therefore it is mechanically necessary for the slope of the occcipital plane to be toward the rear, which it is to a very marked extent. The projection of the lambdoidal crest but accentuates this feature. Surely the head and neck of a horse is not as strong as of a large bull, but the horse has an occipital plane caudally tilted and prolonged into a large crest, and so exhibits much greater "strength", from the usual but erroneous standpoint, than does the bull. The lambdoidal crest of the horse is largely due to the mobility of the head, and the backward inclination of the occiput to the fact that the head is usually held at a sharp angle with the neck. Even more strikingly illustrative of these points is the condition in man and other primates which hold the body erect for most of the time. Hence it is a law that in those mammals which habitually hold the axis of the head at near a right angle to that of the neck, the occipital plane is found to slope backward, to a greater or lesser degree according to the strength of possible antagonistic stimuli.

(D)   Among the Bovidae conditions opposite to those mentioned above are perhaps best exemplified in the water buffalo. In observing a living specimen one is immediately struck by the fact that it carries the axis of the head in almost a straight line with that of the neck, and that it does so almost continuously. We know that it has far more strength in this region than the horse, but there is no lambdoidal crest and the occiput shows some tendency to slope forward rather than to the rear. In the more strongly modified fossorial rodents this occipital tilting is strikingly marked. Those which can still see never have occasion to look in a ventral direction, but only above, to watch for an enemy while the head is level with the burrow entrance. The head is held on a line horizontal with the body, in rest, when tamping the burrow walls with the nose (as *Spalax*), and even above the body when the teeth are being used as a pick while extending the burrow system. Accordingly the tilting of the occipital plane is most pronounced in those forms which have been subjected to fossorial conditions for the longest time, as indicated by their other adaptations in this direction. These rodents have great bulging occipital muscles but usually a very slight lambdoidal crest. The reason for the lack of the latter is probably partly because movements, although they may be of great strength, are limited in distance, that a good lever arm is secured without the necessity of a high crest, and, as seems probable, that a crest is not nearly so liable to develop upon the apex of an obtuse as of an acute bony angle. From contemplation of the above facts it would seem that when the head is normally carried parallel with the axis of the neck the occipital plane will be tilted in an anterior direction.

(E)   Certain developments that I have noted only in Cetacea would seem to point to the possibility of a fifth set of conditions that could be of import in influencing occipital conformation. In *Balaenoptera horealis* Schulte (1916) showed a single semispinalis capitis muscle as huge, occupying practically the whole of the supraoccipital region, while the insertion of the posterior rectus capitis was reduced and shifted largely lateral to the condyle so as to have almost no function in sagittal movements. In *Neomeris* I found the posterior rectus to be much larger, the attachment covering an area perhaps one-third of that occupied by the correspondingly reduced semispinalis, which was double; while in *Monodon* the rectus was enormous and the double attachment of the long back musculature very thin; and Murie (1873) showed that the same condition obtains in *Globiocephala*. Obviously the rectus aids more in static than in active strength, while the longer muscles to the occiput are of chief use in active movements. The former does not need an occipital with a marked slope, while for proper efficiency in the case of a mammal holding its head parallel with the body axis, the latter does. This will be discussed further.

It will thus be seen that there may have been, and doubtless were, more than one stimulus for a given development. These can be discussed, but we cannot tell which were the stronger or, indeed, if some have not been counteracted by antagonistic influences.

We have no means of knowing the static or tensile strength of the cetacean occipital musculature. The term strength is but relative and indefinite at best. Certainly in this order the unbroken body contour in the cervical region means that the underlying muscles are very robust, but in *Neomeris,* for instance, although the complexus was extremely heavy at the base of the neck, it was very thin indeed at its broad attachment to the skull. Of course a porpoise with skull of moderate size may toss the head for a few inches with some show of force, and it is likely that relatively prodigious power would be needed in order to twist the creatures head far to the side in opposition to its wish, but nevertheless cetaceans with large heads cannot even lift these from the ground if they become stranded. The cetacean's head can be but little heavier than the water it displaces, so levitation is almost entirely by flotation. Furthermore it is inconceivable that a whale while swimming does not adjust the position of the head so that it is in equilibrium, balanced so that there is no aquatic force tending to twist the head below or to one side. Hence, leaving out of consideration the fact that the side muscles are employed only occasionally while the dorsal ones are in use almost constantly, the supraoccipital muscles need not be any stronger than those of the side of the neck, unless complications be introduced by the flattening of the whole head.

In all cetaceans there has been a mechanical stimulus for the quantitative shortening of the neck, and this has been accompanied by a shortening of the muscles extending from the thorax to the head. This acts in anchoring the latter more firmly, which is of decided advantage to the animal. In a massive body which moves through the water with the velocity characteristic of whales of less extreme specialization, any really marked lateral movements of the relatively large head might be as disastrous as is the abrupt turning of the front wheels of an automobile running at speed. It would enable the pressure of the water to snap the head sharply to the side and might result in a broken neck. A whale does not need any movement of the head in a sagittal plane for *steering,* because this is accomplished by tilting the flippers. It swims by inducing vertical curvature of the entire body, usually from a point in the anterior thorax as already explained, and this necessitates slight movements of the head in this plane. This is a constantly repeated action, whereas slight lateral movements of the head for steering in the horizontal plane (also facilitated by flipper action) are only occasional. As already stated, these lateral movements must be moderate in degree and should be, as they are, but little if any more decided than the possible curvature of the body proper. In a right turn, for instance, the muscles of the right side need apply practically no power. What is principally needed is a relaxation of the muscles of the left side of the neck, when the water pressure against the left side of the rostrum will force the head toward the right. Then must the left muscles have great static strength to keep the head from turning too far, and active strength through a short distance only to overcome the water pressure and bring the head back into position when it is wished to progress once more in a straight line.

In addition to essential but short sagittal movements of the head in swimming most odontocetes probably have need for tilting the head downward in securing food. The amount of this is unknown, but at least we know that in almost all sorts the rostrum is carried somewhat more depressed and is not directly on a line with the body axis. We would therefore not expect to find the occipital plane so sharply tilted forward as would be the case were the rostral axis parallel with the body. Furthermore, in most odontocetes the rostrum is relatively not nearly so large as in mysticetes and the occipital musculature does not need such leverage to control its movements. In considering this detail, however, there is encountered a great complexity of possible stimuli which it is difficult to explain with clarity, and almost impossible at the present time properly to evaluate.

The Mysticeti never pursue an individual food item but whole schools of relatively small fish or shrimp. Hence there should be no need for any motion of the head connected solely with feeding save that of opening the mouth. There is thus need only for such slight movements as are involved in swimming and steering—very simple ones which have allowed the occipital musculature to assume a correspondingly simple form. They are unlike Odontoceti also in that the rostrum is not statically held in a depressed posture. In balaenopterids the whole head anterior to the occipital plane presents dorsally a practically straight line. In other sorts the whole dorsum of the skull is curved, but still the "average" rostral axis seems fairly on a line with that of the body. At any rate all mysticetes hold the head in a more elevated position than do generalized odontocetes.

In connection with the occipital tilt of mysticetes Miller (1923) considered as essential the presence of a downward pulling force applied to the head. But the downward force, furnished by water pressure, against the head of a baleen whale could hardly be relatively greater than that experienced by a pig while rooting. A force is present, certainly, but the critical factor is the position in which the head is normally held while the force is applied, rather than the force itself. If a whale habitually held its head at a right angle to the neck it would not have a forward tilt to the occipital, but a backward one. A better understanding of the situation may be obtained by examining occipital details in relation with external details. It will be found that in *Rhachianectes,* the most primitive of living mysticetes, the occipital tilting, while perhaps more pronounced than in odontocetes, is very moderate: at the same time it has the longest neck. Tilting is most accentuated in the right whale (balaenid) type, and intermediate in other sorts (*Sibbaldus, Balaenoptera, Megaptera*). Some qualification is necessary in the statement regarding balaenid whales, however. In these the occiput is most sharply tilted in relation to the cranium proper, but the entire skull is sharply curved and sickle-shaped (fig. 20), and the head so held in normal posture that the occipital tilt in relation to body axis is only moderate, and actually less than in balaenopterids. In reality occipital conditions in living mysticetes may be divided into two categories. In one, comprising *Sibbaldus, Balaenoptera* and *Megaptera,* the occipital tilt to body axis is extreme, and may be referred to as the balaenopterid type. In the other group, consisting of the gray and balaenid whales, it is more moderate, and may be called the balaenid type. The rostra of the latter are always downwardly curved (see fig. 9) in varying degree, and the conformation of the head is such that backward pressure of the water is applied both to the rostrum and the throat, so that when the animal is swimming the upward and downward force of the water could easily be equalized by slight adjustments in the posture of the head. And this position of stability must be regarded as the normal one. In the balaenopterid type the dorsum of the whole head is so flattened and so held horizontally that any downward pressure by water resistance would be relatively negligible. The gular region is more curved, however, offering more resistance to the water when the mouth is closed and the throat expanded slightly by force of the baleen plates. Hence in these whales the only action of water resistance when the animal is swimming at speed with

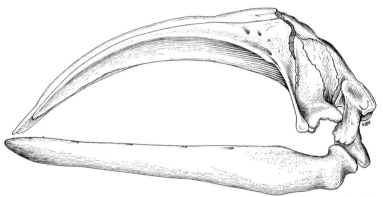

FIGURE 20. Left lateral view of the skull of the right whale, *Eubalaena glacialis*, redrawn from Holder.

the mouth closed is to push upward and backward against the throat, forcing the head up and supporting it without the aid of muscular effort. If one subscribe to the belief, which I do not, that these whales ever move to any extent while feeding with the mouth *wide* open, the chief water force would be applied through the condyles of the mandible. And water force applied to the baleen plates while the mouth is open could be obviated by a slight tilting upward of the snout. In short, all of the above mechanical forces save purely backward pressure at the muzzle may be simply overcome by a slight tilting of the head in one direction or another. The result is that the balaenopterid type of whale is enabled consistently to hold its head in a more elevated position than is practicable for the balaenid type. In fact it *must* do so. In the balaenid type the rostrum is rather evenly rounded from side to side, while in balaenopterids it is flattened and expanded, so that in the latter the rostrum must be kept safely elevated to obviate the danger that when moving at thirty miles an hour the rostrum might become sufficiently depressed for the water to take hold of the broad dorsal plane and force the head sharply downward, with results that might be highly uncomfortable.

In addition to the fact that the occipital should slope more in mysticetes than odontocetes because the former hold the head more elevated, there is the thesis, explained in the next chapter, that because of the larger head the pivot of motion for swimming, normally situated in the anterior thorax, has been shifted somewhat farther forward in mysticetes, which might very likely have a tendency to force the occipital musculature more decidedly rostrad. There is the further theory that the larger, flatter head of the balaenopterids requires a longer power arm for the occipital muscles than the short neck alone can provide; and the possible influence for the same effect suggested by the fact that in the balaenopterid the semispinalis capitis is the one developed for chief control in elevating the head, while in at least some odontocetes the less potentially active posterior rectus is fully as important and may be far more so. In addition the rounded braincase of most porpoises shows that there has been an endocranial opposing force which the occipital has had less success in overcoming than in the much larger mysticetes. Finally, facial migration toward the rear is not so pronounced in mysticetes, and this suggests that in odontocetes this stimulus for facial recession has proved too strong for the occipital tilt to overcome so readily, and that its forward inclination has at least been retarded by this influence. That this latter is not a mere fancy is suggested by the situation in the sperm whale. Although fossils of primitive type (*Diaphorocetus*) show a marked occipital tilt, backward pressure supplied by the developing spermaceti organ has proved the stronger stimulus and overcome it, until now the occipital plane of the living cachalot is practically vertical.

The above are at least some of the reasons accounting for the fact that in Cetacea the forward tilting of the occipital plane is extreme, that it is more pronounced in mysticetes than odontocetes, and more marked, in relation to body axis, in balaenopterid than in balaenid whales.

There are many having their own particular theories who will not agree with me, but to me the evidence as marshalled above is sufficiently strong for the acceptance, at least until many more data are available, of the following thesis. That flotation and water pressure has had an important influence in modelling the *external* shape of the cetacean head. That backward force offered by the water as the animal moved forward has been requisite to the extreme condition of the telescoping of the odontocete rostral elements, but it has been only a secondary factor, the primary one consisting of muscular influence accompanying the extreme recession of the odontocete type of narial musculature. That the moderate forward tilt of the odontocete occipital plane was caused by muscular conditions induced solely by the elevated and static position of the head unaccompanied by the need for cephalic agility, and that as far as the evidence points this region has been totally unaffected by backward pressure of the water. That the elimination or lateral displacement and reduction of the central cranial elements of the Odontoceti have been caused solely by backward crowding of the face and forward crowding of the occiput, and hence by forces that were exclusively or chiefly muscular. That the telescoping and more excessive forward tilting exhibited by the occipital region of mysticeti is due solely to muscle migration finally caused by the static position of the head with need for the minimum of cephalic movement.

Chapter Seven

The Neck

IN MOST mammals the head and neck can develop partly independent of the trunk and the two together may in some respects be compared to an appendage such as the arm or leg. The neck may, and usually does, become long or short according to leg length, so that the mouth may reach the ground; and it almost invariably shortens in mammals which can employ the fore foot as a hand for conveying food to the mouth. The neck must have strength according to size of head and the sort of work that the latter performs, and agility according to the needs of the animal. In the case of a mammal the size of a rat no more need be said, for anything of this size and such light weight could doubtless run as fast and easily were its neck twice as long. But in a large body weighing hundreds of pounds another and vital element is added, constituting adequate equilibration according to the manner of locomotion habitually employed, and this introduces many complex factors which we can only partially understand. In the case of such an amphibious mammal as the sea-lion, the head and neck must act as a balancer, in a gravitational sense, during terrestrial progression. This no longer figures in a mammal that is exclusively aquatic. In such the head and neck usually *must* be of some particular mass and length either so as efficiently to play an important part in swimming, or more rarely the method of progression is such that the anterior end of the animal plays no important part in it, and this portion of the body can then develop in response to some other stimulus.

The conditions which aquatic reptiles seem to have encountered are so at variance with those that have applied to mammals that separate though brief mention should be made of them. Most of the large, extinct, aquatic reptiles seem to have progressed for ages after taking to the water by a rhythmic, diagonal movement of all four limbs that may be compared to the trot, so that the head and neck had no need to act in maintaining equilibrium, or as a rudder. Hence, while this mode of swimming is employed the neck may lengthen or shorten in response to other stimuli. Further, the neck in the two classes can hardly be compared, for mammals exhibit a truly extraordinary conservatism in the number of cervical vertebrae, which limits agility, while the vertebrae of this region in reptiles are as remarkably plastic, and may multiply in some manner or other to a phenomenal number, allowing a length and mobility of neck utterly unattainable in mammals. So an aquatic reptile with a figure suggestive of a wash-tub may have a neck like the half of a huge snake, with as many as 76 cervicals in some plesiosaurs, enabling it to strike at and seize a fish which the body would be incapable of overtaking.

There are two body forms that an aquatic mammal might assume allowing either a lengthening or shortening of the neck according as there are present or absent certain stimuli connected with the acquisition of food. In other words, there are two possible forms in which the proportions of the neck would not vitally affect locomotion. One is the anguilliform or eel-like type, which might be attained by elongation of the body and tail without the length of neck being affected. No living mammal swims by this method. The long-tailed zeuglodonts were definitely modified in this direction and the neck was not elongated. The other type of body referred to above is that in which the legs of one side are separated from those of the opposite side by a relatively wide interval. If this were the original body form, as it likely was in the hippopotamus, we would expect all four legs to be employed in rapid swimming, as is the case in the mud turtle, the ultimate probability being that four paddles of approximately equal size would be developed, and these could perform all functions of swimming and steering, permitting, as far as locomotive factors are concerned, either the lengthening or shortening of the neck. The size of the head in *Hippopotamus* might be expected to inhibit the development of a long neck. The only other mammalian instance for discussion under this head is the platypus. Its body is now somewhat turtle-shaped, and may always have been so, for all four feet are well developed for swimming. The fore feet are now chiefly used for this purpose, which was permitted by the development of an excellent equilibrating organ from the tail. Theoretically it could now undergo a lengthening of the neck, which at present is short; but it would seem that a short neck is advantageous to its habits of feeding. It will occur to the reader that the duck employs somewhat similar methods of feeding and it has a long neck, but this is a necessity for proper equilibration in flight. If it were not then it would not be held in an extended position, but curved as in the heron.

Most of the aquatic mammals listed are not sufficiently specialized for us to tell whether the length of neck is different from that of their ancestors or not. Furthermore, we lack a proper yardstick in that there is no means of telling in a particular case whether an apparent slight shortening of the neck is real or due entirely to a relative elongation of the trunk. Of one thing, however, we can be reasonably sure in most instances. In perhaps a majority of large terrestrial mammals a primary regulator of neck length is limb length, for the neck must usually be long enough to reach the ground. Aquatic mammals lack this stimulus and we can feel reasonably sure that this provides one factor for neck shortening. It seems that the only factor for decided elongation of the neck that could possibly occur for this group would be in the case of one with a slow-moving body which found it of preponderant advantage to secure very active prey by darting movements of a mobile neck. Unless its prey be of marked agility then the aquatic mammal finds it as easy to turn the head by a shift of the entire body as to bend the neck. This is certainly so in the case of Sirenia, and it is a stimulus for cervical shortening, probably to a greater degree than in most aquatic mammals, for these beasts are so sluggish that feeding habits are likely a stronger stimulus for neck length than those arising from locomotion.

The above remarks suggest that the sirenian neck should be definitely but not excessively shortened, and this it is. Additional evidence is furnished by the fact that occasionally in the manati two or three of the cervical vertebrae are fused, the second, third, and fourth then being the ones involved. This is one of the only two living mammals having but six vertebrae in this series. Murie (1872) considered that it is the third cervical that is missing because of the conformation of a slip of the scalenus muscle, while several others have considered that it is the seventh which is lacking. Considering the phenomenal regularity with which just seven cervical somites are laid down in the case of mammals I would be extremely loath to believe, without extremely strong evidence furnished by the conformation of the cervical nerves, that *Trichechus* constitutes an exception to the rule. Rather do I prefer for the present to assume that this genus has but six cervicals for the reason that the thorax long since shifted forward and took unto itself the seventh cervical. In fact it seems that the dugong may even now be undergoing this process, for occasionally if not invariably there is a pair of rudimentary ribs attached to the seventh cervical. This brings to mind the possibility that the stimuli for a short neck in the Sirenia may have been stronger than all facts now indicate, but that this order has inherently lacked the ability to respond to them as readily as have the Cetacea. According to the literature there is some question whether *Hydrodamalis* had six or seven cervicals.

Taylor (1914) stated that the neck of the sea otter is two-tenths of the trunk length, while it is three-tenths in the river otter. With this exception it seems that all aquatic mammals other than the Cetacea, Sirenia and Pinnipedia are either insufficiently specialized to exhibit alteration in the length of the neck, or else they have no terrestrial relatives close enough for significant comparison.

It is from the Pinnipedia and Cetacea that we can tell most about the effect which speedy aquatic locomotion has had upon the neck, and they will be discussed at considerable length. They are amenable to two sets of fundamentally different influences imposed by widely diverse swimming methods. The Phocidae or true seals together with the Cetacea comprise one group, and in this really belong the Sirenia as well. The other group consists of the Otariidae of fur seals and sea-lions. The walrus (*Odobenus*) occupies an intermediate position in some respects. The differences in these two methods of swimming which are of concern to us in the present instance is that in the first group the locomotive impulse is purely from the rear, while in the second it is from the anterior thoracic region.

The cervical complex of the Otariidae is long and sinuous, and the function which it, with the head, plays is two-fold. The animal when on land travels by a shuffling gallop, a gait made necessary by the shortness of its limbs. This is accomplished only by the expenditure of much effort and is greatly facilitated by the violent swinging backward and forward of the neck. Especially is this so in the case of large bulls with huge mass of cervical tissue. This swinging of the neck is an absolute requisite for their terrestrial mode of locomotion. If the neck were but half its length it is likely that travel upon the land would be so difficult that they could do little but wriggle onto a rocky ledge and roll off again.

In swimming, the impulses are from the side in the case of Otariidae, with the center of motion presumably between the two organs for propulsion and near the midthorax. Conditions are thus very similar to the case of a rowboat. A short, tubby skiff is difficult to handle, and one of usual length is best managed when the rower is near the center. Thus, for mechanical reasons, the sea-lion should have considerable mass of body both before and behind the anterior flippers, and this has undoubtedly constituted the chief stimulus for length of neck in this animal. Whether this has been of any influence in the acquisition by adult bulls of their enormous mass of cervical tissue seems doubtful, but must be considered as a possibility, in which case there would have to be present in the female a sex hormone with action of inhibiting in this sex the developing of extraordinary neck size; or vice versa.

This "bull-neck" character in male otariids increases to a quite phenomenal degree at the beginning of the breeding season, when they spend more time on land. In fact, bull fur seals and Steller sea-lions, at least, then spend many weeks on land without feeding or entering the water. Unfortunately those in a position to do so have never reported the conditions involved. Inferentially there is much fat deposited in the cervical region, this being increased as a reserve supply when the breeding period approaches. O. J. Murie has told me that at the time of rut in caribou (*Rangifer*) there is a definite thickening of certain of the cervical muscles, which would be useful in the battles between males. This is probably caused by a hormone released by the awakening sex-glands and is entirely comparable with the lengthening of certain perineal muscle fibers in females at the imminent approach of parturition. I deem it likely that a similar thickening of cervical muscles of bull sea-lions takes place at the beginning of the mating season.

In young males and females the neck is very sinuous with remarkable precision of movement, as all who have watched "trained seals" (usually *Zalophus californianus*) will agree. This is inherent and assuredly of fundamental importance to the economy of these mammals. It is not merely developed by training, as indicated by the statement of Rowley (1929) that when a stone is thrown at a sea-lion cow, "no matter how violently nor how short the range, she will catch the stone with marvelous accuracy in her mouth, often at the expense of breaking off her teeth". Because of the shape and the fact that the propulsive mechanism is situated near the center of the mass the neck may act importantly as a rudder, and yet it may be thrust in all directions for the capture of food without disturbing equilibrium provided that at the same time the hinder end is moved in proper compensation. If the aquatic stimulus were for a shorter neck, however, terrestrial activity would be curtailed, and this would result in still greater independence of the land. So the fact that the long neck which is advantageous for swimming happens also to help terrestrial locomotion should result in a tendency to slow up the rate of aquatic specialization in other directions.

As compared with the seal the cervical transverse processes of the sea-lion are broader, following greater complexity of the longus colli. The spinous processes are also better developed. The most striking additional muscular differences are that the cephalohumeral and humerotrapezius are better situated for twisting lateral movements of the neck and the occipital muscles are evenly distributed along the lambdoidal area to facilitate diversity of head movements.

In the case of an aquatic mammal as highly specialized as the seal or whale, in which the propelling mechanism is situated at the hinder end, an entirely different set of physical laws is introduced. For proper efficiency it is obliged to be of fusiform shape, just as must an airship of the Zeppelin type, the fuselage of an airplane, or (with proper camber) the cross section of its wing. We are entirely justified in accepting it as an incontrovertible fact that with this type of body and of propulsion a long, mobile neck would be impracticable, and that the neck form or its musculature or both must be such that when swimming at speed, this part of the animal is included within the uninterrupted contour of the fusiform body. There must be a point constituting the center of equilibrium, which may also be the pivot of motion, situated anterior to the middle of the body and theoretically this should be at that portion of the thorax having the greatest circumference, which will usually fall at the shoulders. In the Phocidae, especially, the pivotal point of swimming motion is fixed in this region by the fact that here are anchored the lateral muscles comprising the power arm of the tail in one direction, and of the neck and head in the other. Now for most effective results for the muscular power expended it is absolutely essential that there be sufficient mass to the prethoracic part, while if the mass be too great, then efficiency is reduced.

Breder (1926) considered that in most fish the pivotal point lies through the atlas, and this may very well be correct. As discussed in a previous chapter it is deemed that there may be a single pivot of motion coinciding with the center of equilibrium, or in an animal in whose swimming less of the tail is involved there may be two pivots of motion, one posterior and one anterior to the center of equilibrium. In the seal, however, with its relatively small head, it seems that the pivot of motion (or pair of pivots) is situated in the region from which the lateral lumbo-caudal and cervico-cephalic impulses arise. Some of these muscles may shift their points of anchorage to the thorax over the distance of one, two, or three intercostal spaces, or those to the arm may wander slightly, but on the whole their possible migration is very limited.

In general principles the body of the seal (Phocidae) is subjected while swimming to the same laws encountered by whales. We might therefore expect to find in the former the same marked tendency toward a shortening of the cervical series as characterizes the latter group. On the contrary, however, it may seem somewhat surprising to find that in bony details the neck of the seal is little or no shorter than that of the sea-lion, which as far as we can see has no stimulus whatever for a shortened neck. In spite of the latter feature the seal is apparently enabled to meet the condition imposed by its method of swimming that the mass anterior to the thorax be not too great in volume (for proper efficiency) by the fact that its head is relatively small. We cannot be sure, of course, that the length of its neck is absolutely ideal for swimming. In fact it seems likely that there is present some indeterminate amount of stimulus for a shorter neck because in swimming or in resting posture on land the neck is retracted to a marked extent. That this is so is indicated by the fact that if a fish be held above the head of a seal in captivity it will stretch the neck to a phenomenal degree, when it appears fully as long as in the sea-lion. It thus seems probable that in antagonism to the stimulus for a somewhat shorter neck during swimming, there may also be some stimulus, connected with the acquisition of food, for the retention of a moderately long neck. Incidentally the Phocidae may be as yet insufficiently specialized for either one of these to have gained decided ascendency over the other.

Examination of an embalmed specimen does not throw any light on the manner in which this extensibility of the neck is made mechanically possible, but it is doubtless due in part to unusual elasticity of the intervertebral cartilages. The apparent retraction of the neck while swimming is partly real (to as great an extent as the vertebrae will allow) and partly illusory, both because of the unusual breadth through the base of the neck and probably because tension of the lateral neck muscles tends to draw the shoulders forward.

It has been noted in the previous chapter that the muscles of the phocid occiput are so distributed as to facilitate movements in the sagittal and in the horizontal plane. Other muscles of the neck follow the same plan, and are distributed so as theoretically to pull the head up or to the side with less effort than in the sea-lion, and with a minimum of diagonal twisting. If we analyze the swimming movements of the seal it will be seen that practically the entire musculature of either side is concerned. Neither the forward nor the hinder end can be thrown to the side without curving the entire body; hence the muscles concerned in both head-swing and tail-, or pedal-swing really constitute one single group, all the components of which have been specialized toward the single end of efficient aquatic propulsion. Some operate from the anterior thorax, while others—especially the deep and abdominal pectorals on the one hand and the inferior atlantoscapular, cephalohumeral and humerotrapezoid on the other—work in either direction from the anterior limb, so that in such motions the arm acts as a sort of raphe between these two groups of muscles. These muscles concerned with lateral neck movements are very robust and diverge from the head to the sides of the broadened thoracic region in a manner to give them very powerful leverage. They must certainly have an extraordinary amount of normal tone, so that when those of one side contract the muscles upon the opposite side will not relax to a greater extent than is proper. And yet, when the animal wishes to stretch forth its head, the tone is removed and the muscles are relaxed to an abnormal degree. The same may be said of the muscles chiefly concerned with movement of the head in the sagittal plane—the anterior rhomboid and semispinalis for raising, and sternohyoid and tracheovertebral muscles for lowering it. They are powerful and well situated for performing the work which they have to do, which evidently consists chiefly in strong but short depression or elevation of the head to facilitate steering.

As before mentioned it seems certain that because the head of the seal is rather small, instead of relatively large, for the size of the body, the lateral movements of the neck which accompany the act of swimming are distributed over the entire cervical series of vertebrae, and they accordingly would not be expected to depart widely from the usual type. Really the only noteworthy detail is the fact that the spinous processes are reduced and very much smaller than in the sea-lion, which seems an indication that active strength (as contrasted with passive strength) in elevating the head is not of great importance to the animal.

The only ways in which the principle of aquatic locomotion of whales differs from that in seals is that in the former the flattened tail constitutes the primary organ for propulsion instead of the hind feet, and that its plane of movement is vertical instead of horizontal. This means that in the Cetacea the dorsal and ventral neck muscles are required to perfom those acts during swimming which constitute the chief function of the lateral neck muscles of the seal. But the lateral neck muscles are attached to the thorax or arm and belong in entirely different groups from those which swing the tail laterally, while the dorsal neck muscles of chief use are an integral part of the spinal musculature of the entire animal. This permits the fundamental differences that in the seal the center of swimming motion *must* be in the anterior thorax, from which the neck muscles arise, while in the whale there is no muscular inhibition to its shifting either forward or backward, according to requirements introduced by external body form.

The whale has no external neck, or at least there is no constriction at this point, save to a barely discernible extent in *Platanista* and *Delphinapterus,* in both of which the series of cervical vertebrae is relatively longer than usual in this order. This is for the reason that the cervicals have become much shortened in all Cetacea, but in some more than others, and there is variation in other respects, this being best illustrated perhaps in the Odontoceti. The individual vertebrae may be distinct with the centra either fairly thick or wafer thin, all may be fused into one bony complex, or intermediate conditions may obtain, save that invariably, with the exception of the cachalot, any fusion which occurs is first manifested anteriorly. It is probable that this series is shortest in those sorts in which all the elements are fused. Among odontocetes it is likely that *Platanista* has the longest neck of any genus now living, but the skeletons available are disarticulated and unsatisfactory for determining this point. The neck of *Delphinapterus* is also relatively long, as mentioned. Among mysticetes the longest neck occurs in the gray whale *(Rhachianectes),* which is famous for having this member unusually mobile, and thereby hoisting whaleboats several feet above the water by a thrust of the head.

It is difficult to determine the exact percentage of the cervical series to body or even total skeletal length. The cervical length may be taken with ease for the elements are either fused or were in life separated by an insignificant amount of intervertebral substance; but this dimension is almost invariably omitted from existing osteological reports, or else one does not know if the intervertebral disks of the remainder of the skeleton were allowed for. In *Neomeris* I found that the cervical length constituted about four and one-half per cent of the skeletal length, but the specimens were disarticulated so that this percentage is greater than is actually the case in life. From rather rough measurements of the mounted baleen whales in the National Museum, having artificial intervertebral disks of an unknown degree of error, I found that the percentage of cervical length to skeletal length was in *Rhachianectes,* 4.2; *Sibbaldus musculus,* 3.7; *Megaptera nodosa,* 3.6; *Balaenoptera physalus,* 3.4; and in *Eubalaena,* with its fused cervicals, but 2.4. The latter figure also obtains in *Physeter,* according to measurements furnished by Doctor Stone of the specimen in the Philadelphia Academy of Sciences.

Shortening of the cetacean neck has always awakened the liveliest interest and has caused much speculation. The most popular belief may be summed up in the words of Winge (1921): "The head, during swimming is held directed as firmly as possible forward. The neck is not moved, and for this reason it becomes short and stiff. During motion through the water the head is pressed from the front; it is forced back-ward against the cervical vertebrae, which are thereby squeezed exces-sively together and pressed back against the anterior dorsal vertebrae."

The thesis that water pressure against the head has reduced the length of the neck by squeezing it between the head and the thorax hardly seems sound. In the first place it is not alone the neck that has experienced the sort of squeezing meant by Winge, but all vertebrae anterior to the flukes, and the skull as well. In the second, if such a mechanical squeezing would have had any effect this should be just the reverse of a shortening of the cervical complex. According to modern concepts of morphology this should rather have consisted of a narrowing of the skull and elonga-tion of the neck. The whale has of necessity assumed a short neck in order to fulfill certain mechanical requirements contributing to effi-ciency in swimming and I seriously doubt whether water pressure, in the sense referred to above, has entered into the question at all.

Almost if not quite invariably in cetacean literature the shortness of the neck is discussed as though this detail had in itself suffered reduc-tion without reference to the thoracic elements, and this has resulted in much misunderstanding. The cervical vertebrae have not alone been re-duced, but all the vertebrae anterior to just forward of the center of the thorax. In many mammals, including man and cetaceans, the lumbar vertebrae are the most robust of the column. In man the centra of the anterior thoracic vertebrae are progressively thinner and this, to an al-most insensible degree, may be said to continue to, and culminate in, the second cervical. In whales the reduction in centra length is more marked anteriorly. The exact number of vertebrae involved is variable, but perhaps in most sorts the fifth thoracic is the first to show really ap-preciable shortening. The fourth is slightly shorter than the fifth, the third than the fourth, and so on. By the same slight degree the seventh cervical is slightly shorter than the first thoracic and this shortening may culminate in the third or fourth cervical in such whales as have these vertebrae unfused. This is but a general statement and details fre-quently differ, however. In the zeuglodont *Basilosaurus* the neck is clearly shortened, but the transition between the cervical and thoracic series is not gradual. On the contrary, the last cervicals are somewhat abruptly shorter than the first thoracics, and this might be an indication that the cervicals are more responsive to the stimulus for shortening than the thoracic vertebrae; but I believe that rather it is the result of its anguilliform conformation. In some porpoises, such as *Tursiops,* al-though the anterior thoracic vertebrae are considerably shorter than the middle or posterior ones, they are quite a bit thicker than the last cervical. This may be explained, I believe, by the probability that the posterior cervicals have become as thin as possible without fusion, and thinner than any rib-bearing vertebra could become.

Not only is the skull relatively larger in all whales than in the seal, but almost invariably that of the former has accessory equipment in the way of special fatty deposits or baleen armature, which makes the head still larger in comparison to body mass. The head alone therefore furnishes almost or quite the prethoracic mass and weight requisite to high swimming efficiency. In other words the head alone is so large that if there were a neck of respectable length the part of the animal anterior to the thoracic "pivot" of motion in swimming would either be too massive for the part posterior thereto, or else, if the pivot could be shifted forward, and therefore from the thorax to the skull, the neck would really constitute a mechanical part of the body.

In scrutinizing various cranial and cervical conditions there seem to be recognizable two possible factors which might show a tendency for retarding maximum shortening of the neck and fusion of its elements. One of these is the condition imposed by the presence of a tweezer-like beak, or of the comparable tusk of the narwhal. It is readily seen that this equipment would necessitate at least moderate shifting about of the head with a consequent amount of flexibility of the neck, and this should delay fusion of the vertebrae. At any rate the narwhal has all the cervical vertebrae free, as have such long-beaked forms as *Platanista* and the extinct *Eurhinodelphis*. The other is the fact that in those forms with long or slender beaks the cranial part of the skull is relatively small in relation to body size. Hence the head has less mass and this might be compensated for by a slightly longer neck. But non-beaked odontocetes such as the beluga *(Delphinapterus)* may also have the cervicals free, and this fact introduces an element of serious doubt.

At the other extreme are such odontocetes with head of moderate cetacean size as *Hyperoodon, Ziphius, Mesoplodon, Grampus,* and at times *Pseudorca* and *Phocaena,* in which all the cervicals are fused into one solid bone. Still other sorts exhibit every intermediate condition between complete fusion and complete freedom of the cervical elements, but in all existing Delphinidae excepting the narwhal and beluga at least the atlas and axis are fused. At one time I endeavored to reach some conclusion regarding this point by making a list of cervical conditions in all species of whales which I could examine, and attempting to correlate them with any one of numerous other osteological features; but with results that were utterly discouraging. It may be put down partly to the variation in the cervical needs of different sorts, and another possible explanation is furnished by the fact that a very few species (as *Neomeris*) develop a slender, short pair of accessory ribs attached to the seventh cervical vertebra. It thus seems possible that this porpoise has found it less difficult to begin a shifting forward of the thorax than further to reduce the length of the cervical series as a whole.

Certain additional reasons for cervical variation in the larger whales may also be considered, bearing in mind the thesis already explained that in whales with heads of even moderate size the head alone furnishes practically all of the prethoracic mass necessary for efficient locomotion;

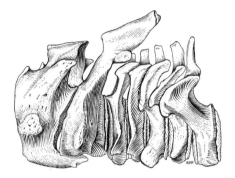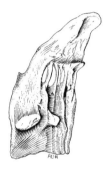

FIGURE 21. Extremes of cervical vertebrae among toothed whales, illustrating a series in which the seven components are separate, and one in which they are all fused. *Delphinapterus* on left, *Grampus* on right.

and this matter of head size is really more pertinent to the present chapter than the last.

One is so used to reading of the prodigeous size of the head in balaenid whales that he is apt to take for granted that their heads are also of unusual *length*. Because of the question of the intervertebral disks it is difficult to find the precise proportions, but I have secured what measurements I could from the literature, and these indicate not only that the skull of *Megaptera* (25 to 31.4 per cent of the total length) may average longer than in the balaenid whales (26.8 to 28.8 per cent), but that there is more individual than generic difference in Mysticeti. *Rhachianectes* (22) and *Balaenoptera acuto-rostrata* (22.6 per cent) both have rather short skulls and the others are intermediate. But there are important differences elsewhere than in *length* of skull. In *Balaenoptera, Sibbal-*

dus, Megaptera, and *Rhachianectes* the whole head is either flattened dorso-ventrally or moderately in transverse direction *(Rhachianectes).* In these the vertebrae are either all free, or, according to Andrews (1916) "two or more of the cervical vertebrae usually become ankylosed," although I personally have seldom noted this condition. This lack of complete ankylosis in mysticetes whose heads average larger than in odontocetes may be attributed to the common belief that the former group is not as far advanced along the evolutional road or as "specialized" (ambiguous term!) as the latter, and consequently that the stimulus for fusion has not been operative for as long a time. In balaenid whales *(Balaena* and *Eubalaena),* however, although the skull is no longer, the head is of prodigeous depth. This increase in total size and mass of head, necessitated by the hypertrophical development of the baleen equipment, has apparently obliged these whales to adopt a makeshift for the accomplishment of swimming. This is even more the case in *Physeter,* whose great spermaceti organ has enlarged the head to a degree where the snout projects several feet beyond the rostrum. In these it would seem that if the pivot of motion occupied its more natural position in the anterior thorax, the increase in the relative size of the head would tend to place this too far to the rear—or too near to the lineal center of the animal—for best efficiency. Whereas in the usual fusiform method of swimming at high speed both the tail and head are curved from some point in the anterior thorax, a disproportionately large head would disturb the proper balance, and it is suggested, partly by inference and partly by the shift forward of the greatest circumference of the body, that in such whales the center of motion has, to as large an extent as it was able, migrated from the thorax to the posterior part of the skull. Theoretically this is rendered possible by the increase in the inertia of the head and the essentially homogeneous nature of the dorsal cervical and the dorsal thoracic musculature. But this condition is in principle less efficient and has resulted in the definite reduction of speed. The stimulus for excessive head size has simply proved stronger than any forces acting in antagonism and under the conditions which the animals have encountered it has not been too great a handicap for continued existence.

Thus the neck, body and tail of these whales seem to all intents to comprise the lever arm for swimming and the head can more properly be compared to a fulcrum (as in 4, figure 1). It might thus be said that while in porpoises the body wags the head, in the sperm and right whales the tendency is for the head to wag the body. The neck is functionally a part of the body and as a fused complex virtually constitutes a thoracic vertebra without ribs. The series has probably become as short as it can while at the same time accommodating essential muscles which are attached at this point, as well as allowing for the emergence through foramina of the cervical nerves. The stimulus has been for the thorax to engulf the neck and to this extent the latter has experienced squeezing, but from a morphological or evolutional rather than a mechanical sense. Although there is no reason for considering the balaenid as any older than the balaenopterid stock the cervicals in the former group have fused more completely probably because the larger mass of the head has increased the strength of the stimulus for fusion. Normally in the former they are completely fused but Flower (1876) has stated that occasionally the seventh is free. The cervicals of the cachalot *(Physeter)* are also usually fused, I believe, but this is not established for the reason that in some of the few specimens known the atlas is free; and this is a unique situation among Cetacea, for in all other whales fusion of the cervical elements first takes place anteriorly and then progresses toward the rear. In those specimens of the manati which show fusion of two or three cervicals the atlas is said always to remain free, however.

In the above discussion account has not been taken of the pygmy right whale *(Neobalaena)* or pigmy sperm whale *(Kogia)* for the reason that these are depauperate, freakish forms. The former is anomalous in that the first thoracic vertebra lacks ribs. In other words, it has eight cervicals, which indicates an exactly opposite trend from that exhibited by all other whales. This circumstance is very puzzling, but it is felt that one exception does not overthrow the arguments that have been advanced.

Before closing this chapter brief mention should be made of other cervical muscles not yet discussed. During swimming the majority of the subcervical muscles of whales act in antagonism to the supracervical ones, but many of them have other functions not connected with locomotion. Thus the hyoid complex is very powerful in odontocetes and these doubtless assist in depressing the head, but whether this is now their primary stimulus cannot be told. They must also be used in the deglutition of bulky food, but this is hardly to be reckoned with in Mysticeti, especially the balaenopterids, because their esophagus is relatively so small, but in this group the depression of the head can probably be largely brought about by flexion of the superficial gular musculature (previously discussed) operating upon the mandibular tip. The hyoideal as well as other small muscles of the prevertebral and lateral parts of the neck are more diverse in their affinities than are the cervical elements of the spinal system and they could never become as simplified as the latter.

To recapitulate briefly; it is believed that the long, flexible neck of the otariid is essential to its special method of swimming. The different method of swimming employed by the phocid and cetaceans necessitates a different cervical conformation. Although feeding habits of the seal may require a neck of considerable length, its effective length while swimming is reduced by the small size of the head and great breadth of the neck base, as well as by as much contraction as possible. The larger size of the head in whales requires marked reduction in neck length, in the porpoise so as to bring the head nearer the thorax, and possibly in balaenid and sperm whales so as to bring the thorax nearer the head and mechanically to make of the cervical series a single element comparable to a thoracic vertebra without ribs.

Chapter Eight

The Trunk

THERE are numerous factors affecting the precise conformation of the mammalian thorax, but it is invariably of such shape that only very minor adjustments are necessary in order to bring it to an ideal stream-line form. Perhaps in the majority of quadrupeds the broadest or deepest part of the body contour is just back of the anterior limbs, although this point not infrequently is situated in the posterior thorax, or even through the abdomen, in the more paunchy sorts. But no mammalian thorax is so unsuitable in form that it will offer any real resistance in swimming and in consequence there is no very strong stimulus for a definite change in its shape even when sufficient time has elapsed for there to be very advanced aquatic specialization in other respects. I deem the Pinnipedia, Sirenia and Cetacea to be the only aquatic mammals to have reached this point. It is commonly accepted as fact that the great thoracic diameter of *Hippopotamus* is a result of aquatic habits. I doubt this and although I may be entirely mistaken I believe that the primary reason that the thorax is larger in this animal than the rhinoceros, for instance, is that the more succulent food of the former necessitates an alimentary tract of prodigious capacity, with stomach more than a dozen feet in length.

We can be sure, however, that in pinnipeds, sirenians, and cetaceans the aquatic life has been lived for sufficient time for there to have been modifications in the external form of the trunk—especially in the first and last orders, which are more speedy. The requirements for external conformation are simple and necessitate merely that the greatest girth be approximately at the pivot of motion for swimming. In pinnipeds, sirenians and most cetaceans this is in the region of the anterior thorax, a possible exception being in the case of adult bull sea-lions, where the greatest girth may be at the base of the neck. In balaenid and sperm whales, where it has been argued that because of the huge size of the head the pivot of swimming motion has tended to shift forward, the greatest girth seems to be through the posterior part of the head, as would be expected. For practical purposes, apparently, this is all that is necessary, for it would make very little difference whether the cross section of such a body be slightly flattened in the vertical or horizontal plane. We can be sure, however, that if the aquatic influence were uncomplicated by any muscle pull or lifting action of the lungs the tendency would be for the cross section of the anterior trunk eventually to assume a shape that was exactly circular, for this is the ideal both for locomotion and for such purposes as retention of body heat, as well as that external pressure may be distributed evenly to all the internal organs.

The point of greatest bodily circumference may be, and is, shifted according to aquatic requirements without especial reference to the form of the thoracic cavity itself, as fleshy or fatty tissue may easily be deposited in the required region. The actual external shape of the anterior trunk may be influenced by deposits of fat, by the conformation of the shoulders and their muscles, by the height of the vertebral spines and consequent thickness of the spinal musculature, and by the form of the thoracic cavity.

In the fetal state the shape of the cross section of the thoracic cavity, in all mammals without exception, I believe, is slightly broader transversely than in a sagittal direction. In terrestrial sorts its shape in after life will alter according to posture, the fundamental forces concerned being gravity and muscle pull. In a strictly quadrupedal form, and even such a primate as the baboon, gravity will tend to pull down the sternum. This is doubtless assisted in some obscure way by muscular stress, and by the requirement in swifter, larger mammals that the anterior limbs be as near together as practicable to reduce any propensity to waddle. The result will be a thorax narrow transversely. That muscle pull need not be of great influence in the attainment of this result is indicated by the situation in the existing sloths, which spend their lives either hanging by their limbs or curled up in sleep. We might expect to find that the constriction of the chest between the pendent arms had resulted in its becoming narrower but on the contrary it is much broader than thick, indicating that muscle pull has not been an important factor in the expected direction, but that the chief influence was the removal of ventral, gravitational pull, which has permitted the ribs to spring outward. In mammals of erect posture, as man, there is also no gravitational pull in a ventral direction upon the sternum, but here the throwing back of the shoulders (ostensibly) has permitted a broadening, usually, of the thorax chiefly by tension of the muscles extending from the arm to the midline, both dorsally and ventrally. Occasionally, in man, a "pigeon-breasted" individual is encountered, in which the thorax is almost circular, so other and obscure factors are indicated. Or there may be some such unusual situation as is found in bats, which require a broad chest to accommodate a huge mass of pectoral musculature.

In the case of mammals which are exclusively aquatic gravitation does not enter the question. It is true that at least most whales are slightly heavier than the water which they displace and that for equalized flotation they need a lung-full of air. Hence there is some slight force of gravity experienced; but when distributed over the entire animal it is absolutely negligible. In consequence the conformational stimuli experienced by the thorax are those supplied by muscle pull, levitation of the lungs within the cavity, such amount of tendency for a circular trunk as progression through the water may determine, the influence which the location of heavier bones may introduce, and at times a possible fifth factor. In regard to the latter, certain odontocetes indulge in a rocking motion during progression. The killer whale especially is in the habit of swimming slowly and taking a fresh breath every few seconds. Accordingly the head is first elevated above and then depressed below the surface, and the back rocks to and fro in the sagittal plane. Undoubtedly this has had some effect, however slight, upon the conformation of the entire animal. But it is impossible to know the relative importance of any of these influences.

It is popularly believed that aquatic life augments the lung capacity and in consequence that there is an increase in the diameter of the thorax. This is entirely logical and may be conceded. Certainly the thorax of the sea otter seems to be considerably more capacious than in its river cousin. But usually there is no sure yardstick whereby this can be measured and it must be inferred. For one thing there are no thoroughly aquatic mammals sufficiently close to terrestrial forms for adequate comparison; and an expansion of the thorax may merely indicate that there has been a shortening of the abdomen, or an increase in the size and capacity of the alimentary tract may have crowded the thoracic cavity.

Unfortunately I have neglected to examine the cross section of the chest of any fresh pinniped or cetacean, and a preserved specimen soor. becomes so distorted by its own weight it is valueless for this pur pose. Nor can great reliance be placed in a mounted specimen.

Among the smaller aquatic mammals the only one (so far as I know) having details of the thorax that merit attention in the present connection is the insectivore otter *(Potomogale)*. Dobson (1882) showed that in this animal the pectoral muscles are surprisingly unlike the general pattern that is so characteristic of insectivores. Rather is it suggestive of conditions in the sea-lion. I believe, however, that this resemblance is fortuitous and that there is shown merely an intermediate stage in eventual convergence toward a similarity in appearance to pectoral conditions in the seals (Phocidae). In certain respects the swimming of *Potomogale* and the Phocidae is similar. In neither are the fore limbs used in swimming, and in both the hinder end is swung from side to side. Therefore, as discussed more fully in the case of the seal, there is a tendency for the posterior part of the pectoralis and the latissimus dorsi to develop toward the end that they may assist in pulling the posterior part to the side. In *Potomogale* there is a clear tendency in this direction, and the two muscles are continuous along their borders. The anterior part of the pectoralis unites over the arm with the inferior margin of the trapezius, insertion being upon the humerus distad of the greater tuberosity. By means of these two muscles, at least, the arm may easily be held stationery, acting as a base from which other muscles may operate. Deserving of mention in this connection is also the sternocleidomastoid, which is apparently of much importance in lateral movements of the head that are requisite to swimming. This muscle arises from the midventral line for a surprising distance beneath the pectoral mass.

In mounted skeletons of seals the anterior thorax is usually slightly broader than high, or it may be markedly broader *(P. groenlandica)*; in sea-lions it is definitely narrowed; and in the walrus fairly intermediate between the two. These facts need not be of great significance in the present connection, however, for there is equal variation in terrestrial carnivores. At any rate this difference in pinnipeds is accentuated, in the case of preserved specimens, to the point where a seal can hardly be balanced upon its side, while one has difficulty in so balancing a sea-lion upon its sternum. The sea-lion spends considerable time on land, supported partly by the anterior limbs. The thorax thus experiences a ventral gravitational pull for at least some of the time, and during this time there is the same static muscle stress as acts upon the usual terrestrial quadruped. High spinous processes and accompanying musculature in this region further accentuate the depth of the trunk as compared to its width. What effect the levitating influence of the inflated lungs or of the swimming actions of the anterior limbs have had while the animal is in the water we do not know, although it is not improbable that the constant adductive movements of the flippers have tended to make the chest narrower.

At least most seals of the genus *Phoca* never use the anterior limb on land for supporting the body save for the briefest periods, although some others of the Phocidae, as *Mirounga,* habitually do so. Furthermore, gravitational force can very rarely act upon the sternum. On the contrary the sternum presses upon the ground both when the animal is resting, and wriggling along the surface, so there is at such times actually as much force pressing *upward* against the strenum as downward. Presumably these factors have been of importance in shaping the thoracic cavity. There are other elements, however, which further modify the shape of the trunk. Because the spinal musculature has become greatly broadened for the purpose of lateral movements of the hinder end and because there is no need for much movement in the sagittal plane, the spinous processes of the anterior thorax are very low and the musculature concerned very thin. This further reduces the sagittal dimension of the trunk. The transverse dimension is increased by the fact that the lateral muscles of both trunk and neck need as great a lever arm as possible more effectively to accomplish the lateral movements used in swimming. The muscles not only are very robust but they are overlain by a thick layer of fat, and between them and in all interstices there is much connective tissue surrounding networks of blood vessels. The broad thorax is of distinct advantage to the animal in swimming. Whether flattening in the sagittal plane actually increases the swimming ability of the seal or whether, not of advantage, it is yet of insufficient disadvantage to be of consequence, is unknown.

In comparison with a typical terrestrial carnivore the chief characteristics of the vertebral column of the Pinnipedia are the looseness of the articulations, the elasticity of the intervertebral disks, the latter being difficult to investigate after death, and the fact that the spinous processes exhibit no definite change in slope, or anticline. The latter is a character shared by all essentially aquatic mammals. The usual quadruped has a definite center of motion in the vertebral column, which it would be more logical to call the center of suspension for the reason that the back-bone of a quadruped may with propriety be compared to a double-pier, cantilever bridge. From the above center the anterior spinal and some of the shoulder muscles operate in one direction and the posterior spinal and some of the pelvic muscles in the other, resulting in a backward slope of the spines in the anterior thorax and their forward inclination in the extreme posterior thorax and lumbar region. And there is usually a rather abrupt alteration in the character of the spines where the slope changes. The position of this center depends upon the stress encountered, as pointed out by D'Arcy Thompson (1917). If the chief weight is borne by the fore legs the center will be farther forward, and if the hind limbs bear all the burden, as in the kangaroo, then the center will be shifted far to the rear. In a completely aquatic mammal the skeleton has no resemblance to a double-pier, cantilever bridge, but may better be

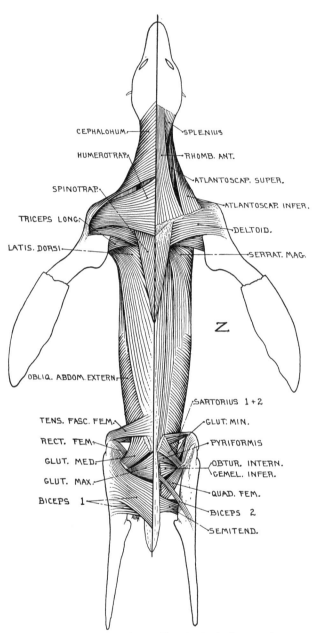

FIGURE 22. Dorsal musculature of a sea-lion (*Zalophus*): superficial layer upon the left, and much of the next deeper layer to the right of the medial line.

[158]

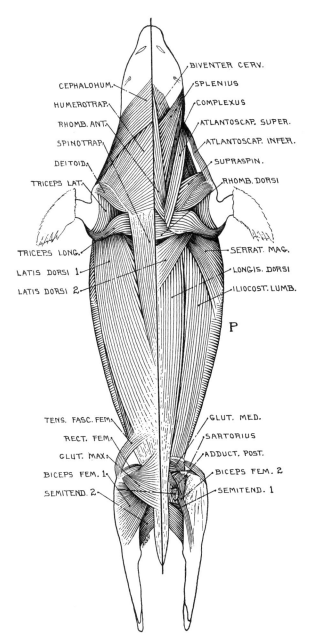

FIGURE 23. Dorsal musculature of a seal (*Phoca*): superficial layer upon the
left, and much of the next deeper layer to the right of the medial line.

[159]

likened to a pontoon bridge, for the body is supported in the water throughout its entire length, and in theory each individual vertebra bears its proportionate share of the load, although muscle stress introduces a disturbing element. The curve of the vertebral column is always rather gradual, but there must be some one point upon the arc from which the column curves in either direction. In the anterior thorax as well as the posterior neck the spinous processes of otariids are moderately high, which is chiefly for supplying an efficient lever arm for the muscles raising the head and neck, but spinous height rapidly decreases toward the posterior thorax and in the lumbar region, indicating that the lumbar spinal musculature need not be very strong for such actions as the animal finds necessary. In phocids, on the other hand, the thoracic spines are very low, with no better definition than the lateral processes. Their height is slightly greater in the lumbar series, however, and here the centra are also more massive.

As in all aquatic mammals the articulations of the pinniped vertebrae are reduced, for as stresses are applied while swimming to a large extent throughout the columnar length rather than at particular points there is little need for local strength, while there is increased need for flexibility of the column. This is attained (especially in the Otariidae) by virtual abandonment of an interlocking type of zygapophyses, and at the same time reducing all other processes which might offer mechanical restriction to limberness. Anapophyses are absent as such in the lumbar series. In otariids (at least in *Zalophus*) the zygapophyses of one side are very close to those of the other and in the lumbar region the articulations are such as apparently to prevent any marked concavity in the outline of the dorsal surface, but permitting an unusual amount of convexity. In a seal *(Phoca)*, relative to size, the zygapophyses of opposite sides are about three times as far apart, and in the lumbar region the articulations are such as to allow very definite concavity in the outline of the dorsal surface. Especially is the latter the case between the lumbar and sacral series, to the degree where the sacrum may be elevated to a quite surprising extent. And although the cleaned skeleton does not show it there is possible a very remarkable amount of concavity in the posterior thoracic and anterior lumbar regions of at least some phocids, for photographs of *Mirounga* show that this animal can bend the column to quite a right angle at this point. Convexity of the lumbar series cannot be determined from osteological examination, for this would depend mainly upon the amount of play allowed by the zygapophyseal articulations. Presumably the possible amount of convexity is not very great and somewhat less than in otariids, but intervertebral flexibility would render it easy for the seal to develop this ability did it have occasion for doing so.

The number of ribs in the most generalized mammals is thirteen pairs, while in the Pinnipedia they number fifteen, except in the walrus which has fourteen pairs. This is accompanied in the former case by five, and in the latter, by six lumbar vertebrae. So it is seen that the members of this order have responded to a stimulus for a longer thorax, although it is conceivable that this result was attained before aquatic habits were adopted. Ten (usually) pairs of ribs are attached to the vertebrae by both capitulum and tuberculum, and this, I believe, is the situation in at least the majority of terrestrial carnivores. So in the latter detail there has apparently been no alteration. The ribs are not otherwise note-worthy.

In pinnipeds the sternum is usually composed of six or seven bony elements, while in terrestrial carnivores it averages eight or nine. So the sternum in the former group has suffered reduction, this being more accentuated in the Phocidae than the Otariidae because in the latter the individual elements have a greater length. In all pinnipeds there is a presternal extension of the manubrium, partly cartilaginous, which is markedly well developed, especially in the Phocidae, and this is of very definite importance to the animal.

With all of the above bony details of the pinniped thorax in mind we can proceed to scrutinize some of the muscular stimuli involved. In otariids the cervical and anterior thoracic muscles are developed not only for agility of neck muscles for enabling the head and neck to act as a balancer both on land and in the water, but to assist in movements of the anterior limb in both situations. The spinal musculature retains to a large extent its regular function of bending the vertebral column in the sagittal plane. The lateral and ventral thoracic and the abdominal muscles are mostly subservient to the functions of pulling the anterior limb posteriorly and medially, and pulling the sacral region down and forward during terrestrial progression. The musculature controlling the anterior limbs is discussed more fully under the latter heading, but it may here be noted that the absence of a clavicle in the Pinnipedia (as is usual in Carnivora), allows much freedom in movement of the shoulder, and the specialized functions involved may or may not have had some influence upon the conformation of the anterior thorax. In both the sea-lions and seals the noteworthy forward extension of the manubrium, partly bony and partly cartilaginous, has undoubtedly been brought about in response to the need for more powerful action of the pectoral muscles

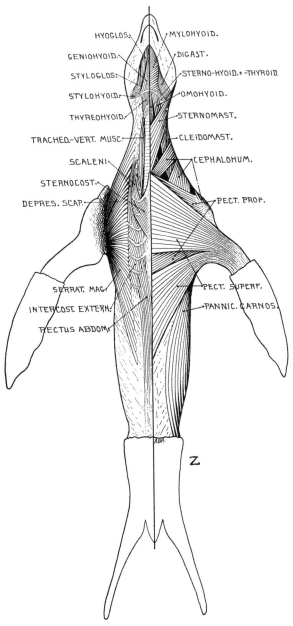

FIGURE 24. Ventral musculature of a sea-lion (*Zalophus*): superficial layer upon the right, and much of the next deeper layer to the left of the medial line.

[162]

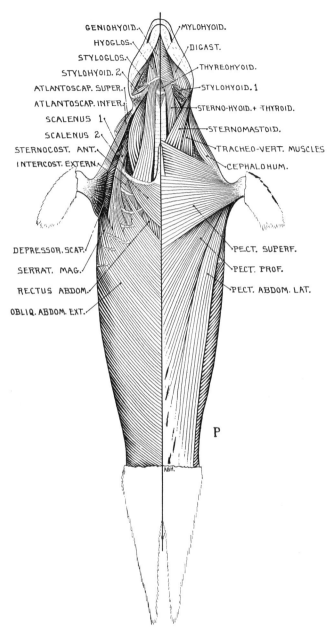

GENIOHYOID.
MYLOHYOID.
HYOGLOS.
DIGAST.
STYLOGLOS.
STYLOHYOID. 2.
THYREOHYOID.
STYLOHYOID. 1
ATLANTOSCAP. SUPER.
ATLANTOSCAP. INFER.
STERNO-HYOID.+ THYROID.
SCALENUS 1
STERNOMASTOID.
SCALENUS 2.
STERNOCOST. ANT.
TRACHEO-VERT. MUSCLES
INTERCOST. EXTERN.
CEPHALOHUM.

DEPRESSOR. SCAP.
PECT. SUPERF.
SERRAT. MAG.
PECT. PROF.
RECTUS ABDOM.
PECT. ABDOM. LAT.
OBLIQ. ABDOM. EXT.

P

ABH.

FIGURE 25. Ventral musculature of a seal (*Phoca*): superficial layer upon the
right and much of the next deeper layer to the left of the medial line.

during adduction combined with extension forward of the limb. In the sea-lion this correspondingly affects the cephalohumeral, part of which arises from the anterior border of the pectoral.

In the seal the panniculus carnosus muscle does not converge markedly to the arm pit, but the fibers have an even cranio-ventral inclination. This condition may have been purely an ancestral inheritance or may be useful in the wriggling motions accompanying terrestrial locomotion, but could hardly be of any help in swimming. Contrasted with this is the situation in the sea-lion, in which fibers of the postbrachial part of the panniculus all converge strongly to the arm pit. In this animal the sheet of muscle covers the knee and extends quite to the base of the tail, and as a result it is of great help in movement upon the land. Contraction of this panniculus assists in flexing the lumbar region so that the hind feet may be placed flat on the ground, and in galloping, contraction of the panniculus after extension of the anterior limbs helps in pulling forward the entire hinder end of the animal. In swimming the panniculus can act from the other end and lend power to backward thrusts of the fore flippers. This, combined with adductive motion, is also the function of the posterior part of the pectoralis.

In the seal lateral movements of the hinder end are prerequisite to swimming, and one would imagine that for it a panniculus of the sea-lion type would be very useful, but as already said this has not been developed, possibly because any purely brachial stimulus for it has been lacking. It is logical however, to expect lateral movements, accomplished mainly by the spinal musculature, to be markedly assisted by ventral muscles, and this is brought about by an extraordinary development of the pectoral muscles. Midventrally the latter extend from the sternum (deep pectoral part) practically to the pelvis, although muscle fibers do not occur quite so far caudad, while more laterally the abdominal pectoral virtually reaches the knee. This abdominal division is extremely heavy and thick in its anterior portion. The whole postbrachial part of the pectoral thus can operate to pull the hinder end sidewise while the dorsal musculature counteracts a downward pull by the posterior pectoral. But such flexion tends to pull the arm to the rear and this must be counteracted by antagonistic action of those lateral cervical muscles that are attached to the arm. Thus the arm operates mechanically as a sort of raphe, from which cervical muscles act in pulling the head to the side while at the same time the posterior pectoral is doing the same for the hinder end.

As the above is the case one might expect to find a somewhat similar situation in the dorsal muscles, and this is so to a modified degree. In the seal the spinal musculature is the chief agent in lateral movements of the posterior end. This has phenomenal breadth, as well as considerable thickness, in the lumbar region. The sublumbar musculature is well, although not remarkably, developed, but the iliocostalis and longissimus have expanded and where the latter is attached to the ilium this bone has turned laterad to provide greater accommodation for the muscular connection. As the iliocostal is the most lateral division of the erector spinae it naturally is of most consequence in the lateral movements employed for swimming. It might very well be expected to maintain its robustness as far as the occiput, but this it does not do. As it continues onto the thorax from the lumbar region it gradually thins and virtually disappears over the anterior thorax. The function of the iliocostalis is therefore almost exclusively for operating the hinder end in swimming. The forward end of the animal must consequently be controlled by a different group of dorsal muscles. As with the more ventral anterior muscles concerned in swimming, this is accomplished partly by muscles extending from the arm to the head or neck. The splenius, humerotrapezoid, cephalohumeral and both atlantoscapulars are better situated in the seal than the sea-lion for purely lateral motion of the head, and for antagonism the spinotrapezoid projects farther back in the former, while the latissimus is more extensive and is double. The conformation of the latter indicates that it might be of distinct aid in sidewise curving of the hinder end, but it seems probable that the chief stimulus for its development was to act in antagonism to prebrachial musculature.

By the above statements it is not meant to imply that only those muscles mentioned are used by the seal in swimming. There are present extremely heavy subvertebral cervical muscles which I suspect are used in antagonism to translate into lateral movement the action of the semispinalis capitis, which otherwise would largely result in raising the head. Similarly almost every muscle of the body and neck should have some use in swimming, but the actions of many of them in this connection are obscure.

The sirenian skeleton is remarkably heavy and dense, especially in the manati. This is popularly believed to be for the purpose of enabling these animals more readily to sink from the surface to their pastures upon the bottom. This may be the proper explanation but it should not be accepted without considerable reservation. Sirenians descend to very moderate depths. Why should they be better fitted for doing so than whales, which descend to great depths? Also it would seem of at least equal importance that they should be able to rise to the surface with celerity. As a matter of fact they have doubtless adopted a middle course and their specific gravity is probably almost the same as that of the water which they displace. If this be the case then their bones are particularly dense to compensate for the unusual lightness of the rest of the body.

As sirenians never leave the water the thoracic stimulus of fore leg support usually present in the terrestrial mammal is entirely lacking, as it is in cetaceans, with this difference, that in sirenians the anterior limb is much more mobile. Appreciable gravitational pull upon the thorax is also lacking, and there remain as discernible influences only the stimulus for an aquatic mammal to assume a circular thoracic cavity, and the effect that levitation by the inflated lungs may have. As a matter of fact in cross section the chest seems to be definitely broader than high. Levitation by the lungs might be expected to raise the curve of the ribs well above the vertebral column, but this is no more marked than the condition encountered in many terrestrial mammals, both quadrupedal and bipedal.

Sirenians have experienced quite a remarkable lengthening of the thorax which has operated to shorten the lumbar region. There seems to be considerable variation in the number of ribs. Stannius recorded a manati with 15 ribs while Murie encountered individuals with 16, 17 and 18 pairs. *Halicore* may have 18 or 19. In the latter genus the conformation of the ribs is not unusual, but in *Trichechus* they are very remarkable. In this animal the individual thoracic vertebrae are much longer and therefore the rib centers are farther apart. Presumably in compensation the ribs are phenomenally broadened so that the intercostal spaces are not particularly wide. Furthermore in this genus (at least in the species *latirostris*) the distal ends of the first twelve ribs are on a line virtually parallel with the vertebral column, the remainder of the series becoming successively shorter. It is said that always in this order all ribs have both capitular and tubercular attachment to the vertebrae. This character increases the rigidity of the thorax and might be expected to accompany a long sternum with strong and well calcified sternal ribs. On the contrary the sternum is much reduced and in the manati especially about two-thirds of the costal series apparently have no costal cartilages at all, their atrophy leaving a relic in the shape of nodular bony growths upon the distal extremities of the true costae. Flower has illustrated the sternum of a young *Halicore* in which four pairs of costal cartilages join the manubrial-xiphoid interval (figure 3). In a mounted skeleton of this genus in the National Museum, with sternum completely

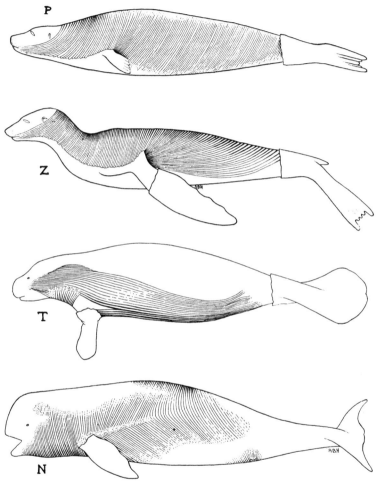

FIGURE 26. Panniculus carnosus and cranial continuation of (P) seal (*Phoca*), (Z) sea-lion *(Zalophus)*, (T) manati *(Trichechus)* redrawn from Murie), and (N) porpoise (*Neomeris*).

ossified, there are but three pairs of ribs extending to this area, and also three pairs in a *Trichechus*, the first, however, apparently attached to the posterior part of the manubrium. In the Sirenia, therefore, there is encountered the condition of a thorax extremely strong above and extremely weak below. It accordingly seems likely that as the Sirenia do not descend very deeply the aquatic pressure upon the thorax is relatively uniform. The thorax does not need to adapt itself for fluctuating compressibility and the attachment of the ribs to the vertebrae is therefore less elastic. A necessity in this connection would seem to be that the animal breathe almost entirely by the diaphragm.

In Sirenia the articulations of the vertebrae in the lumbo-caudal series are definitely reduced, this being most marked in the manati. Anapophyses are lacking as are also well defined postzygapophyses from this series. Metapophyses are present in the manati although they are fairly fused with the prezygapophyses. In the dugong there are no metapophyses occurring as processes in the thoracico-lumbar series and the zygapophyses do not project laterad but are flush with the bony laminae of the spinous processes. In the manati the spines and zygapophyses are rugose with numerous small but well defined prominences. In the dugong the spinous processes are of moderate and approximately equal height in the thoracico-lumbar series. In the manati the spines are not quite so long. Murie (1872) has stated that in the manati the intervertebral substance is very limited in amount, being no more in thickness than a tenth of an inch, and this seems quite surprising in view of the fact that in other aquatic mammals this substance is unusually generous in amount, and in consideration of the extent to which this animal can curve its body and tail. The latter is often used as a prop (see figure 7) to keep the body from resting on the ground while feeding, and the curvature of the lumbo-caudal region is then excessive.

From Murie's excellent illustrations it is seen that the panniculus carnosus of the manati is a most extraordinary muscle. Although some of its fibers stretch to the fore limb, the main body of the muscle, which may be as much as an inch and a half in thickness, extends from the pelvic region to below the eye, in a broad, powerful sheet (figure 26). It reaches quite to the midventral line, and passes both laterad and mediad of the arm, the sphincter colli then becoming more complex over the neck. Clearly this great muscle should have two functions; one as an accessory belly strengthener, as pointed out by Murie, to support the viscera and partly to make up for the absence of costal cartilages of the usual sort. The other function is to help bend the body in swimming. When one side of the panniculus is contracted there evidently follows a twisting, more or less lateral motion, but when both are flexed together,

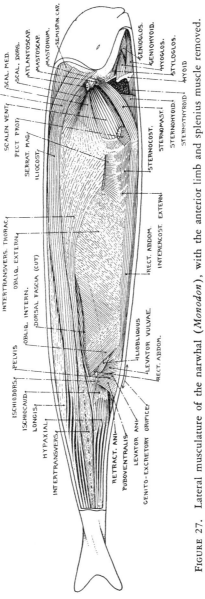

FIGURE 27. Lateral musculature of the narwhal (*Monodon*), with the anterior limb and splenius muscle removed.

[169]

curvature is in the vertical plane, the result being that the tail is depressed.

The manati thus effects swimming by flexion of the hypaxial caudal mass, and of the panniculus and rectus abdominis to depress the rear half of the animal, and this may be assisted by some of the neck muscles acting to depress the head. The hypaxial muscle is less extensive than in the Cetacea because the lumbar region is shorter. It is made up of a massive superficial and a robust deeper division, both caudad of the last rib, and in addition, a smaller muscle which reaches within the thorax and which probably represents a quadratus lumborum. The antagonist of this ventral group is, of course, the erector spinae. The divisions of this seem to be well fused, as one would expect. A spinalis dorsi continuous with a levator caudae internus was said to have been distinct, while a broad sacro-lumbalis was confined to the thorax.

As previously indicated one cannot be sure whether the theoretical mechanical stimulus encountered by aquatic mammals is for an exactly round thoracic cavity, or a round trunk as a whole. Personally I consider it likely that this stimulus alone is rather feeble, or else that it is overcome by much stronger ones. Just as a combination of factors has resulted in the assumption of a trunk laterally flattened by speedy pelagic fishes such as the mackerel, so might one expect to find that the swiftest whales are so flattened but in the sagittal plane, to correspond with the different plane of tail motion. As a matter of fact these two conditions are complicated by at least two factors that should influence tail shape, as discussed in the next chapter.

The thoracic shape in whales is so variable that all our arguments fail of application. It is impossible to ascertain the precise thoracic cross section of a living, adult mysticete. Mounted skeletons indicate that the cavity is either slightly broader than high, or else is approximately circular. Allowing for the spinal musculature this would make the entire trunk either approximately circular or else higher than broad. And the latter seems to be the situation in at least the majority of porpoises, to a really quite marked extent in some. Levitation by the lungs may have been of influence in shaping those with broadened thorax, but this argument takes us no farther. We might expect to find that the very strong levitation of the lungs during deep diving had elevated the dorsal curve of the ribs well above the vertebral centra, but this is no more marked than in many terrestrial mammals.

As with the shape of its cross section there is really extreme variation in the length of the cetacean thorax, it having become elongated in some

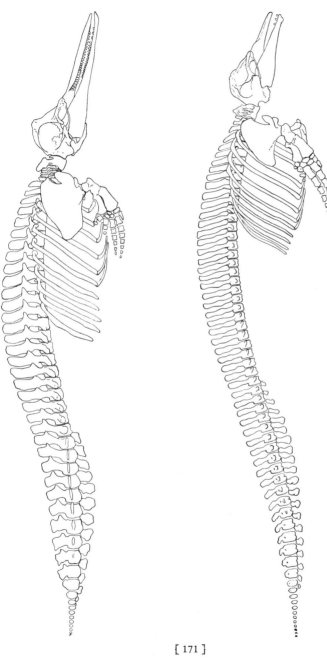

FIGURE 28. Skeletons of toothed whales (*Delphinus sinensis* above, and *Grampus griseus*, redrawn from Flower) to illustrate approximate maximum and minimum development of the interlocking of the vertebrae (prezygapophyses).

sorts and shortened in others. *Neobalaena* has no less than 17 pairs of ribs, while *Hyperoodon* has 9, which is the least number in any living mammal. Other sorts vary between these two extremes. There is indicated some instability in the anterior thorax by the fact that a few cetaceans have a pair of rudimentary cervical ribs, and that occasionally in mysticetes the first rib is bifurcated, one head going to the seventh cervical and the other to the first thoracic vertebra. It is perhaps usually stated that some whales exhibit this condition, which gives the impression that it is a character of certain species. As far as I can ascertain, however, it is an individual though perhaps fairly common development. Thus no mounted mysticete skeleton in the National collection shows any bifurcation of the first rib, while the Sei whale obtained in Japan by R. C. Andrews (1916) does show it. In view of the fact that the more primitive fossil whales all had seven cervical vertebrae, so far as known, the two details mentioned above might well be interpreted as an indication of morphological effort to attain, in a somewhat different manner, a forward shift of the thorax. But against this reasoning there is the totally antagonistic fact that *Neobalaena,* anomalous in so many ways, has the first eight vertebrae without ribs, so that in this genus there seems to have been a shift of the thorax toward the rear.

The thorax of whales varies in the strength with which the ribs are attached to the vertebral column. Ribs with both tubercular and capitular attachment vary in number among toothed whales from at least eight (and possibly more in some sorts not examined) in *Berardius, Kogia, Monodon,* and *Phocaena* to as few as four in *Stenodelphis* and some individuals of *Tursiops.* It has been stated in the literature that the ribs of existing mysticetes are all single-headed, but this is erroneous. Conditions are variable and the articulations of the ribs are certainly reduced, but all individuals of this group which I have examined show at least a few ribs that cannot be called single-headed, possibly with the exception of *Megaptera.* It seems normal for the first rib to have no capitular projection toward the centrum. In *Eubalaena* this is also lacking from the second rib, it is slight upon the third, and upon the fourth reaches half way from the tuberculum to the centrum, thence gradually shortening in caudal sequence. In *Rhachianectes* the second to sixth ribs inclusive have capitular projections that almost reach the centra. In *Sibbaldus* the second, third and fourth ribs show this character, while in *Megaptera* there is some capitular projection upon the third rib, which, however, falls considerably short of the centrum, and this is still less defined in the more posterior ribs. Thus, although there may not be bony connection between capitulum and centrum, a process representing the former undoubtedly indicates that there is strong ligamentous connection.

Accompanying the increased elasticity of the dorsal thorax is a comparable development of the ventral thorax brought about by a reduction in the costal attachments to the sternum. This is attained partially by an increase in the number of the so-called floating ribs through the elimination of the cartilages which more usually are attached to the series of costae often but unfortunately termed false ribs, and by a reduction in number of the true sternal or cartilaginous ribs, which in odontocetes seem invariably to become calcified with age. Reduction of the sternal ribs is inevitably accompanied by a corresponding reduction in the length of the sternum. Atrophy of the latter complex does not follow any one rule but may be attained by elimination of any of the posterior elements. The xiphoid and posterior sternebers may disappear, but whales apparently never show the condition encountered in *Halicore,* in which the sternebral series has become so reduced that this has no individual centers of ossification (apparently), several costal cartilages thus having attachment to the xiphomanubrial interval, although this later experiences general calcification.

That a reduction of the cetacean sternum is now in progress is indicated by the statement of Beddard (1900) that in *Phocaena* it shrinks progressively from the young to the adult state. Among Odontoceti there may be seven pairs of ribs with sternal attachment *(Berardius),* six pairs *(Monodon),* or the sternum may be much more reduced, the elements finally fusing in the adult, and resulting in a single bone that may be broader than long (as *Neomeris).* Even when the sternum consists of several elements complete ossification of the whole often occurs in old age. In all porpoises without exception, I believe, the manubrium is broader than any other sternal element, is indented antero-medially, and produced in a process antero-laterally (chiefly for the sternohyoid muscle), with a lateral expansion for the pectoralis minor attachment. The sternum of *Physeter* is of peculiar form and Flower (1876) considered that it consists of a manubrium and two sternebrae; but Owen stated that it has four elements. In all odontocetes at least the ossification of the twe centers of each sterneber is delayed. Occasionally (as in *Monodon,* figure 31) two pairs of ribs have attachments well forward upon the first sternal element and there thus seems to be some question regarding the homology of the latter. Either all or most of the ribs have shifted forward, thus placing the second pair upon the manubrium, the latter and first sternebra have experienced first crowding and then fusion of their ossification centers, or, more likely, the two centers of the manubrium were first separated by an antero-medial indentation, then becoming atrophied, and the first sterneber finally became hypertrophied to take the place of the true manubrium. This point should be more thoroughly investigated when adequate material is available.

In mysticetes the sternum is still further reduced to a single bone, evidently the manubrium, to which is attached but the first pair of ribs. It is frequently stated that this bone is of some particular shape in one species and of another shape in a second, but as a matter of fact the shape appears to be almost as variable individually as specifically. It is probably always more or less roughly triangular or heart-shaped in the balaenid whales, while in the others it may be either U-shaped, T-shaped, cross-shaped or rather irregular. Usually, however, there is a posteromedial projection forming with the lateral process a concave border into which fits the end of the first rib. Certainly in these whales the sternum has suffered as much reduction as is possible without a change in the attachments of (chiefly) the smaller, more fleshy pectoral (when there are two divisions) and of the sterno-mastoid.

Winge (1921) stated that when there is a tendency for the costal cartilages to disappear the sternum loses an essential stimulus and becomes reduced. This is self-evident, whether it be considered that the sternal elements are morphologically derived from the costae, or (according to a later thesis), that the derivation of this complex was originally from the medioventral part of the shoulder girdle. Schulte (1916) ascribed to J. C. Vaughan the verbal opinion that in descending to great depths the pressure of the water on the abdomen would force rostrad the diaphragm, which would in turn force the ribs out, accounting for the reduction of rib connections and of the sternum. Müller (1898) believed that the reason the mysticete sternum is more reduced than in odontocetes is that the diaphragm of the former is less muscular, indicating a greater use of the thorax in breathing.

The sternum as a whole can hardly suffer reduction in the number of its elements after the fashion characteristic in Cetacea without corresponding reduction of the costal cartilages. That the latter is not necessarily dependent upon decrease in strength of the costal-vertebral articulations is suggested by the fact that in the manati these articulations are strengthened, while the sternal ribs are considerably reduced. While not absolutely dependent one on the other it is but logical to infer that increase in the elasticity of the dorsal thorax has usually gone hand in hand with the same adaptation ventrad. Then there is the matter of the

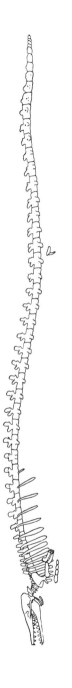

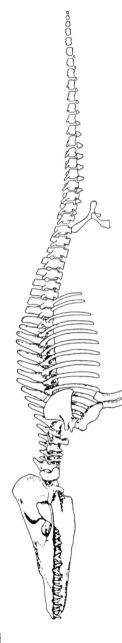

FIGURE 29. Skeletons of zeuglodonts. *Basilosaurus* above (redrawn from Gidley), and *Zeuglodon osiris* (redrawn from Abel).

function of the ventral thorax as a scaffold for muscular attachments. The costal cartilages cannot be reduced unless conditions permit of alteration in the costal attachments of the transversalis abdominis, and in the more ventral of the intercostal muscles. The sternum cannot be reduced without alteration in certain aspects of the rectus abdominis attachments and reduction in origin of the pectoralis. Chief of these are the latter two. The major sternal attachment of the rectus may easily shift forward anyway, and an extensive origin of the ectal pectoral may not greatly involve the sternum proper but only the linea alba. Where there is a heavy anterior pectoral mass attached broadly to the bone, which is the usual character of the minor division when this is present, or of the major if this be greatly reduced, this affects the sternum mainly in a broadening of the manubrial part. Hence it appears likely that muscular adjustments permitting the shortening of the sternum are easily made. In this connection there should also be mentioned the possibility that the extraordinarily developed gular musculature of mysticetes may well have been instrumental in reducing the sternum in this group.

In fine, all we can logically infer regarding the reduction of the sternum in the Cetacea may be summed up in a single paragraph. There is lacking the brachial and muscular stimulus for a sternum of moderate size. Vaughan's opinion that water pressure at great depths would force the diaphragm forward hardly seems well taken. Such pressure is naturally the same over all parts of the animal at any given depth and we know that at great depths external and internal pressure *must* be equalized, for no thorax could otherwise withstand a pressure in excess of a ton to the square inch. It therefore seems logical to infer that the increased elasticity dorsad and the reduction of the sternum in this order is to allow for the amount of compressibility of the thorax needed to prevent water pressure from cracking the ribs.

The vertebral column of the Cetacea is noteworthy for the very marked reduction (least pronounced in *Platanista*), which really amounts to entire abandonment, of interlocking or articulation of the vertebrae, and increase in size of the intervertebral disks, both of which are modifications to increase the uniformity with which the column may be curved without tendency to bend at one or more particular points. It is permitted by the fact that the body has been supported by flotation only for a very long period of time and, incidentally, is a step in retroversion toward the primitive chordate condition. As support is by flotation acting upon all parts of the body it might be expected that the static position assumed by the vertebral column would be a straight line. The column, however, is affected by the fact that it is located above the longitudinal body axis and by tension of the muscles that extend toward the head on one hand and the tail on the other, and in life is slightly curved to a greater or lesser degree depending on muscular differences and body shape.

The terminal epiphyses of the cetacean vertebrae are very distinct and become thoroughly ossified into thin disks which fuse with the bodies very late, especially in balaenid whales. Flower (1876) has stated that whales appear to differ from all other mammals inasmuch as the neuro-central suture is always placed a little above the junction of the arch with the body. Anapophyses and postzygapophyses are absent as true processes. Prezygapophyses may be well developed and present upon all vertebrae, as in Mysticeti, *Stenodelphis, Monodon, Mesoplodon*, etc., indicating that there is definite zygapophyseal articulation (at least by ligaments) throughout the entire column; or on the other hand they may be totally absent from the central lumbar series, becoming gradually differentiated craniad upon the first few lumbar and caudad on the last few, or even entirely eliminated (*Grampus griseus,* figure 28). This shows that in the area where they are not well defined zygapophyseal articulations do not exist, which increases the limberness of the column by just this much. Metapophyses are developed and in caudal sequence gradually arise from low upon the neural arch to the base of the spinous process. In the anterior thoracic series the di- and metapophyses usually occur unseparated, I believe, the diapophyses in most sorts of whales gradually separating and descending from high on the arch to the centrum, thus becoming parapophyses. In the Physeteridae, Kogidae, and the Ziphiidae, however, there is a different condition, for the diapophyses do not gradually change to parapophyses. On the contrary the former diminish and disappear, while at the same time the latter become defined well ventrad of the diapophyses and gradually increase in size.

As previously remarked most cetologists refer to the phenomenal shortness of the whale's neck as though this character were entirely disassociated from thoracic conditions. As a matter of fact this shortening process merely culminates in the neck, but also involves the anterior third of the thorax. In mysticetes there is an almost insensible, gradually increasing shortening of the centra from about the fifth thoracic to the third cervical (in those sorts with free neck vertebrae). In most porpoises which have the posterior cervicals unfused these may be relatively thinner, to a considerably greater extent than would be possible for any thoracic vertebra bearing a rib. Consequently there is usually a more abrupt transition in the thickness of the centra between the thoracic and cervical series. The proper interpretation of these facts seems to be not merely that the cetacean neck has become markedly shortened, but that all vertebrae anterior to the thoracic pivot of motion have experienced strong stimuli for shortening, to which they have responded to various degrees according to their capabilities, this shortening having been necessary in order that the animal might assume the fusiform shape best fitted for swimming.

Winge (1921) has stated that the abandonment by the fore limbs of the function of supporting the body results in the reduction of the height of the anterior thoracic spines. This character, however, is only secondarily dependent upon function of the anterior limb. If in quadrupeds the fore limb be of greater importance during locomotion than the hind, then not only must the spinal musculature have corresponding strength in this region but the spines must be higher to give better leverage for muscles that elevate a heavier head and neck (as in *Bison*) (Thompson, 1917). When the fore limb is no longer used for support the anterior part of the erector spinae usually becomes weaker (phocids, bats, etc.), or it may undergo an independent modification for increased strength, resulting in spinous processes that are phenomenally developed, as in most Cetacea. The height, and to a lesser extent the character, of these varies considerably, however, reflecting corresponding variation in the spinal musculature. In *Mesoplodon* the height of the spines of the entire column is quite phenomenal. In mounted skeletons of *Delphinus* and *Lagenorhynchus* the spines are relatively broad sagittally and so close together that it is difficult to understand how the requisite amount of movement would be possible; but perhaps in life the intervertebral disks are unusually thick. In *Globiocephala,* on the other hand, there is sufficient distance between the spines for great mobility of the column. Winge (1921) has said that "bending of the column in the vertical plane—is reduced or abandoned." This is an extraordinary statement in view of the fact that all the modifications of the vertebrae with the possible exception of spinous height are for increase of mobility and all locomotor movements in whales are instigated by motion of the column in this plane.

In some respects the lumbar region of Cetacea really constitutes a part of the tail, from a functional standpoint, but certain of its characteristics may be discussed in the present chapter. Flower and Lydekker (1891) have stated that *Neobalaena* and *Inia* have but three lumbar vertebrae. At the other extreme *Grampus* may have 21, *Lagenorhynchus*, 23, and *Delphinus,* 24 (21 in another skeleton). This of course, is figuring on the

basis that there are no sacrals present, the length of the lumbar series being determined by the position of the first chevron bone of the tail. As
a matter of fact we have no assurance that in such a mammal without a
sacrum there can not have been great alteration in the chevron compliment so that the first of these bones might now occur anterior or far
posterior to its ancestral position, meaning, in the latter case, that the
first few of the series have been eliminated. Either this has been the case,
lengthening of the lumbar region having been brought about by absorption of the anterior caudal elements, or else this has been due to the lay-
ing down in embryo of accessory lumbar vertebrae anlage. It is, how-

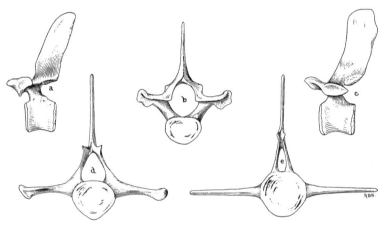

FIGURE 30. Vertebrae of the porpoise *Phocaena*, illustrating differences in the
positions of the various details: (*a* and *b*) fourth thoracic; (*c*) eighth
thoracic; and (*d* and *e*) lumbar vertebrae.

ever, probably not astonishing that there should be such lumbar variation in a mammal in which there is no attachment of the pelvis to the
vertebrae and in which the apaxial and hypaxial musculature have each
experienced such a complete degree of fusion. It seems that this would
involve merely a shifting forward or backward of the pelvic region by
shortening or lengthening of certain muscles with little regard for the
bones, and this should not entail any great difficulty. Presuming that
this has been the case, I have no idea regarding the stimuli involved or
the advantages gained. There is just one point that might throw some
light on the question. In at least most cetaceans having short or fairly
long (as many as about a dozen vertebrae) lumbar regions, prezygapophy-

ses are continuously and uniformly present as far as the peduncle and in these, as far as can be told from bony details, the post-thoracic apaxial musculature is relatively homogeneous, so that the lumbar region seems to constitute the base of the tail. In porpoises with an excessive number of lumbar vertebrae it seems that all spines except those near either end lack zygapophyses, so that in these sorts there are involved two regions, in some manner separately specialized, of the lumbo-caudal part of the erector spinae, one in the lumbar and the other in the caudal area. Conditions suggest that a need experienced for longer lumbar muscles may have been instrumental in lengthening this region in certain porpoises. This in turn suggests that there may be some difference in the exact muscular action by which these two sorts of cetaceans accomplish swimming.

In Cetacea the spinous processes and parapophyses, constituting simple transverse processes, of the lumbar series are greatly developed, while other bony protuberances are either much reduced or entirely absent. This corresponds to the simplification of the spinal musculature. The whale swims by movements in the vertical plane of the caudal appendage or flukes, and the spinal muscles are called upon for little else. Naturally, with almost perpetual use these have acquired phenomenal thickness. In the lumbar area the dorsal muscles are imperfectly divisible into two series (iliocostalis and longissimus) while in the posterior thorax these may be even more homogeneous, and continue onto the head with little change either in mass or character. The muscular action concomitant to swimming has already been discussed to a considerable extent so that here little need be said, save to repeat that in whales, as contrasted to seals, the back muscles can act more uniformly throughout the entire vertebral length, permitting more latitude in possible shifting of the pivot of motion.

The erector spinae operates to raise the tail. Consequently in whales there must be an antagonist to depress the tail with potential force that is approximately equal. This is provided by the extraordinary modification of the infravertebral or hypaxial musculature of the tail base and lumbar region. In the latter situation there are presumably psoas and quadratus lumborum elements, but it is entirely out of the question to decide whether all of these have become hypertrophied or some have increased at the expense of others, for anteriorly they are so simplified that only one superficial separation is possible. In what may be termed the posterior lumbar area the hypaxial and apaxial musculature are of approximately equal mass, but whether they are each capable of exerting

[180]

precisely the same force is unknown. If there be some inequality involved, then this must be compensated for by tilting of the flippers. Beyond question, however, there is a tendency for perfect equalization,

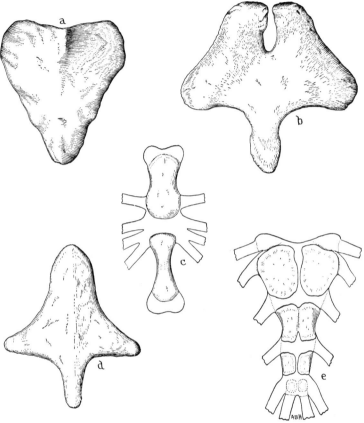

FIGURE 31. Sterna of whales and sirenians: (*a*) *Balaena* and (*c*) young dugong (*Halicore*) (both redrawn from Flower); (*b*) *Sibbaldus* and (*d*) *Balaenoptera borealis* (both from specimens in the U. S. National Museum); and (*e*) young narwhal (*Monodon*) (from a photograph in the American Museum of Natural History).

which presumably has been accomplished as fully as the muscular conformation will allow. In respect to the latter point, the hypaxial muscles continue robustly only to the last rib, and thence disappear within two or

three costal spaces. Hence, while the apaxial muscles operate throughout the entire length of the vertebral column, the hypaxials can bend
only the lumbo-caudal series. Anteriorly what downward bending of the
posterior thorax and of the head is necessary in swimming must be taken
over chiefly by the powerful rectus abdominis on the one hand, and the
ventral neck muscles on the other. In this connection a point should be
mentioned that may already have occurred to the reader. In theory the
swimming motion of whales has been discussed as though the curve assumed by the body were a perfect arc. In practice, however, this may not
be so. Not only may the costal equipment prevent as much possible curvature of the thoracic vertebrae as of the lumbo-caudal series, but this
may further be reduced by the greater mass in the former region. It
might therefore be more correct to consider that the neck and tail each
bend from the thorax to a considerable degree, while the curve of the
thoracic vertebrae is more moderate. This, at the present time, is almost
impossible of determination.

Theoretically lateral motions by the tail are unimportant to the Cetacea,
but actually there must be considerable strength in this plane for thrashing about when the need arises. Such movement may be accomplished
chiefly by the flexion of the apaxial and hypaxial muscles of a single side,
and also by the intertransversarii, which occupy the space between the
transverse processes of the lumbo-caudal vertebrae. In cetaceans these
are unusually modified and upon the thorax, of at least most odontocetes,
spread out into a thin sheet covering a large part of the ribs deep to the
latissimus dorsi layer. In *Monodon* there is a convergence of the fibers
from each direction to (about) the sixteenth lumbo-caudal vertebra, indicating a center in this region for lateral movement. It is likely that the
intertransversarii act fully as much in preventing too much curvature as in
instigating it. In other words the function is probably as much static as
active.

Chapter Nine

The Tail

THE QUESTION of the physiological development of the hinder end of an aquatic mammal for the function of primary propulsion is an involved one. Of certain points one can be sure, while regarding others there is some uncertainty and only probabilities may be advanced, this for the reason that there are indications that in particular cases the development has not been in a straight line but has been somewhat by trial and error. As a fundamental concept, however, I have no hesitation in making the unequivocal statement that the evolutional tendency when a mammal takes to a free-swimming type of water habitat is always for it eventually to develop the rear end into the primary, oscillating organ for its propulsion through the water. If it does not do so it is a sign that it originally made a wrong start or that it has encountered antagonistic stimuli of such strength that it was diverted from the most efficient evolutionary development. In theory it makes no difference whether this ideal propulsive force is furnished by the flattened tail as in the whale or by the adpressed hind feet of the seal. In practice, however, it is probable that the muscles of the seal that are involved in swimming can never become as homogeneously specialized for a single function as are those of the whale's tail. I do not know and consider that speculation on this point would be well nigh useless.

The important point is that although swimming by the whale and the seal entails widely different muscular action, the principle is in theory the same, and this principle is the only one that a mammal can adopt which is thoroughly economical in practice, for it is the only one by which there is no lost motion or energy. In the case of other sorts of swimming motion by the hind feet, or by the fore feet, either recovery motions are necessary or a part of the flippers must overcome water resistance—alternatives which detract by just so much from the propulsive strokes.

For a proper consideration of the situation some brief recapitulation is necessary. When a mammal first takes to the water it has a caudal equipment that may be divided into three categories. It is either without a functional tail, like the bears for instance, it has a long tail, or this member is of intermediate length. If the tail originally was short it is likely to remain so and never be of economic value to the animal, in which case the limbs will be the members developed for propulsion. If the tail be long it is extremely likely that it will eventually become the prime means of swimming, *unless* it be diverted at a rather early stage in its development by the necessity for fulfilling some other function. If it be of intermediate length one of two things may happen, depending upon its exact size originally and the other conformational features of the mammal. In this must be included inherent capacity for change in the desired direction. It may become shorter and cease to be of consequence or it may become longer and of greater importance. In illustration of the possibilities in this line it may be mentioned that in the case of a mammal with moderate length of tail we can never predict what the development will be because the antagonistic stimuli may be too nearly equal. Although the modification is entirely according to law it appears to us as fortuitous whether the tail or the hind limbs gains the initial ascendency and implied evolutional velocity of the chief swimming organ.

In approaching this subject of caudal evolution one must clear his mind of all idea that this member changed to the form in which it may now occur in the most specialized of aquatic mammals for some single fundamental reason. True, its development has followed definite laws and in *most* cases it is predictable what form the tail will finally assume merely by observing the methods of swimming which any mammal now employs, but the development is, nevertheless, step by step. Thus the current belief that the cetacean tail is flattened vertically so that the animal may more readily ascend to the surface for breathing is not only erroneous in practice, but involves an improper mental approach to the entire subject of aquatic specialization. Not only can an animal with horizontally flattened tail ascend as easily for the reason that elevation of the body is accomplished by the equilibrating rather than the propelling mechanism, but no animal could start on its evolutional career with any such particular end in view. The direction in which the tail will finally be flattened is dependent not upon the tail itself but upon the direction in which it is involuntarily (usually) moved during swimming movements of the unspecialized ancestral form.

For ease in discussion the tails of aquatic mammals may be divided into two classes, which involve two fundamentally different principles: Those which are narrow in the horizontal plane or which will eventually become so; and those which are flattened in the vertical plane or will eventually become so. In these two classes are included even such tails as are too short to be of any possible use for propulsion, and they thus comprise all aquatic mammals except the capybara, which is entirely devoid of a tail.

1. Tails narrow in the horizontal and broadened in the vertical plane. In the first group below are included all those genera in which the tail has been flattened in the horizontal plane, or else is provided with a ventral keel of stiff hairs. In the second table are those genera in which the tail is still terete but which may be expected eventually to develop a narrow tail rather than a flat one. There is some doubt in this regard respecting *Chironectes*, as explained later.

Tail narrow	Tail round
Desmana	*Chironectes*
Galemys	*Neosorex*
Neomys	*Atophyrax*
Chimarrogale	*Myocastor*
Crossogale	*Dasymys*
Nectogale	*Nilopegamys*
Limnogale	*Arvicola*
Potomogale	*Neofiber*
Crossomys	*Rheomys*
Hydromys	*Anotomys*
Parahydromys	
Ondatra	
Ichthyomys	
Hippopotamus	

It will be noted that all the above mammals having horizontally flattened tails are either insectivores or rodents with the single exception of the hippopotamus, and in none except *Potomogale,* and possibly *Limnogale,* does the tail constitute the principal organ of propulsion. Also that none, except *Potomogale* perhaps, is really very highly specialized in an aquatic direction.

From a study of the question I am led to believe that in order that an aquatic mammal shall finally acquire a tail that is horizontally flattened its terrestrial ancestor must have had the following characteristics: a normally cylindrical body not particularly elongated; a tail, preferably of considerable length, which was not much enlarged at the base, thus showing an abrupt transition in size between the hinder end of the body proper and the base of the tail (of the character occurring in typical rodents, as the rat); feet of the normal rodent or insectivore character, in which the hind feet are sufficiently larger than the fore feet so that there will be no doubt but that the former will be of greater importance to locomotion while the animal is swimming "dog-fashion".

Given the above characters the course of aquatic development will ordinarily be as follows: When the animal first takes to the water, swimming will normally be accomplished by the movement of all four feet. The hind feet, however, being larger than the fore feet will be depended upon more and will gradually become larger, acquiring webbing or a fringe of bristles. At this stage the fore feet are only an incidental aid to locomotion. Swimming will normally be by alternate strokes of the hind feet. I accept this as the most efficient method that will almost invariably be employed merely because it is the rule, in birds as in mammals. Swimming by alternate strokes of the hind feet involves wriggling the hinder end of the body, which will cause a sinuous motion of the tail. This latter will aid forward locomotion in degree according to the area of the lateral tail surface. I have no hesitation whatsoever in stating that this horizontal lashing of the tail, so well illustrated in the case of the muskrat (*Ondatra*) constitutes a strong stimulus for lateral flattening of this member. Why this is so no one knows, but the evidence is overwhelming that just such stimulation will initiate development in a useful direction. It may be purely natural selection, it may be chiefly because the friction of the water against the upper and under sides of the tail tends to develop a dorsal and ventral thickening, or a complex of unknown factors may be involved. At any rate, as the flattening of the tail progresses it will be of greater and greater proportional importance in swimming, for it is theoretically much more efficient than can be the alternate kicking of the hind feet because the latter necessitates recovery motions. Finally this specialization of the tail will increase possible speed to the point at which movement of the hind feet would be more of a hindrance than a help and use of the feet as a primary, or even secondary, means of speedy aquatic locomotion will be abandoned. The final step in this direction would be the assumption of a fish-like tail comparable to that of the whale, but with flukes vertical instead of horizontal, and presumably the elimination of the hind feet. No mammal is thus developed, for none with horizontal flattening of the tail is independent of the land. All are of rather small size, are inhabitants of streams more properly than of large rivers, and are not yet very highly specialized in an aquatic direction, *Potomogale* being the most modified of the lot.

The above thesis is perfectly consistent with the facts except in two instances, one of these being *Hippopotomus* and the other *Potomogale*. The size of the tail in the former animal is entirely too small to be of the slightest use either in swimming or in steering, and yet it is laterally flattened to a phenomenal extent. The only explanation for this condition would seem to be that in spite of its small size it has been wriggled from side to side by the alternate strokes of the hind limbs for a sufficiently long time for it to have responded to the same stimulus that would have affected it were it sufficiently large to be an aid in propulsion. The case of *Potomogale* is more difficult. Its tail is remarkably efficient as a flattened propulsive organ and yet the feet are small and so unmodified that we are justified in assuming that they are not used while the animal is in the water. It seems that the only logical theory is that originally the terrestrial ancestor of this mammal was equipped with small feet while its tail, although round, was unusually heavy, especially at the base, so that in connection with a slim body, it was even then a more efficient swimming organ than the small feet. If it came more natural for this terrestrial ancestor to move the tail from side to side, as a dog does, rather than in the vertical plane, flattening in the horizontal plane would be the inevitable result.

The insectivores and rodents listed above as having terete or round tails can be dismissed with the statement that they are either not sufficiently specialized for there to have been caudal change, or else this member has been unusually conservative. The case of the water opossum *(Chironectes)* is somewhat different. The webbing of its hind feet is quite extraordinary in degree, and it is difficult to understand how this was brought about so completely without some flattening of the tail, *providing* that the animal swims by alternate kicks of the feet— movements which would of necessity involve the wriggling of the tail from side to side and this should be easy of accomplishment because of the robustness of the tail base, as in so many marsupials. There is no authentic statement, I believe, of the exact method of swimming employed by this animal and it does not seem impossible, although it is improbable, that it might propel itself by means of kicking the feet in unison, and if this should ultimately prove to be the case, then the roundness of its tail would be expected.

It seems that at least in insectivores and rodents the first change in the tail brought about by their aquatic life is usually the acquisition of a border of hairs, above or below or both, that are longer and stiffer than those upon other parts of the tail. *If* there be anything to the theory of the eventual inheritance of acquired characters throughout lengthy periods of time, then it seems that a hairy border was in some way stimulated to growth by the endlessly repeated lashing of the tail from side to side during propulsion through the water. This motion results in considerable friction by the water along both the upper and under side of the tail and this may have been the activating agent, throughout a very long period of time in order to have final effect.

A similar possibility may explain the development in insectivores and rodents of a tail flattened transversely and expanded in the vertical plane. Any stimulus that increased growth of hair upon the dorsal and ventral borders of the tail would presumably result, at a later stage of aquatic specialization, in the deposition of more subcutaneous tissue along these borders. And this is mainly fatty or even partly glandular, rather than muscular or bony. But this is pure speculation, for as we cannot even find out what really causes baldness in man one can hardly speak authoritatively on the reasons governing hair growth upon the tail of a shrew. A rather baffling coincidence in the tails just considered is the fact that although hairy caudal keels evidently precede (usually) flattening of the member itself, in those forms in which the latter is the case the tail is no hairier above or below than upon the sides.

Of those insectivores and rodents whose tails are flattened in the horizontal plane *Desmana, Potomogale* and *Ondatra* are the only ones in which this character is really pronounced, although *Galemys* and *Limnogale* approach the conditions existing in the first two genera respectively. The tails of *Desmana* and *Galemys* are somewhat swollen posterior to the base, as is usual in the family Talpidae, this being due to the presence of bodies that are largely fatty but which may also be glandular in character. In the muskrat (*Ondatra*), however, the tail is unusually slender, the skin not only being rather thin but without any sign of subcutaneous fat and so closely adherent to the muscular and tendinous tissue that this member is quite difficult to skin. In these aquatic moles and in the muskrat the base of the tail is not enlarged but is abruptly smaller than the hind quarters, approximately as in the common brown rat. In the muskrat at least the tail describes sharp sinuous movements throughout a considerable arc during aquatic propulsion. This motion follows oscillation of the hind quarters while the hind feet are kicked alternately and is, undoubtedly, chiefly involuntary. The flattening of the tail, however, has progressed to the point where this member is a definite, although still a secondary, aid to swimming, and but a slightly greater specialization in this direction would enable it to constitute a propulsive organ equal in importance to the hind feet.

In fact I feel sure that I have seen a muskrat swimming slowly by means of the tail alone, while the feet were trailed, but at speed the feet must still be employed.

In the above mammals the musculature controlling lateral movements of the tail is not as yet appreciably specialized, although it must be inappreciably so. In other words the muscles must be considerably more fitted for this work than is the case in the nearest terrestrial relatives, but not to the extent that this has as yet had much effect upon gross conformation. All that can now be said is that as this specialization becomes more marked, the muscles controlling lateral motions of the tail will become larger, broader, and there will develop a tendency for muscular simplification and partial fusion.

I regard it as doubtful if the tail of *Potomogale* ever presented quite the appearance that we find in most terrestrial insectivores. As already mentioned the case of this genus is somewhat puzzling, but it seems to me probable that its tail was already quite robust at its base when it first took to the water. The tail of this genus is now as perfected a swimming organ as it can probably ever be unless the animal become less dependent upon the land. Swimming is evidently accomplished not by a violent wriggling of the tail, as in the muskrat, but by more circumscribed, although stronger, motions, enabling the animal to slip through the water with remarkable ease and speed. I have never had a specimen for dissection but Dobson (1882) has illustrated its musculature to good advantage. The surprising detail shown is the way in which the gluteal complex has become specialized. One division of the latter has expanded so as to stretch from the knee posteriorly and far over the base of the tail. Clearly this gives the latter added lateral leverage, so that a number of the muscles of the posterior limb can assist oscillations of the tail, either directly, or indirectly as antagonists. The muscles from the pelvis, consisting of Dobson's ischio- and ileo-caudalis, are both large and long, and indeed there is marked increase in the robustness of all the muscles of the tail proper. Unlike most mammals of this size the muscles of the base of the tail are entirely fleshy, rather than chiefly tendinous, and this is the reason for the gradual extension onto the tail of the body contour.

Before abandoning the subject of mammalian tails that have become flattened transversely it is justifiable to discuss future potentialities and successive steps in their eventual development which seem most likely. This, of course, is pure speculation, but is of great interest, and possibilities can at least be mentioned.

The probabilities certainly are that the tail of the muskrat, for instance, will continue to expand vertically for some time to come, for we have every reason to believe that the stimulus that started it on its present course of development will continue. It is therefore improbable that the present direction of flattening would change through an angle of 90 degrees, so that eventually its tail would ever resemble that of the beaver. Evidently the stimuli for the two sorts of caudal flattening exhibited by these two mammals were always different. At least this thesis can be accepted as a likely one, and also that the tail of the muskrat will continue throughout the ages, to grow higher.

Bilateral symmetry among vertebrates is almost invariable, while dorso-ventral symmetry, in its exact meaning, is very rare, but still, after a general fashion, is sometimes encountered. Thus, the dorsal and ventral halves of some sorts of fish are often exactly alike in shape. The difficulties arising from the situation were the tail of a whale longer and wider on one side than the other are too apparent to need discussion. But asymmetry in the *vertically* expanded tail of some vertebrate is a different matter. In a lop-sided whale of the former character all manner of gyrations would be performed because the asymmetry of the tail would throw the body to the side and there is present no pair of vertical rudders, one above and the other below the body, to counteract the uneven influence. But were the tail expanded vertically instead of horizontally asymmetry need have no disturbing influence that could not easily (theoretically) be overcome by compensating tilting of the fore limb rudders, or proper curvature of the body. Tails dorso-ventrally asymmetrical with which their owners get along very well indeed are seen in certain newts and sharks.

But asymmetry of the sort exhibited by the newt's tail might be passively developed, as far as concerns swimming. Could there be any active stimulus connected with swimming that could cause asymmetry in a tail that is vertically expanded? Undoubtedly yes. There might be sufficient dorso-ventral asymmetry of the body, advantageous for feeding or other habits, so that there would eventually and naturally follow a compensating asymmetry of the tail to counteract it: or there might be asymmetrical influences introduced by an organism swimming habitually near the surface or near the bottom. Obviously, if a shark swam at the surface it would be disadvantageous to have one lobe of the tail projecting into the air, or conversely, a dependent tail-fork dragging over the bottom would be a handicap. From another aspect, swimming near the surface or bottom respectively introduces uneven degrees of water friction operating upon the upper and lower parts of the body.

Although it is not denied that the vertically expanded tail of an aquatic mammal might eventually develop into perfect vertical flukes without ever showing any definite asymmetry, it is deemed more likely that there will be marked asymmetry exhibited during some stage of the process. If this be so the form of the tail might be either epibatic (the upper lobe longer and larger than the lower) as in most sharks, or hypobatic (the lower lobe longer than the upper) as in some of the extinct marine reptiles. The latter condition would probably be the more likely, for I believe an epibatic condition of the tail in any air-breathing vertebrate is unknown.

Before abandoning the subject of tails that have expanded in the vertical plane it may be well briefly to discuss conditions in some of the extinct aquatic reptiles. Always in the flukes of the Cetacea and Sirenia the caudal vertebrae extend straight toward the rear and pass through the center of the tail to the vicinity of its medial notch. Thus the caudal tendons are enabled to operate practically from the posterior border of the tail and the force exerted upon each of the two lobes is symmetrical. Always in reptiles having a bilobed tail, however, the vertebrae followed the lower lobe to the tip, the caudal axis bending sharply ventrad at the tail base (peduncle). This clearly indicates a fundamental difference that from the very beginning has underlain the evolution of these two sorts of tails. Fraas has offered a reconstruction of what he considered to have been four of the stages in the attainment of this reptilian development. The precise shape of each tail is largely speculative, of course, and it is questionable whether in *Mixosaurus* the upper lobe should not be placed farther back (fig. 33), because the greater height of the spines near the tail base may really indicate that an augmented muscle mass existed at this point; but the principle seems sound, for only this sort of gradual, asymmetrical development of the tail could account for the situation of the caudal vertebrae within the lower lobe.

Obviously in the case of sharks the development of the epibatic tail has been the most favorable for bottom feeding habits, so that the shorter lower lobe would not drag upon the bottom. Equally obvious is the advantage of a markedly asymmetrical hypobatic tail for swimming near the surface, so that no high upper tail lobe will project above the water. As deeper swimming was habitually indulged in a higher upper lobe could develop, this ultimately attaining the size of the lower lobe.

Von Huene (1922) mentioned that in the latipinnate ichthyosaurs (as *Mixosaurus*) the tail seems to have been a very poor propeller and that locomotion was probably accomplished chiefly by the paddles. In longipinnate sorts such as *Ichthyosaurus,* however, he claimed that the upper lobe of the tail was immovable and together with the dorsal fin functioned as a rudder. It is impossible that in such a highly specialized form one-half of the tail could have acted efficiently as a rudder and the other half as a primary propeller. Undoubtedly as the upper half of the tail approached the lower half in size and form there was a stiffening of the caudal tissue, just as in whales, toward the ultimate goal that the caudal tendons operated from the base of the tail rather than its lower tip, enabling the upper lobe to impart almost, if not quite, as much propulsive force as the lower half. Any asymmetry in this force could then have been equalized by tilting of the flippers.

The second class into which I have divided the tails of aquatic mammals is as follows:

II. Tails flattened in the vertical and broadened in the horizontal plane. In the first group below are included all those genera in which the tail has been flattened vertically, and in the second, those in which the tail is still terete but which may be expected eventually to develop a vertically flat tail.

Tail flattened	Tail round
Ornithorhynchus	Mustela (the aquatic forms)
Lutrinae (part)	Lutrinae (part)
Enhydrinae	
Pinnipedia	
Castor	
Sirenia	
Cetacea	

There are two members of the above groups which I am unable to discuss with any great feeling of certainty. These are *Ornithorhynchus* and *Castor*. It is not difficult to determine the economic value to them of their caudal equipment, but it is very puzzling to envision the process by which the specialization was initiated. The case of the platypus is of lesser moment, perhaps, It is such an anomalous beast in so many ways; we are entirely ignorant of the ancestral type, and it is the only aquatic mammal with a sizable tail that swims chiefly by means of the fore feet. Hence we must be satisfied with assigning the proper function to its tail without attempting to visualize the successive steps through which it originally passed. According to all reports the tail is now used as a rudder, and for the purpose of keeping the head depressed when the animal is nuzzling about in the mud for its food. The latter function would seem to be the critical one.

Externally the tail of the beaver is phenomenally broadened and the integument is scaly in appearance, although cornification is not marked. Subcutaneously the tissue is fatty, for the most part, but the lateral muscles at the base of the member are very broad and powerful. The caudal vertebrae are unusual in that they exhibit marked broadening, especially those proximad, and this character is shared by the sacral series. In the caudal compliment there are about 23 to 25 vertebrae, which are short, wide and depressed, with very wide transverse processes which become double at the middle.

The reason for the sort of flattening shown by the tail of the beaver is puzzling. The function, once popularly believed, as a vehicle for carrying mud, was long since proved to be erroneous. Many people still consider that the flat tail was developed so the animal might slap it on the surface of the water and thus more quickly submerge, but this hardly seems to be sound logic, for the muskrat can disappear with an abruptness that is equally startling. Its present function as an organ for giving warning signals (by surface slappings) is undoubtedly incidental and secondary, as is any use to which the tail may occasionally be put for tamping mud while building dams. It is obvious, however, that without a tail flattened in *some* manner the beaver would progress in circles while towing logs and sticks. This, then, must be listed as a primary need for a very broad tail, but one flattened in the horizontal plane should be much more effective for this purpose, hence, while the rudder function may have very materially assisted in the expansion process, it could hardly have initiated the present direction of flattening. Similarly with the fact that the tail is occasionally used as a scull. I have watched a beaver swimming slowly by the tail alone, pulling the appendage latero-ventrad first to one side and then the other. In this also a laterally flattened tail would be more effective. As the beaver swims by alternate strokes of the hind feet it seems that this almost certainly introduced some original stimulus for flattening the tail in the transverse plane. Hence there appear to be at least three factors for which a narrowed tail would be more favorable than a broadened one, and it accordingly appears likely that the broadening stimulus would have to be an unusually strong one to overcome them.

The beaver often walks erect with an armful of mud, and also uses the tail as a prop while cutting trees (Bailey, 1923), or it may make more use of this member for keeping near the stream bottoms than is suspected. Of these three circumstances it seems that the second would be more critical in possibly causing a flat tail, but I do not feel assured that this is the case and I prefer to make it plain that I have no conviction on the subject.

The Pinnipedia may be said to lack a functional tail and it seems highly probable that they have descended from an ancestry with tail much as in the living bears. In the walrus the "crotch" stretches uninterruptedly between the heels caudad of the bony tip of the tail. In a sea-lion (*Zalophus*) the external tail constituted but six per cent (60 mm.) of the total length, while in a seal (*Phoca*) this percentage was seven (72 mm.). The tail can have not the slightest use as an active aid either for propulsion or steering, but in the sea-lion the tail was slightly thicker in horizontal than in vertical dimension, while in the seal this was very much more pronounced. In the latter at least it was plainly to be seen that the shape of the tail permitted this member to fit perfectly into the cleft between the heels so as to effect an uninterrupted body contour. The stimulus for this specialization is unknown. It was probably of an entirely different nature from that which has caused a broadening of the tail in aquatic forms in which this member is the primary propulsive organ, and for the present all that can be said is that it constitutes another instance of the fact that when an animal experiences the need for a modification, the latter will often appear eventually.

The tail of the otter is of particular interest in the present connection for the reason that it seems to me likely that the original ancestors of the whales had a largely similar conformation of the body. The tail of this animal is exceedingly thick at base, so that there is a gradual taper from the hindquarters to the tail tip. At the same time the body is rather long and sinuous and the legs are short. The feet are often used in swimming, but as propulsive organs they seem to be of decidedly secondary importance. Swimming is mainly accomplished by curvature of both the lumbar and caudal regions in the vertical plane. Extension of the vertebrae, after flexion, is often accompanied by a kick of both hind legs, not in unison nor yet in alternation, but with a sort of galloping action, but the animal appears able to swim with equal ease and speed when the feet are folded against the body. The shape of both body and tail base, together with great muscular power in the latter region, is largely accountable for this, and as the feet are relatively small and little changed for swimming (in the common *Lutra*), it seems either that the feet of mustelids have been unusually resistent to modification or else, more probably, that the tail has always been of such character as to lend itself readily to swimming. This is also indicated by the fact that the feet of the mink and sumpfotter are not in the slightest modified for swimming and the tail, although much less specialized in this direction than that of the otter, plays a very important part in propulsion. In this connection the feet of the otter will be more fully discussed in a future chapter.

In but one genus of river otter is the tail other than round, at least to an extent that has proven noticeable. In *Pteronura* this member has a sort of fleshy keel upon either side, and hence, is flattened vertically and expanded horizontally. This is a development which other otters may be expected to follow.

In the sea otter (*Enhydra*) the tail is also expanded laterally, evidently to a more marked degree than in any river otter, but it is relatively shorter and does not reach beyond the tips of the toes when the hind limbs are extended backward. But one cannot discuss the development of the tail of this animal with any degree of assurance until more is known about its swimming habits. Certainly the hind feet of the sea otter are of greater importance in swimming than is the case in the river otter, and therefore, by analogy, the tail is of less importance. But the tail is more specialized, and it is therefore likely that it has experienced aquatic influences for a longer period of time. It may once have been relatively shorter and have experienced a stimulus for elongation, or it may always have been approximately of its present length but the use to which the hind feet were put in swimming enabled these members to gain the evolutional ascendency. It appears very likely, however, that the tail of the ancestor of the sea otter when it first took to the water was not *equally* as long and robust as of the river otter at the same stage of its history, else the tail, being theoretically a more ideal organ for propulsion, would have gained the evolutional ascendency, as it has in the Lutrinae, before the hind limbs had gotten well started.

The future course of enhydrine development is uncertain. If the tail were longer I should predict as a matter of course that this appendage would increase in importance and size, gradually supplanting the hind feet as the chief organ of propulsion, and that the latter would dwindle in size. Actually the tail now seems to be at a critical stage. The hind feet are so large and (apparently) efficient that the animal may be unable to alter its present course of development, or the tail may be unable to catch up in the race of aquatic adaption and it may well be that it will henceforth play a rôle of increasing unimportance. Incidentally any function of steering which it may now have would be insufficent to save it from atrophy, in the latter course of events, for it is situated too close to the primary propulsive organ to act efficiently as an equilibrator.

The tail in the Sirenia varies. In *Halicore* and *Hydrodamalis* it is quite whale-like, there being two pointed flukes with a partly defined medial notch between, and a constricted peduncle. Little more than this can be said, for the muscular anatomy of neither has been investigated; but in main features, consisting chiefly of simplification of the apaxial, and fusion and hypertrophy of hypaxial elements, it undoubtedly resembles the Cetacea. In *Trichechus* the tail is usually stated as shovel-shaped, but in its exact conformation there is probably specific variation. Thus Murie (1872) figured a specimen (figure 6) of what he calls *Manatus americanus* (=*T. latirostris*) from the West Indies in which there is a medial indentation of the posterior tail, while elsewhere (1880) he showed an individual from British Guiana with tail tip somewhat pointed (figure 6). But even yet it is not known whether the animals from these two regions really represent two species. The illustrations will give a better idea of the sirenian tail than can a description. It will be noted that in this respect the manati is much less specialized than the dugong, but that the former represents a stage through which it is not improbable that the latter at one time passed. In *Trichechus* there is no well defined peduncle, but merely a slight taper of the posterior part before the lateral expansions of the tail begin. The latter is readily seen to be fairly intermediate between a stage on the one hand wherein there was either no lateral broadening of the tail, or one comparable in degree to that existing in *Enhydra,* and on the other, conditions as now to be found in *Halicore*. In a mounted skeleton of *Trichechus* in the National Museum there are 22 lumbo-caudal vertebrae, while in a specimen of *Halicore* this series numbers 26. In the latter the transverse processes gradually diminish in width from the thorax to the region of the peduncle, thence widening once more until near the caudal tip. This is an interesting occurrence for the reason that it is a character which the Cetacea do not show to the slightest degree. It seems to be an additional instance of the fact that two animals frequently respond differently to the same stimulus.

The caudal conditions in the Sirenia are not readily analyzed. It is currently believed that the sirenian ancestor was of proboscidean stock and we would therefore picture the caudal stimuli to have been somewhat similar to those that the hippopotamus has experienced. One would naturally envision this ancestral type as swimming by alternate strokes of at least the posterior, if not all four, limbs, while an inadequate tail trailed behind. If this had been the case the stimulus would have been chiefly for a transversely flattened tail. In order for the tail to have attained the shape that we now find it, it is probable that either one of two things was an original requisite: (a) That the terrestrial ancestor of the sirenians had a sinuous body with long tail that was especially robust at base, and was of a rather active disposition. These premises seem necessary in order that it could swim with the sinous movements in the vertical plane that the otter now employs in swimming. But we surely have no justification for considering that any member of the proboscidean stock was ever this sort of mammal. (b) The other alternative is that the sirenian tail developed along the lines that it followed (vertical flattening) for the same reason that the tails of the platypus and beaver did. Of these I regard the platypus' tail as the more significant, for this animal and the sirenians (at least the manati) utilize the tail extensively for keeping near the bottom. Murie's illustrations, reproduced in figure 7, were drawn from life and show this in an interesting manner.

The subject of the sirenian tail should not be abandoned without calling attention to another possibility. As with most highly specialized aquatic mammals it is likely that the more distinct genera are of great antiquity. It is very likely that the manatis on the one hand, and the dugongs (and possibly *Hydrodamalis*) on the other began to diverge as separate groups soon after the sirenian ancestor took to the water, if not before. It is not improbable that this took place before the tail had experienced any aquatic modification at all, in which event the tail of the manati and of the dugong has each followed an entirely independent course of development, diverging in details to a greater or lesser extent according to varying conditions which each has experienced. Presumably the tail of the dugong is of the higher aquatic type, but this does not necessarily mean that it ever passed through the exact stage which is now exemplified by the manati. Each type of tail may now be perfectly fitted to the needs of the animal and the manati may never develop the more specialized tail of the dugong. There is the possibility that the flukes of the latter have developed from skin folds of rather limited extent near the tail tip, while the tail of the manati may have been evolved from an appendage having lateral keels along its entire length. There is little evidence either for or against this possibility.

Caudal conditions in living and known fossil Cetacea seem to be of three sorts; one represented by the zeuglodont *Basilosaurus,* the second by *Zeuglodon osiris,* and the third by all odontocetes and mysticetes. *Basilosaurus* and its ilk had an anguilliform type of body, the tail being excessively long. Posterior to the thorax the zygapophyses did not articulate and the transverse and spinous processes were short, indicating that the spinal musculature was not developed to a degree where it could handle flukes of the modern cetacean sort. These vertebral details show that there was great mobility of the tail to allow for serpentine motions and it seems certain that the propulsive mechanism must have been in the nature of some sort of continuous fin fold running in a fore and aft direction, although this may not have been of entirely uniform width. Presumably there was a symmetrically placed pair of these extending for most of the length of the tail. As the caudal expansion of living cetaceans is horizontal it is certainly reasonable to assume that this was also

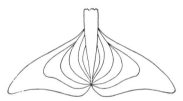

FIGURE 32. Six stages in the ontogenetic development of the cetacean flukes, after Ryder.

the case in zeuglodonts, and that it progressed by undulations in the vertical plane after the manner of the traditional sea serpent.

The second sort of cetacean caudal condition is represented by *Zeuglodon osiris.* In this the tail length was comparable to modern porpoises, and although the high spinous processes of the thorax and anterior lumbar region indicate that there was powerful musculature in these areas for control of the tail, the spines of the posterior lumbar region and tail were remarkably weak, showing that here there were no heavy muscles. It is therefore doubtful if this animal had abruptly expanded flukes; and the tail was too short to have anguilliform keels upon either side. It therefore appears not unlikely that it was provided with a caudal equipment more nearly resembling the manati than a living porpoise. Incidentally there were likely at one time or another zeuglodonts which had a caudal propulsive apparatus fairly intermediate in character between that of *Basilosaurus* and *Zeuglodon osiris.*

[198]

As far as I know there is no reason for believing that any odontocete or mysticete whose remains have yet been discovered had flukes very different from what we are accustomed to consider as characteristic of this order, although it is true that I have not examined the details of the caudal vertebrae of a great many fossils.

The caudal equipment of the whale has been a subject for the liveliest controversy, as has been the case with so many cetacean details. Thus Gray held the belief that the flukes were derived from the entire hind limbs, while Ryder (1885) considered that they represent the pedes only. The latter's lengthly defense of this belief is a curiously artless combination of established fact and fancy. He argued that the original development in the Cetacea was along somewhat the same lines as that followed by the seals, and that the hind limbs were at one time used in oscillating movements for propulsion. He believed that the feet and tail were later inclosed in a single fold of integument and that finally the bony and muscular part of the hind limbs became atrophied and shrank toward the pelvis, leaving the integumentary part of the pedal expansions attached to the side of the tail. There were some converts to this view, chiefly among those who were unwilling to relinquish the thesis that pinnipeds, sirenians and cetaceans all represent different stages of development from a common derivative. It seems hardly necessary to point out that if this were the actual situation there would not only be clear and incontrovertible proof of it in cetacean embryos, but if the flukes had been derived from anything except fibrous, dermal dilations of the lateral tail the adult would necessarily exhibit some cartilaginous or muscular relic of the fact in this region. On the contrary I have no fear of contradiction when I say that in the light of present knowledge the evidence is conclusive that the posterior termination of the whale is composed of caudal elements only.

In attempting to explain the asymmetry of the odontocete skull Steinmann (1912) made the claim that the Cetacea originated from the ichthyosaurs, and hence that the whale's flukes have become horizontal from an originally vertical position. This theory is also untenable. Had this been the case the whale would have experienced a period of (say) from one to several millions of years during which its flukes would have been at an angle of 45 degrees (more or less) to both the vertical and horizontal. This would have obliged one-half of the erector spinae musculature to have become atrophied and the other half hypertrophied, while a similar fate, but in reverse order, would have overtaken the hypaxial muscles. In truth we would then have a cetacean of astounding asymmetry.

Also in trying to explain the asymmetry of the skull in toothed whales Kükenthal (1908) made the claim that the flukes are asymmetrical. He was led to this belief because in many of the embryos which he examined the caudal lobes were set somewhat awry, and from this he argued that a similar condition in the tails of baleen whales was accompanied by slight though recognizable asymmetry in the mysticete skull. His thesis was that an asymmetrical tail, used in sculling movements, tends to turn the animal to the left, when the unequal pressure of the water upon the two sides of the head will have resulted, throughout long ages, in an asymmetrical skull. At the present time this theory is given very slight credence. It is true that Kükenthal figured the transected tail of a rorqual which showed the flukes in somewhat oblique relation to the vertebrae, but even if this accurately depicted the conditions in life this individual may have been pathological. The caudal conditions in preserved embryos are entirely without significance, for the preservative acting upon soft tissue in a cramped position invariably distorts the flukes, and almost always they will remain fixed in the shape of an S. I have examined numerous cetaceans, both odontocetes and mysticetes, with possible asymmetry of the tail in mind and have never found the slightest indication of such being the case.

Ray seems to have been the originator of the belief that the horizontal direction of the whale's flukes was attained because the animal is an airbreather, and such provision enables it more easily to seek the surface for a fresh breath. In the literature this is often repeated without comment. A little reflection, however, will convince one that there can be nothing to this reasoning, as Beddard (1900) and a few others seem to have concluded. In the first place no such function as ease of ascent from great depths could have had the slightest influence upon the initial stages of tail change in the cetacean ancestor. Caudal development must have started in conformity with the manner of swimmng then employed without regard to any final use to which the tail would be put. In the second place, when the tail is used as the primary means of propulsion it will become specialized so as to drive the animal forward in a straight line, regardless of the direction of flattening, while other details of the body will take over the function of steering. At speed the whale evidently elevates or depresses its line of progress by the flippers alone, save in the case of an abrupt turn, and it could with the greatest ease swim either straight up or straight down were the flukes vertical instead of horizontal.

As already mentioned it seems to me highly probable that whales have descended from an essentially active ancestry, and that some method of swimming has always been employed which involved movement of the tail in the vertical plane, originally somewhat after the style now to be seen in the river otters. Successive steps in caudal development then would be a lateral flange upon either side, next the broadening of the terminal part of this flange without a corresponding increase in the width of the proximal portion. Following an increased tendency toward segregation in the extreme distal portion of the tail of the lateral expansions, there would finally result the graceful, bilobed flukes of the rorquals, which are probably more specialized than those of any other cetacean.

That an immense length of time has elapsed since the first step in the development of the cetacean flukes is indicated by embryological evidence. Ryder showed that in a *Delphinapterus* fetus of about an inch and a half in length this expansion is apparent, and of spear-shaped form. Indeed, the literature is replete with such evidence. Contrary to what one might expect this caudal expansion does not occur upon the peduncle proper but apparently is confined to the area of the tail tip which will later support the fully-grown flukes (figure 32). This fact may contribute some evidence to the theory that the cetacean tail was not comparable to that of the present manati during any stage of its evolution, but that the lateral expansion had always been confined to the position which it now occupies.

Naturalists are not entirely in accord regarding the precise manner in which the flukes of whales are used, for opportunities for observation are infrequent and porpoises may move the flukes too rapidly for the human eye to follow. A number of observers have marvelled that when looking down upon a porpoise swimming at speed, as just in front of the bow of a boat, no movement was appreciable, and yet the animal not only maintained its position but easily darted ahead when it so wished. From this it must be inferred that the fluke movement of at least some porpoises is through a short arc, is rapid, and correspondingly very powerful. It has been claimed from time to time that the movements of the cetacean flukes are not strictly in the vertical plane but are slightly oblique, first to one side and then the other (scull-like) or even somewhat twisting (partly screw-like). It is possible that both are used at times, in addition to strictly vertical thrusts, this depending upon the speed, the sort of whale concerned, and the conformation of the flukes. I think it more likely that a cetacean with relatively narrow flukes would employ a sculling motion, so as to reach out first on one side and then on the other for undisturbed water, more readily than one with very broad flukes.

Breder (1926) has stated that in twelve high speed fish the width (or height) of the tail averaged 21 per cent of the total length. The rorquals are the speediest of the large whales, and one which I measured had a length of 63 feet, with flukes 15 feet broad, so this same proportion in this animal was almost 24 per cent. The tips of the flukes were therefore of sufficient length to reach well laterad into undisturbed water as the animal swam, and I have been assured by observers, including trained naturalists, that in this sort of whale there is no lateral or sculling motion to be noted when the animal is swimming.

The shape of the flukes differs in various types of whales. Especially in less speedy sorts the posterior border may be straight, or this edge of either lobe convex in outline and with a median notch. In the rorquals, however, the flukes are more falcate in shape, the tips extending farther back and the posterior border being more suggestive of an S. Hence, the tail is thus inclined to be forked. Apropos of this, Breder (1926) has stated that fishes with squarish or spatulate tails are comparatively slow but capable of extremely sudden short bursts of speed, while those with deeply forked or lunate tails are capable of long continued swimming at high velocity, the more lunate the tail the faster being the fish. In discussing this fact Nichols (1915) has pointed out that during speedy locomotion the water displaced by either side of the body of a fish should, directly the fish has passed, meet again with a minimum of disturbance by the median part of the vibrating tail. Breder considers that in the main this is true, but that there are other factors involved is shown by the fact that when he cut a prominent fork in the tail of a fish normally having this member of spatulate shape, no greater speed was attained nor was this reduced, but the motions of the body were different. It is therefore justifiable to presume that in the faster sorts of whales such as the rorqual, the broad, falcate, slightly forked flukes are of prime importance in attaining and maintaining speed, but that there are also other factors involved, perhaps of equal import, including, at least, shape of body and muscular conformation. In other words if a gray whale could temporarily be equipped with the tail of a rorqual there is no reason to believe that it could swim any better for the reason that it probably has not the equipment to operate such a tail in the most advantageous manner.

The structure of the whale's flukes is truly remarkable. The smooth slope of their contour and nice variation in thickness, as the situations of the various stresses dictate, are very suggestive of their evident fitness for function. The epidermis is, of course, very thin as it is over the remainder of the body, and beneath this is a pure white, fibrous tissue, somewhat fatty and very elastic in texture, which has phenomenal toughness and yet may be sliced with a knife without difficulty. The strength of this tissue is forcibly impressed upon an observer as he watches a whale of 75 tons being drawn up the slip tail first. If the flukes catch beneath some obstruction either the latter is torn from its moorings or the flukes snap into place with an abruptness and strength that shakes the entire animal, but very seldom is any serious damage suffered by the

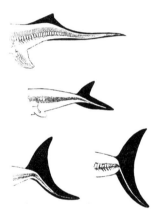

FIGURE 33. Suggested restoration of caudal outlines, illustrating progressive development, of the Ichthyosauridae: (a) *Mixosaurus;* (b) young and (c) adult stages of *Stenopterygius;* and (d) *Ichthyosaurus* (after Fraas).

flukes themselves. It is difficult to understand why, during the time that it took for cetacean flukes to reach their present stage of perfection, the caudal vertebrae themselves did not experience profound alterations in the way of broadening, and as a result we must believe that any unyielding stiffener of such a sort would not prove advantageous. Of course there may ultimately be acquired within the caudal lobes an equipment of supporting cartilages, in which ossification centers might appear, but as there is as yet no sign of any such development there is no reason to think that there ever will be.

In all the Cetacea the transition from peduncle to flukes is very abrupt, the latter always diverging at almost a right angle. The cross section of the peduncle varies greatly according to the habits—to a large extent the speed—of the animal. In some of the slower sorts of whales (balaenids, gray whale, etc.) this may be almost circular and fairly elliptic. I believe that the height is invariably greater than the width, however. At the other extreme in peduncular shape are the rorquals. In a finback *(Balaenoptera physalus)* of 65 feet the peduncle directly anterior to the flukes measured approximately one foot in width and four feet in height. Instead of being purely elliptic the peduncle was quite sharply keeled above and below. It is obvious that such a shape would prove of very great economic benefit in reducing water resistance to the minimum as the animal elevates and depresses the tail while swimming. Sections cut from the peduncle show that in a specimen of this size the dorsal and ventral 8 inches of the respective keels are composed of the same sort of fibrous tissue as the flukes. So it is obvious, I think, that these narrow, angular, peduncular keels have been built up above and below the great spinal tendons for the sole function of reducing water resistance during swimming, which is just what the flukes have done to *increase* resistance. But such contradictory situations are frequently encountered in any study of specialized organisms.

Save for the fibrous keels as described above, the shape of the peduncle of the Cetacea is dependent upon details of the vertebrae of this region, and therefore upon the character of the musculature concerned. It is clear that in those sorts of whales with relatively broad peduncles locomotion is retarded by just so much, and a broad peduncle can be of use only to give the caudal muscles greater leverage in lashing the tail laterally—a movement that cannot be of very great importance to the animal. With a peduncle that is relatively very high or deep a whale is equipped not only to elevate and depress the peduncle with the minimum of water resistance, but also to secure, by means of long spinous processes and chevron bones, increased leverage by which the tail may be raised and lowered with greater power or greater ease.

The caudal vertebrae of living Cetacea vary in number from 16 in *Neobalaena* to 32 (or possibly more) in *Steno* and *Lagenorhynchus,* and the variation in height of spines and width of transverse processes is great. Prezygapophyses are present in mysticetes, and in most odontocetes, but in some forms (as *Grampus*) of the latter these are suppressed. Chevron bones are present, taking the place below that is filled above the column by the spinous processes. Their development corresponds to that of the spines, the latter invariably being slightly longer. As Flower has said regarding the caudal vertebrae, in passing backward the arches and processes gradually disappear, and the bodies become compressed and elevated vertically. At a point corresponding to the posterior part of the peduncle there is an abrupt change in the character of the vertebrae, and thereafter those which in life were situated within the confines of the flukes become smaller and broader, the column ending in a series of vertebrae that are no more than bony buttons. To the vertebrae of the flukes pass bundles of large tendons, both from the apaxial mass above, and the hypaxial musculature below.

Chapter Ten

The Pectoral Limb

It is a well established belief that all terrestrial vertebrates were originally derived from a fish-like aquatic ancestry. The anterior limb, or ichthyopterygium, of fish has certain well defined characteristics, among which is the lack of clear distinction between the proximal elements ("brachium" and "antibrachium"), the ocurrence usually of more than five series of distal components ("digits"), and the indeterminate number of elements ("phalanges") of which each of these are composed, according to the exigencies of individual cases. In mammals the anterior limb, or cheiropterygium, is composed of distinct brachial and antibrachial segments, the digits never number more than five, nor the phalangeal ones (including metatarsi) more than four except in whales. What is more logical than to presume that some of the steps taken by the mammalian stock as it arose, by whatever process and by whatever path, from a fish-like ancestry, should eventually be retraced in more or less complete degree as some of its representatives once more become completely fitted for an aquatic existence? Working with such a thesis as a tentative basis certain of the trends which are found to be exhibited take on added significance.

When a mammal first takes to the water the fore limb is usually, if not always, used as an aid to locomotion, and there will be a lengthy period during which its function is very inefficient for the reason that it is ill fitted for the part that it plays. The final fate of the fore limb in respect to aquatic modifications undoubtedly depends upon a great number of factors, but it seems that the chief determinant rests upon the question of whether or not the hind limbs or (and) the tail gain evolutional ascendancy over the pectoral appendages. The chances of the hind limbs gaining the lead are much more than even. In the first place, with very few exceptions (some sorts of bovines, as *Bison,* the hyenas, etc.) the hind limbs of terrestrial mammals are larger, more powerful, or both, than the complementary pair. They would then be more vigorously kicked, and would gradually play an increasingly important part in aquatic locomotion. Second, the fore limbs may be used for other purposes besides propulsion when the animal is in the water, as in helping to introduce food into the mouth or in prying up stones upon the bottom. Third, the most efficient method of applying the propulsive force in swimming is from the hinder end, and it is likely that the animal very quickly discovers this fact. Especially if there be present a tail of respectable proportions and length I regard it almost as a certainty that in mammals the fore limbs will not constitute the primary means of propulsion. It seems likely, then, that unless there be some special feature of body conformation (as a short tail) or special feeding habits that might introduce disturbing elements, the final, primary, propulsive organ of an aquatic mammal will not be the pair of pectoral limbs.

If a highly specialized aquatic mammal swim by means of oscillations of its hinder end—either the tail or hind feet—it should have some anterior provision for steering and equilibration. A slow baleen whale, with its enormous head, might possibly get along very well without a rudder, but for an odontocete or pinniped, pursuing individual food items, inability to steer (i.e. make sharp turns) would spell speedy death by starvation. If the head be of sufficiently small size so that a neck of considerable length be possible (see chapter 7), then this part of the animal can take over most of the function of rudder. It can swim along in a straight line with neck retracted until such time as it desires to turn abruptly, when the head can be thrust sharply to the side and the body will follow. But at times the head has other duties to perform. Perhaps a speedy fish is being pursued and the endeavor is to seize it as quickly as possible. Then is it advisable that there be a separate apparatus for steering, and this is supplied by the fore feet. Thus in the true seals (Phocidae) there is such nice interaction between the head and the neck on the one hand, and the anterior limbs on the other, that it is impossible to tell which is of the most importance in steering. But it is important to note that in this animal it is likely that if the fore limbs were held immovable against the sides steering would be accomplished with just as much effectiveness, at least for a short while, although it is probable that exhaustion would follow more speedily.

In the Cetacea matters are somewhat different. The whole body may be curved moderately in any direction and thus effect turning, but the neck is so short that it is capable of little more curvature than the thorax and it is likely that when the animal is progressing at speed the entire vertebral column is so occupied with oscillations concomitant to swimming that other motions of the body concerned purely with vertical steering are not often attempted, this preferably being accomplished by simple tilting of the flippers. Being of importance in this regard to a mammal capable of high aquatic speed it is to be expected that they would readily become specialized in a variety of ways according to the needs of the particular species or genus concerned.

There here may arise the question of whether the pectoral limbs of the Cetacea ever constituted a primary swimming organ. Beddard (1900) considered this to be quite likely and others have entertained the notion. No one can deny this conclusively but in view of what evidence there is it is not at all likely. There is nothing in either the osteology or myology of the Cetacea to support such a theory, purely mechanical reasons would prevent it from being a likelihood, and it is not probable that any aquatic mammal capable of developing such a perfected propeller as the cetacean flukes would have so used the flipper to any marked extent. On the other hand, the osteology of the dugong does render it possible that at some comparatively recent time its flippers could have been used for propulsion in quite efficient fashion; but it seems more probable that osteological development has been merely convergent and that its brachial musculature has been put to some other use than as even a moderately important aid to swimming.

It is seen from the above that in a mammal that is highly specialized for an aquatic life the anterior limb should normally have either one of two functions; that of steering and equilibration more often, their importance in this function depending upon other bodily details as well as habits, and that of primary organs for propulsion more rarely.

If the fore limb be not of great importance in steering (or swimming), it may never become essentially fin-like. If it is constantly used for this purpose, or to apply the propulsive force in swimming, it will finally assume the characters of an efficient paddle or fin. In either case efficiency demands that there should be a surface of broad area which may be brought into action against the water, and a relatively thin border, which may cut through the water with the least amount of resistance that is practicable in connection with requisite strength. As an ideal, however, the cross section of this paddle or flipper will not show two lines that are perfectly parallel, but rather will such a cross section be of a fusiform shape, as is the wing of a bird or of an aeroplane. For one thing the anterior border can not be too thin because of needed strength. The posterior not only can be but should be thinner to allow for greater limberness, and to reduce suction or partial vacuum as the flipper passes through the water.

Whether for propulsion or as a specialized rudder the anterior limb of an exclusively marine mammal may be expected to have become relatively stiff, although elastic. In effect it becomes either a horizontal rudder or an oar. As such it should have a single joint, analogous to a row-lock, and this should be situated just within the body contour. There should not be additional joints and those already existing will tend to become immobile, unless there be need for them because of feeding habits.

In attempting to determine the ideal position for the flipper of a marine mammal that uses this member for steering there are encountered a number of interacting factors which render the question a difficult one. It is apparent that in the whale the most effective position for simple steering fins would be in the vertical plane, one above and another below, so that by slight tilting the body would be thrown to the right or left: but such an alteration in fore limb posture would be impossible. The same act could be accomplished by carrying the limb horizontal but tilting it so as to offer as much resistance to the water as possible and at the same time pressing the opposite limb against the body. This concerns only rudder action. But some equipment for pure equilibration should also be advantageous. If a whale swim but languidly with only the tail proper involved then this may not be necessary, but if it swim so violently that there is no real interval between the cephalic and caudal amphikinetic parts, there will be a greater or lesser tendency for the thorax to move upward or downward at each stroke of the flukes. In order to reduce this lost motion to a minimum, large flippers widely extended would be of the greatest aid. In addition, an animal provided with flippers of this character so held would find it easiest to depress or elevate the head by slight tilts of these members. Thus it is seen that in an animal such as a whale a pair of horizontal flippers should best accomplish elevation and depression of the body (equilibration proper) and by operating one at a time these might be just as efficient in steering from side to side as if they were vertically placed.

The above argument sounds logical but it may be complicated to a large degree by yet another factor. Back in the early history of the world when the first vertebrates were being evolved these developed an equipment of primitive fin folds. Opinion differs as to the ideal arrangement of these and there are proponents of two different theories, but all are agreed that there was a pair situated latero-ventrally from which were derived all four vertebrate limbs. The point which is of interest in the present connection is that there was evidently an elemental stimulus connected with aquatic requirements which operated to develop pectoral fins at a certain angle to the body axis. It is by no means unlikely that there is now the same stimulus, but in reverse order, operating in the case of aquatic mammals, to place the axis of the fore limbs at some particular angle. Such a stimulus is impossible of analysis, and the question seems too speculative to follow further.

In the case of a mammal such as the sea-lion which swims exclusively by means of the anterior limbs, there are certain principles of efficiency which must govern the motions employed. This does not concern low speed, during which several sorts of makeshifts may be employed, some for relaxation and some in pure fun. During leisurely progress the flippers may be advanced and then brought rearward with a broad sweep, this corresponding to the overhand stroke employed by human swimmers. It is true that during action of this sort there is an active recovery motion which, during the forward stroke, must substantially increase water resistance, but what resistance is added at this time is subtracted during the backward stroke and after much thought I have come to the conclusion that as far as concerns only resistance it makes no practical difference whether the flippers, during slow progression, are held with their forward borders in the same transverse plane, or whether they are alternately advanced and retarded.

The above "overhand" movement is not employed by sea-lions for speedy progression, so far as my experience shows. The reason for this seems to be that when a certain degree of speed has been passed, the animal is incapable of operating the long flippers on the backward stroke sufficiently fast to supply any propulsive force. The principle is somewhat the same as that which correlates running speed with the celerity with which the legs may be moved. Another deterent is the fact that during such a backward stroke the broad surface of the flippers would be presented in an antero-posterior direction. Hence, even though the tip of the flipper could be retarded sufficiently quickly to supply some propulsion, the broad part of the wrist would be so situated as to act as a strong brake. By this method of overhand swimming, therefore, there appears to be a certain unknown speed limit which sea-lions are incapable of exceeding. So something more efficient must be evolved—some way of utilizing an oblique, rather than a direct, thrust against the water.

As a result the Otariidae now swim at high speed not by flexing and extending the arm, but by adducting and abducting it, with hardly any antero-posterior action at all so far as I can determine. As the flipper is adducted it is turned obliquely or "feathered," so that the thicker an-

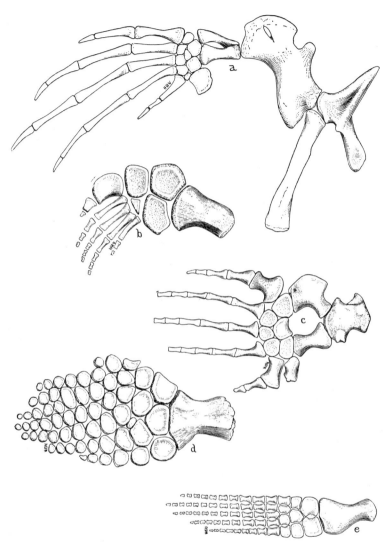

FIGURE 34. Left pectoral limbs of some aquatic reptiles: (*a*) leatherback turtle (*Dermachelys*); (*b*) *Geosaurus;* (*c*) *Clidastes* (a moasaur); (*d*) *Opthalmosaurus* (an ichthyosaur); and (*e*) *Elasmosaurus* (a plesiosaur). The last four are redrawn from Williston.

terior border really supplies the force, and precedes the posterior border, the motion of which is more passive. By this method of swimming the possible speed is very great indeed, and in practice depends directly only upon the power with which the flipper is adducted. Another advantage is that this need not entail any lost motion, for abduction, although necessarily much weaker because of muscle conformation, consists of the same motions as adduction but in the opposite direction, and can contribute at least some propulsive force. Thus in swimming by adduction and abduction in the transverse or largely vertical plane all movements can be utilized for forward propulsion and the only resistance is offered by the anterior borders of the flippers. It is undoubtedly by just this method, or one substantially the same, that aquatic birds, including penguins, which habitually pursue speedy prey by "flying" under water utilize their wings; but the case of the marine turtles is different.

I have had but limited opportunity for observing the actions of penguins, but so far as my experience goes the motions of their pectoral appendages differ in no important respect from those of sea-lions. Movement is almost entirely in the abductive-adductive (transverse) plane, the wings are "feathered" during the stroke so that the force applied is oblique, and although the abductive or upward stroke is too fast to follow satisfactorily it seems likely that it is so performed that it furnishes at least a slight amount of forward propulsion. There is the difference, however, that in the penguin the static posture of the paddles is almost horizontal, while in the sea-lion it is more adducted, and the arc of movement varies accordingly.

The swimming movements of the marine turtles are rather hard to describe. The elbow protrudes from the body contour and the enormous humerus is worked chiefly up and down. The forearm segment, however, is to all intents a mechanical part of the paddle and the whole is *extended* (see figure 34) with respect to the humerus in a manner never encountered in the Mammalia. By virtue of this alteration at the elbow the turtle's flipper is given a definitely caudal inclination, so that although the humerus is chiefly worked dorsad and ventrad from a transverse position, the flipper, with its axis almost parallel to that of the body, operates by thrusting the water chiefly backward by means of the palm. Perhaps this situation has some bearing upon the fact that hyperphalangy is not met with among the turtles.

By the above it is not meant to imply that no other swimming motions are indulged in by either penguins or turtles. Those described are merely believed to be the most efficient ones by means of which their best speed is attained.

This outline of the known factors responsible for fore limb conformation and action in highly specialized aquatic mammals has been given first in order that the reader may have a better understanding of what follows. The details to be discussed in the present chapter have such great interdependence that it is difficult to arrange the subjects properly and to avoid some repetition.

Clavicular conditions vary greatly among all sorts of mammals and it is not easy to distinguish just the critical factor that determines the presence or absence of this bone. In general it may be said to be lacking in those mammals which use the pectoral limb for support only, and present in those which are in the habit of using the hands for grasping; but there are many exceptions, to the latter statement especially. The clavicle should be considered not in the nature of a strengthening member, but rather as a strut to prevent an undesirable degree of adduction of the shoulder. In heavy mammals that bound about, landing solidly upon the fore legs, it would be in danger of breakage, and is accordingly absent; nor does there seem to be much need for it in the case of the more narrow-chested mammals.

The clavicle is lacking from all the more highly specialized aquatic mammals—pinnipeds, sirenians and cetaceans—it is functionally absent in the carnivores, usually present in rodents but absent in a number of heterogeneous sorts (including the capybara), and present in all insectivores except *Potomogale,* and presumably *Limnogale.* The latter is really the only significant fact. For all we know the terrestrial ancestors of pinnipeds, sirenians and cetaceans may have lacked a clavicle. It seems reasonable to suppose, however, that the way in which whales use the flippers would introduce a stimulus for the elimination of the clavicle were one present. On the other hand, one would surmise that the otariids might find a clavicle of advantage to the way in which they constantly adduct the flippers during swimming.

It is perhaps unsafe to attempt to analyze the muscular factors underlying the attachment of the shoulder to the body of aquatic mammals, but existing conditions may be mentioned and a few possibilities advanced.

A trapezius is lacking in whales but is present in both pinnipeds and sirenians, and it seems fully as likely that this muscle has always been absent from the cetacean stock as that it has been eliminated by aquatic habits. It is not known what scapular motions a whale finds of advantage to its well-being, but there appears to be a rather uniform plan of scapular suspension employed in this order and it is likely that very few, or very circumscribed, movements of the shoulder are indulged in. There is a tendency apparent for reduction of the width of the shoulder attachments, which are more segregated into three areas than is the usual case: below and slightly to the rear there is a narrowed pectoralis (either single or double) ; above and slightly to the rear a broad rhomboid, which may be continuous along its border with a narrowed latissimus dorsi; and anteriorly there is anchorage to the atlas and mastoid region by two or three narrow muscles (atlanto- and mastoscapular, and masto-humeral). The serratus magnus may be either very narrow, or very broad as as to function as a very efficient antagonist to the extensive rhomboid.

In pinnipeds the suspension of the shoulder is according to a different principle. The dorsal, ventral and cranial anchors are all spread to a remarkable degree, allowing powerful movement in any and all directions. There are three widely-spreading trapezius divisions, the anterior rhomboid reaches the head, and other shoulder muscles are specialized accordingly.

There is no movement of the shoulder involved in the act of swimming by the true seals, and yet it is inconceivable that such specialized shoulder musculature could constitute merely a phylogenetic inheritance from a more specialized ancestor. If we look upon it as having developed for the purpose of lending all possible assistance to the act of swimming, then is it more understandable. The action is entirely too complex for simple analysis, but after much study of the question I must believe that the end toward which the shoulder muscles of these animals have striven is for the purpose of accentuating the lateral movements of the hinder end of the body in one direction, and the forward end in the other. The muscles anterior to the shoulder would in this case act largely as antagonists to those posterior thereto, and the shoulder is thus comparable functionally to a sort of raphe between the two groups. The chief muscles concerned are the phenomenal pectoralis and latissimus dorsi on the one hand, and the cephalohumeral and humerotrapezius on the other. Movement of the shoulder muscles other than those of importance to swimming may be relatively incidental in this animal.

In the Otariidae or sea-lions the functions of the shoulder muscles are very different from those in the seal. The shoulder proper does not play a largely passive part during swimming, but an essentially active one. Some of the extrinsic muscles (chiefly the pectoralis, but the cephalohumeral and latissimus are also of major import) are directly concerned with flipper movement, while others have the function of adjusting the position of the shoulder itself. It is certain that these latter adjustments are extremely frequent and extensive. The part which they play in actual swimming is unknown, but during terrestrial activity the scapula slides about beneath the skin in a quite surprizing fashion. The vertebral borders of the scapulae may appear to meet well above the back bone, or they may be slid far ventrad. The pectoralis is of course largely responsible for the latter act, and the humerotrapezius is peculiarly fitted for elevation of the scapula, inserting upon almost the entire length of the humerus and with only incidental attachment to the spine of the scapula.

It is only from the pinnipeds, sirenians and cetaceans that we can hope to learn anything regarding scapular tendencies in aquatic mammals. Beyond any question this bone has undergone some degree of broadening in both sea-lions and whales. There may be considered to have been some tendency in this direction in the manati, and in some seals, while in the dugong and the majority of seals the scapula assumes a more falciform shape. It may safely be assumed that the variation in the general shape of the scapula, including the position of the spine, is due entirely to muscle stress. In Cetacea this bone is usually quite broadly fan-shaped, more so in some and less in *Physeter,* in which the flipper may be presumed to be less efficient as a rudder because of the unwieldly size of the head. Invariably the infraspinous space occupies practically the entire lateral aspect of the bone, while the bony area of the supraspinous fossa is insignificant and perhaps but one-hundredth as large, or occasionally it actually does not exist, as in *Platanista* (figure 36) and *Megaptera.* The supraspinous muscle is, then, of decreased importance. But the infraspinatus is not relatively larger so as to fill the infrapinous space. On the contrary this muscle covers but a half or two thirds of the latter; but its details are variable. We must therefore seek other muscles that have had need of a greater angle of leverage and have accordingly stimulated the broadening of the scapula. Judging by conditions in *Monodon* and *Neomeris* it is likely that in toothed whales this stimulus has been supplied either by the subscapularis, which covers the entire medial surface of the bone, or (and) the deltoid, which in the latter genus arises from the entire, and in the former, almost the whole, vertebral border. Contributing to the situation may also have been a stimulus for extension of the glenovertebral angle by the teres major and (or) serratus magnus, which latter in *Monodon* is especially extensive. In at least some of the Mysticeti the stimulus for broadening of the scapula has resulted in an extreme development at the glenovertebral, and to a lesser extent at the coracovertebral, angle of the suprascapular cartilage (figure 36). The chief reason for this extreme cartilaginous extension in a posterior direction is seen in figures given by Schulte (1916) to be probably the serratus magnus muscle, attached to the posterior part. This muscle is thereby given an especially efficient lever arm for operating as a depresser anguli scapulae, ostensibly of great use in tilting the flipper for equilibration. It should also be mentioned that in Mysticeti the deltoid has become differentiated both in origin and insertion for assuming in even fuller degree the function normally held by the supraspinatus.

In the Cetacea the condition of the acromion is quite curious and there is apparently no good reason for its existence. It does not increase the leverage of any muscle, for nothing is attached to it save incidentally. The same may be said of the coracoid process. Both may be developed to an extraordinary degree, as in *Sibbaldus* for instance, or both may be entirely obliterated, as in *Megaptera* (a southern species of this genus is said to have a short acromion), apparently with an equal lack of reason for both conditions.

It is interesting to note that the scapula in the zeuglodont *Basilosaurus* is exactly what one would expect it to be supposing that it represents a stage through which modern whales have passed. In shape it is entirely whale-like. The supraspinous fossa is relatively smaller than usual but still much larger than in any whale, being about a third or two-fifths of the infraspinous fossa. Along the glenoid border there is also a fossa for a strong teres muscle. The acromion is greatly developed.

To a large extent scapular conditions in the sirenians resemble those in the pinnipeds, but Murie did not give sufficient detail regarding this part of the manati to make it advisable for me to pursue the subject further. Noteworthy in the manati is the irregularity of the spine, apparently attributable to some detail of the deltoid, and the pointed acromion. In the dugong the acromion is strongly distinct but it projects at a right angle to the scapular plane.

Ostensibly the medial muscles of the sea-lion scapula must be those which have been instrumental in causing the broadening of this bone, for the lateral ones are very different from those in whales. The supraspinous fossa is considerably larger than the infraspinous space, while in the Phocidae it is smaller. In both groups the origin of the infraspinous muscle occupies about half the space posterior to the spine, but the muscular conditions over the remainder of this area are so involved and

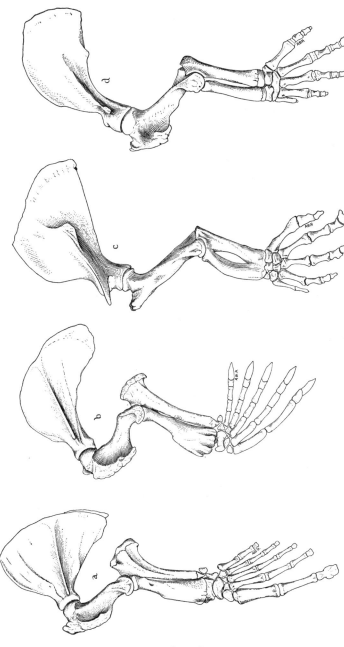

FIGURE 35. Left pectoral limbs of pinnipeds and sirenians: (a) sea-lion (*Zalophus*); (b) seal (*Phoca*); (c) manati (*Trichechus*); and (d) dugong (*Halicore*).

diverse that but little more can be deciphered. There seems to be but little doubt, however, that the more falcate shape of the scapula in *Phoca* signifies a more caudal extension of the glenovertebral angle for the chief purpose of furnishing increased leverage for the enormous triceps, rather than to any extrinsic limb muscle.

In all aquatic mammals the ball-and-socket character of the shoulder joint is retained. In the Pinnipedia, Sirenia and the Zeuglodonts (at least in some of them) the synovial character of the elbow joint is also retained, and this is true also of the carpal bones of pinnipeds, save that in the Otariidae the mobility of the latter region is reduced. In modern whales, however, the joints distad of the shoulder have all lost their synovial character and instead are entirely fibrous. That part of the limb that projects beyond the body contour of whales is thus incapable of bending movement save that the fibrous interosseous tissue gives considerable elasticity to the flipper. This character of resilience is apparently of much value and it is to be expected that it will increase to a certain ideal optimum. One way in which this might be accomplished is in the reduction in length of the interfibrous (i.e osseous) elements.

The advantage of an elastic but non-jointed paddle to a whale is too obvious to need discussion. Apparently it might be of equal desirability to a swimming sea-lion, but this animal is obliged to have adequately bendable joints in the arm if it is to continue movement on land. Although the latter function is of no use to a sirenian, it does need to bend the elbow during feeding, for the flipper is then employed for drawing herbage toward the mouth, and Murie mentioned that in the manati the joints are very lax and their ligaments simple.

But a fore limb in the shape of a true paddle is needed by whale, sirenian and sea-lion—not a spatulate enlargement of the manus upon the end of a long arm. I think this is obvious and that it may be accepted without argument that the normal tendency in such aquatic mammals as whales and sea-lions is for increase in relative size of the paddle part (the manus) and decrease in relative size of the non-paddle part (the antibrachium and humerus). There undoubtedly are several different factors that may help this development, and likely some unknown ones that tend to hinder it.

In such a non-jointed paddle as we may presume to be ideal for propulsion or equilibration in the water the only essential limb muscles are those which operate the arm as a whole, bending it upon the shoulder joint. Not only must there be adequate provision for flexion and extension, abduction and adduction, but also for tilting or rotation, in order that a whale may elevate or depress its line of progress. It is clear that small muscles inserted near the head of the humerus would be capable of but feebly waving about a long arm with broad paddle upon its end. For efficiency the arm should be shortened, as already argued, the critical muscles should be strengthened, and their effective leverage increased by a migration of their insertions distad from the head of the humerus. Such alteration in muscle attachment will effect alteration in the bones, and this in turn may greatly change the functions of the muscles involved.

As already discussed there is no longer any reason for aquatic forms to hold the limbs in vertically dependent posture and the tendency, if uncomplicated, is probably for these members to be held at an angle of about 45 degrees. We have no means of knowing the exact angle favored by various sorts of cetaceans, but the osteological evidence would indicate either that in most porpoises the habitual posture of the flippers is more abducted than in Mysticeti—which seems unlikely—or that *the chief work performed* is instigated from a position with the flippers more elevated or abducted in the former cetaceans. Theoretically this chief work should consist of strong downward movements of the flippers after they have first been elevated, ostensibly for elevating the anterior end of the animal.

Incidentally it should be mentioned that at least most cetaceans probably cannot extend the flipper forward to an angle greater than 90 degrees to the body axis, if indeed even this much extension be possible. The point at issue, however, is that whales abduct (elevate) the limb at the shoulder joint to a considerably greater amount than does the average terrestrial mammal, the degree depending upon the sort of whale. Thus, whereas the limb movement in most mammals is fore and aft, or by extension and flexion, in whales it is rather in the transverse plane, involving abduction and adduction. This change is accompanied by certain definite alterations in the shoulder, which are reflected in the humerus.

It is frequently stated that there has been rotation of the cetacean humerus. I do not altogether approve of this term as it is somewhat misleading, and the process has evidently been entirely different from that experienced by man, in which the upper arm has been rotated by a shift in the usual position of the elbow joint. In whales the elbow is in the plane usual in Mammalia. In the Mysticeti the elements of proximal humerus have not undergone much alteration in position, but in the Odontoceti they have, apparently not by any twisting of the bone but by individual shifting of the items. In the whalebone whales the humeral head may be said to occupy a caudo-lateral position in respect to the axis of the shaft indicating that there is not such definite or else such habitual flexion of this segment as in most mammals, and that it is held somewhat more abducted. The tuberosities have not altered their positions. The lesser, situated mediad, is practically undeveloped as a process and is indicated merely by a pronounced rugosity and slight elevation of the bone. The determinant in developing this region into a true process is the subscapularis, although for *Balaenoptera borealis* Schulte (1916) showed the coracobrachialis and mastohumeralis as also attached upon this area. This lack of development of the lesser tuberosity indicates that either the subscapularis is unusually weak, which its extent belies, that it normally operates when the arm is much adducted, or else that it has a somewhat altered function, for instance to effect rotation of the humerus when the arm is considerably flexed. I regard the latter as the most likely of these three possibilities, although there is nothing else to indicate that this is the case.

In the Mysticeti the greater tuberosity is fairly well developed and may be practically as high as the head. According to Schulte's figures the reason for this is partly, though perhaps unimportantly, the infraspinatus attachment, and (chiefly) the deltoid, which here inserts. We must presume for the present that this muscular condition is also found in other mysticetes. It will thus be seen that although this process appears from an exclusively osteological viewpoint to be homologous with a greater tuberosity it is hardly so except in position, and is actually a deltoid process. The inference then, according to Schulte's figures, is that in at least one mysticete the deltoid has gradually been altered so that its origin occupies the scapular area normally held by the supraspinatus, and that its insertion enables it to function in the same way, except that the latter also stretches far distad and onto the forearm in a way that no supraspinatus ever does, thus not only effecting moderate extension from a more flexed posture of the arm (shown by the height of the process) but also effectively aiding to *maintain* with a minimum of effort such moderate extension as far as an angle of 90 degrees with the body.

In odontocetes the details of the proximal humerus are quite different. Instead of the head being located somewhat toward the rear of the shaft it is usually situated to the side, and the lateral side at that. Conformation of the tuberosities is essentially variable, undoubtedly reflecting important differences in the muscular equipment. On the whole the lesser tuberosity may be said to occupy its normal position, at least in those sorts examined, but because of the shift of the head this process is so located as to allow the subscapularis, which inserts upon it, to act even more directly as an adductor than usual. If the habitual posture of a humerus be vertical (to the body axis) then the subscapularis can operate efficiently upon a lesser tuberosity that is either low or poorly defined. If the humerus be held markedly abducted then equal efficiency will demand a lesser tuberosity of great height and projection. The latter is the case in such an odontocete as *Tursiops,* in which the tuberosity is just about as large as, and higher than, the head, indicating in connection with the lateral situation of the latter that the normal posture of the flipper may be a pronouncedly abducted one. But this process is not so high in a number of other toothed whales, and there is even some variation in its precise situation.

As there is no occasion to extend the arm beyond an angle of 90 degrees with the body axis there is not only no muscle corresponding in function to a clavoacromiodeltoid, save a weak mastohumeralis poorly placed for this purpose, but the supraspinatus, normally an extensor of the humerus, has not only suffered enormous reduction but its function has changed. In those sorts dissected its insertion has shifted mediad to the anterior border of the lesser tuberosity and it accordingly acts as a rotator to elevate the anterior border of the humerus. The infraspinatus, also normally inserting upon the greater tuberosity, has shifted its attachment distad and slightly laterad and now is inserted chiefly into a fossa, very characteristic of most odontocetes, situated upon the humeral shaft, so that it acts not only as an upward rotator of the anterior border but also effectively as an abductor of the arm. Usually in mammals the greater tuberosity has developed into an eminence because of the stimulus supplied by the insertions upon it of the supraspinatus, infraspinatus, and teres minor, but in the toothed whales the first two muscles have shifted elsewhere and the last does not occur as a distinct division, so the greater tuberosity has ceased to exist as a process strictly homologous with the eminence to which this term is applied in most mammals.

In a number of toothed whales which I have examined there is some variation in the conditions of the proximal humerus as recounted above. Slight eminences, situated in this or that direction from their situation in the dissected specimens bespeak corresponding muscular variation, and there may actually be a prominence which might be mistaken for a greater tuberosity of slight definition, but it is believed that this is only analogous, rather than homologous, as discussed below.

In those odontocetes dissected the extremely robust deltoid inserts upon the whole lateral face of the distal humerus, acting as a powerful abductor of the arm, but this insertion is thicker and stronger craniad and it likely has some rotating action as well. The latter fact is not shown osteologically in all toothed whales but in fully adult *Tursiops,* for instance, it seems clearly indicated by a very pronounced process upon the distal third of the anterior aspect of the humerus. This strongly impresses one with the fact that if the latter protuberance were shifted more to the proximal end of the bone, as is largely the case in some individuals of *Kogia,* it would form precisely the same sort of "greater tuberosity" as appears to have been caused in the Mysticeti by the deltoid. There is a similar process, evidently attributable to the deltoid, in zeuglodonts, but located just distad to the middle of the shaft.

The above muscles taken together seem to show that in the Cetacea there is a reduced power of extension of the arm, and increased power of abduction, and perhaps a slightly greater power of both adduction and rotation upward of the anterior border. The latter act is also markedly assisted in mysticetes by flexion of the serratus magnus. There is not apparent any special provision for *depression* of the anterior border of the arm. The general conformation perhaps makes this motion not so necessary, or it may be effected by special action of the latissimus dorsi or other muscles.

The distal humerus of the Cetacea lacks the stimulus supplied by functional antibrachial muscles and a synovial joint. Accordingly the lateral and medial elevations of the condyles have atrophied. Even were there absent some undefinable stimulus for a broadening in an antero-posterior direction of all the arm bones, which indubitably exists, this entire lack of reason for condyle definition would likely be sufficient to allow the distal humerus to broaden out in an antero-posterior direction to conform to the extent in this plane of the antibrachial bones.

It is unfortunate but nevertheless true that when I was preparing a previous paper on the pinnipeds (Howell, 1929) I was unable clearly to observe all the motions followed by a sea-lion when swimming at speed. I gained a satisfactory understanding of movements during more leisurely progress and made the mistake of supposing that these were used at a faster gait, for I had not then encountered the correct conditions of light and clarity of the water to see them dart at speed deep below the surface. This course seemed entirely justified for the reason that the posterior part of the pectoralis and the latissimus dorsi are so well developed for executing powerful backward thrusts. The statements

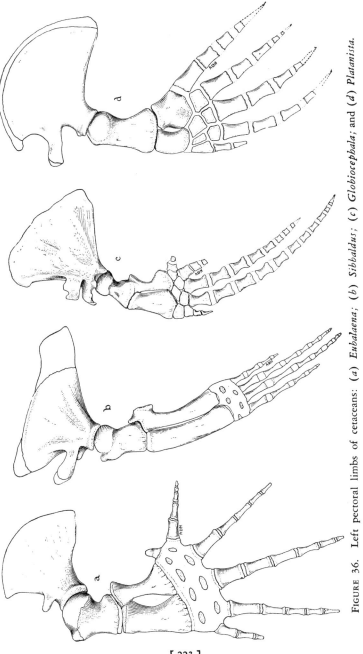

FIGURE 36. Left pectoral limbs of cetaceans: (*a*) *Eubalaena*; (*b*) *Sibbaldus*; (*c*) *Globiocephala*; and (*d*) *Platanista*.

which I then made were that the sea-lion swims by advancing the flippers, not in unison nor yet in alternation but with a sort of galloping movement, and then by partial rotation so as to present the broad aspect of the flipper, progressed by means of strong backward sweeps of these members. More recently, however, I have succeeded in obtaining clear views of the whole process. Undoubtedly the muscles mentioned *were* developed for backward thrusts at a time when the animals were less specialized and as yet incapable of the speed that they now attain. At present, however, they are too speedy for just this method of propulsion to be efficient, as already discussed at the beginning of this chapter, and for mechanical reasons rather than purely muscular ones, they are obliged when progressing as fast as possible to do so by means of adductive and abductive thrusts of the flippers, with the minimum of extension or flexion. The anterior border rather than the posterior is where the force is applied, while the latter is allowed passively to follow through, so that there is an oblique thrust and both adduction and abduction are utilized for propulsion, although the latter undoubtedly is productive of much less power, possibly an insignificant amount.

Unlike the case of the whales the flipper of a sea-lion is not stiffened by fibrous joints. The elbow does not cut much of a figure in the bending of the external arm for it is close to the body contour, but although possible flexion at the wrist is not great, extension at this point is through an angle of 90 degrees to allow for terrestrial progression, and this must be overcome by mechanical means during swimming so that the wrist will not bend backward during the adductive stroke, and so that muscles will not be wearied by continual effort to prevent such bending. Extension of the arm is no more needed than in whales, the degree of abduction need not be greater, but adduction and flexion must, or at least should be, greater. In relaxed posture the radial or anterior border should be presented straight, forward, for this position is the mean from which pronation and supination are instigated. Accordingly the normal terrestrial position is for the manus to extend from the wrist in an exactly lateral direction rather than somewhat craniad as usually shown in mounted skeletons.[2]

In the seals (Phocidae) the pectoral limb is not used in swimming, as already discussed, save during such acts as turning. It is supposed to be employed for such purposes as scratching holes in the ice, and its strictly

[2] This inaccuracy will be noted in the photograph (fig. 14) of the fur seal herewith depicted, which is the same as that shown in my previous paper referred to.

terrestrial uses are very incidental, save in the elephant seal *(Mirounga)*, in which the manus helps support the weight. It is usually kept folded back against the body with the segments markedly flexed so that all of it save the manus is contained within the body contour. Extension is possible, however, and I have seen an animal on the steep margin of its pool stretch forth the manus so that as well as I could judge the arm was extruded from the body contour as far as the elbow.

The normal or static position of the humerus in respect to the scapula, determined to the best of my ability during dissection, differs only slightly in sea-lions and seals. The angle formed by the humeral axis with the spine seems to be slightly less in the latter, showing a greater degree of flexure, but the difference is not sufficient to be of much significance. The position of the humeral head in relation to the shaft is slightly more mediad in the sea-lion. This, I judge, is less marked than one might expect, in consideration of the fact that the most efficient method of swimming is by abduction-adduction movements, but this development may be so recent that it has not had sufficient time to have had marked osteological effect, and after all many other motions of the humerus are of great importance. In phocids the head is more posterior to the axis of the shaft, showing that extension-flexion motions are of more pronounced import, and its conformation is such as to indicate that possible flexion of the humerus in relation to the scapula is more extreme in the seal.

In the sea-lion the greater tuberosity is markedly higher than the head and than the lesser, while in the seal it is much lower, the lesser tuberosity having greater elevation. The reason for the height of the greater tuberosity in the sea-lion seems solely attributable to the supraspinatus muscle, acting chiefly as an extensor, and the strength of this muscle is shown in the scapula also by the great extent of the supraspinous fossa. The infraspinatus, chiefly a rotator, also inserts upon this process, but more proximad, and its position in *Phoca* is quite comparable, but in this animal the supraspinatus is evidently not required to act so strongly as an extensor, for not only is the supraspinous fossa of the scapula much smaller, but the greater tuberosity, upon which it inserts, is hardly higher than the head and very much lower than in the sea-lion, giving this muscle reduced leverage.

The conformation of the lesser tuberosity in the two pinnipeds considered is not altogether what one would expect. In the sea-lion, although robust and subtended by a heavy ridge, it is rather low and not nearly as high as the head; consequently much lower than the greater tuberosity. Its height proximad has nothing to do with the strength of the attached muscles, of course, for this detail depends upon the optimum angle of leverage. Its relatively low height, then, would indicate that the muscles operating from it are chiefly flexed when the arm is in a rather adducted posture. Not only is the subscapularis well placed in respect to the head for leverage in adduction, but insertion of the episubscapularis (absent in seals) is located still farther distad along the ridge, giving added power. What one might not expect in this animal is that the lesser tuberosity is not situated directly mediad of the head (to be most efficient in adduction) but medio-craniad. This situation might easily translate movement that might otherwise be adduction to one largely of rotation. This, in effect, may be actually what occurs at times; or flexion of the same muscles may be productive of pure adduction when other muscles are used as antagonists. It seems that chief among the latter might be the teres major, whose insertion is broadly along the medial shaft of the humerus distad of the middle of the bone.

In the seal these medial insertions are disposed upon the humerus for different action. The elevation of the lesser tuberosity is quite surprising, it being much higher than either the head or the "greater" tuberosity. The attachment of the cephalohumeral thereto can hardly account for its height, and the only reason apparent is that the chief work of the subscapularis, and attendant subscapulo-capsularis (absent in sea-lions), is performed when the arm is pronouncedly abducted or elevated. This might well take the form of a strong downward heave of the manus while swimming for the purpose of quickly elevating the body, or while scratching holes in the ice. Other muscles of this region are less well situated for adduction of the humerus than in the sea-lion. There is no episubscapularis and the insertion of the teres major is farther craniad in respect to the head, theoretically giving a greater rotating action to this muscle.

The deltoid ridge, comprising a proximal continuation of the greater tuberosity, is phenomenally developed in the pinnipeds, and a complex of muscular stimuli is brought to bear upon it. In the seal this ridge with tuberosity is but little more than half the length of the bone, while in the sea-lion it is almost two-thirds, and in the latter especially the concerned muscles accordingly have a phenomenal leverage. While in the seal the humeral attachment of the cephalohumeral is confined to the greater tuberosity, in the sea-lion it extends over the entire length of the deltoid crest. In the seal this is the case with the humerotrapezius, while in the sea-lion the insertion of this muscle is for practically the entire length of the humerus. In both animals the pectoral is double, one division being confined to the deltoid ridge and the other being almost as long as the whole humerus. In the seal the atlantoscapularis inferior inserts upon most of the deltoid ridge, while in the sea-lion this muscle does not reach the humerus but inserts upon the spine of the scapula. Most of these insertions are by fascia and although it would be utterly unjustifiable to say that the muscles mentioned have been only of secondary importance in the development of such a high deltoid crest, still it may be stated that all of them would appear to be equally effective if operating upon a humerus without a crest. The large deltoid, however, which operates chiefly as a flexor by virtue of the shortness of the humerus, is given the function also of a powerful rotator by the lateral definition of this crest.

Little more need be said regarding the pinniped pectoralis, for it has already been considered in preceding chapters. In *Phoca* the main development of this complex is posteriorly, for swinging the hinder end sidewise, but the part medial to the arm is also very powerful, either for quickly adducting the arm for steering, holding it firmly adducted while the muscles anterior and posterior from the arm are used in swimming, or probably both. In the sea-lion although the abdominal pectoral is powerful it is far less so. It can be used to help control flexion of the arm during terrestrial progression or backward sweeps of the flipper when the need arises. In the seal the insertion of the pectorals upon the humerus gives unusually efficient leverage, and in the sea-lion this is even better developed. Not only is there double insertion over the length of the humerus but extension of a superficial sheet onto the forearm, giving the greatest possible leverage for strong and constantly repeated adduction of the appendage. Flexion of the part inserting over the deltoid crest also greatly aids the rotation of the anterior flipper border that is so necessary during swimming.

The significance of most of these extrinsic muscles of the humerus has already been discussed in relation to the neck and other parts of the body, but they must receive further consideration in the present connection. Although these muscles have many significant similarities in seals and sea-lions, yet their functions appear to be very different. It seems almost certain that in the sea-lion the muscles extending from the humerus to the head and neck are used chiefly for extending the arm craniad, both when the need arises while in the water and during progression by a lunging gallop while on land. And similarly that the abdominal pectoral, latissimus, and panniculus may be likewise used for flexing the arm. It appears equally likely that the corresponding muscle groups in the seal have their chief use in pulling the head and the hinder end respectively in a lateral direction while swimming. To me it seems beyond question that if these highly specialized muscles extending from the phocid humerus were actually a relic from a time when seals might have used the arm as a primary means of propulsion the external form of the manus would now show far more indication of this fact.

In the sea-lion the lateral epicondyle of the distal humerus projects scarcely laterad of the trochlea, but is much better defined in the seal, while in the former the medial epicondyle is greatly developed, but slightly so in the seal. This is an indication that in the sea-lion the flexor muscles of the fore arm have more leverage and hence are presumably more efficient than the extensors, and that in the seal the reverse is the case. These muscles will be more fully discussed later.

The musculature of the dugong is almost unknown but it is likely that in brachial details it is considerably different from the manati because there are very important differences in the form of the humerus. In addition, Murie failed to figure many important items of the manati, and in consequence my discussion of the sirenian brachium should be considered as only tentative and subject to amendment.

The humerus in the dugong and the manati show two very different trends, that of the former being very pinniped-like in some general respects, while in the latter this is not the case. The head in both is situated fairly posterior to the shaft axis. This is a rather trustworthy indication that the chief direction of movement is in the sagittal plane, or at least that any other primarily important movement that might have been recently adopted has not yet had time to cause appreciable alteration osteologically. This is also borne out by the fact that the lesser tuberosity is not medially situated. Rather is it continuous with the greater, the two conjoined forming a broad, transverse ridge fairly anterior to the head. The medial part of this ridge, homologous to the lesser tuberosity, is much higher than the latter, indicating, I should think, that the humerus is normally held somewhat abducted from the scapular plane. There is no teres minor in this genus, so the scapular muscles inserting upon this ridge, or broadened tuberosity, are the subscapularis upon the medial part, the infraspinatus upon the outer, and the supraspinatus between them. The first, operating alone, should accomplish some inner rotation with adduction; the second, slight abduction with outer rotation, or flexion when the humerus is also strongly flexed; and the third extension: or by operating the first in antagonism to the second, all three can act in extension. And extension of the flipper is very important in the act of drawing herbage toward the mouth. In the manati the deltoid is inserted not upon a ridge but on a rugosity upon the middle of the shaft. It accomplishes chiefly flexion, this being strongly aided by action of the teres major, which with the latissimus inserts upon a similar but larger and more distal rugosity upon the opposite (medial) side of the shaft.

In the dugong the humeral head is also posterior to the shaft axis, indicating that flexion and extension is the chief movement, the greater tuberosity is higher than the lesser, the former is located craniad of the head as usual, and both are individually distinct instead of continuous as in the manati. In addition there is a heavy and high deltoid crest continuous distad with the greater tuberosity. Humeral conditions are essentially similar in the Steller sea cow.

In the dugong osteological details of the brachium so greatly resemble those in the sea-lion that I have no choice but render the opinion that the musculature of this region must have many points of similarity. The supraspinous fossa is very much smaller in the dugong, but the muscle should have a comparable action and insert with the infraspinatus upon the greater tuberosity. Origin of the teres minor cannot be as extensive and this muscle must be reduced—perhaps absent. The subscapularis is likely used separately for adduction and there should be a strong deltoid inserting upon the deltoid crest for use in rotation of the humerus. This is all that can be said. In view of the fact that we know so little of the way in which the limb of this animal is mostly used and consequently are unable intelligently to compare its real function with that of the manati limb, it would hardly be justifiable to advance further possibilities. It may be stated, however, that the anatomical details of the manati indicate that this animal has never used the pectoral limb as a means of propulsion through the water since this member became specialized, while the osteology of the dugong does show that the limb might once have been used either for this function or some other that necessitated muscle action of much the same sort.

After having discussed the brachium and before taking up details of the antibrachium it is proper to discuss alterations in the length of both of these segments, and consequently in the whole arm, which aquatic mammals have experienced. For the reason that different sorts of vertebrates may respond in such diverse ways to the same stimuli we may not derive much help in this from a scrutiny of brachial conditions in aquatic reptiles, but to do so will at least be interesting.

In discussing the modification which the pectoral limb has undergone in both mammals and reptiles of aquatic habits Williston (1914) stated that the humerus has become greatly shortened in aquatic types having a tail fitted for primary propulsion and even in some having short tails, as seals, and to a lesser degree, sea otters. In those which use the legs for direct propulsion, as plesiosaurs and marine turtles, the humerus is elongated. In all save seals and otters whose limbs are used rather as sculls than oars the lower limb bones are always shortened. These statements are not entirely accurate. In all highly specialized aquatic mammals now living, regardless of the method of propulsion employed, the humerus has become shortened. On the other hand, so far as I am acquainted with the facts, in all aquatic reptile the sequence seems to be for the antibrachial segment first to experience shortening, followed later by a shortening of the humerus. At times, as in the leatherback turtle (*Dermochelys,* fig. 34), the disparity in size of these segments is still very pronounced, the humerus being huge and the antibrachium really of insignificant size. Apparently this is fundamentally characteristic of the order and different from what seems to be the usual sequence in aquatic mammals. The likelihood is that this difference is chiefly attributable to the dissimilarity in the osteological plan and basic muscular equipment of the two orders. For the present it is hardly profitable to pursue the subject further.

In spite of the elongation of the sea-lion manus the total bony arm length from the shoulder joint is, relative to length of body vertebrae, very much shorter than in such a terrestrial carnivore as a cat, while in the seal it is 20 per cent less. This may be said to be due entirely to a diminution in the length of the brachial and antibrachial segments. Relative to length of body the humerus in the sea-lion is but 65, and in the seal but 45 per cent of the length of this bone in the cat. The radius, as representing the antibrachial length, is found to be 105 in the sea-lion and 96 per cent in the seal of their respective humeral lengths, while in the cat this percentage is about 104. It is somewhat surprising to find the proportions of these two segments so uniform in these three mammals, and it is an indication that the stimulus for reduction in the arm length of these pinnipeds has apparently been quite uniform, or at least has had a uniform result, in both segments between the scapula and manus. In the walrus, however, the radius is only about 80 per cent of the humeral length.

That there has been a pronounced reduction in brachial and anti-brachial length in sirenians is apparent and yet it is difficult to compare its degree with this detail of the Pinnipedia. It is clear, however, that in relation to arm length there has not been as great reduction in this dimension of the sirenian humerus, while there has been a greater amount in the radius and ulna. But it is unexpected to discover that in both genera considered the radius is just 68 per cent of the humeral length.

In living cetaceans there is invariably a phenomenal shortening of the humerus to the extent where this bone is at times two thirds as broad as long, but its proportion to the radial length is extremely vari-able. In almost all sorts it is definitely shorter than the radius, but in *Physeter, Kogia, Stenodelphis* and *Platanista* it is longer—in the latter over 200 per cent of the radial length. In other odontocetes it is slightly shorter, while in mysticetes this is considerably more pronounced. Thus in *Rhachianectes* the humerus is about 68 per cent of the radial length, and 60 per cent in *Sibbaldus*. Speculation of the above cetacean facts is, however, sadly complicated by the circumstance that the arm of *Basiolosaurus,* more primitive than any recent whale, shows a much reduced antibrachium and a rather long humerus.

As with the carpus and digits of the Cetacea the elbow joint in this order is entirely fibrous in character. Apparently this is due to the non-development of the articular structures characteristic of a synovial joint, rather than to the alteration of these. It seems that the fibers of the existing joint develop directly from the perichondrium or periosteum to the degree where they supply the amount of stiffening that the animal finds essential.

Whereas the external axilla of the sea-lion falls midway of the anti-brachium and of the seal opposite the wrist, it is apparently situated just proximad of the elbow in the Sirenia, at about the proximal third of the antibrachium in at least some odontocetes (porpoises), and at about the same point in Mysticeti. The optimum position of the elbow in respect to the body contour in aquatic mammals thus seems to be somewhat variable, depending upon individual requirements, and of course upon the amount of specialization.

In no other mammals than those discussed above does it seem safe to say that there has been alteration in the osteology of the fore limb in-duced by life in the water. Taylor (1914) has made the observation that in the sea otter this member is relatively smaller than in the river otter, and the short legs of the hippopotamus may be considered either in this light or as having been caused by decreased terrestrial activity. But such alteration has been too slight in degree to make discussion profitable.

As the anterior limb of aquatic mammals has always been a subject for the liveliest speculation there will undoubtedly be expected of me some statement regarding my conviction on this question. Unfortunately the facts do not warrant any very strong convictions, for each case is different and seems to constitute a law unto itself. In the first place it is likely that the paths followed by aquatic reptiles are throughout most of their course so different, because of a different equipment to begin with, that they can certainly not be compared with any intelligence, at least during their earlier stages. At long last, after a staggering length of time, the paddle of an ichthyosaur and that of a cetacean, if used for the same purpose throughout ages, may show a convergence of characters to the point where they have essentially similar details. The Cetacea already show promise of this, but no other aquatic mammal is sufficiently specialized for this to be apparent. Diverse sorts pursue individual paths and even though these trend in the same general direction they are often far apart and have become deflected from the straight line by numerous obstacles. All that I feel convinced of at present is that in aquatic mammals there is a stimulus for the shortening of the length of the part of the arm situated between the scapula and the manus. It further seems likely that the control of the arm as a swimming paddle or equilibrator is facilitated by a humerus shortened to some degree and with the insertions of some of the critical muscles shifted farther distad. In addition it appears logical that after the disappearance of a synovial elbow joint and marked atrophy of the musculature of the lower arm the shortening of the antibrachium should be enabled to progress at an accelerated rate. But any opinion as to whether the brachium or antibrachium should be shorter in some hypothetical aquatic mammal would be merely fanciful.

Regarding the course of future development, we are justified in considering that eventually a flipper of the character of that found in *Platanista* (fig. 36) will likely assume the essential characters of that of an ichthyosaur (fig. 34), in which the humerus, radius and ulna have been reduced to flattened ossicles of the same appearance as the carpal bones. This may or may not be the final goal of the pectoral limbs of mysticetes, sirenians and non-phocid pinnipeds. It is not improbable but is surely not a certainty. I will even go so far as to say that it seems to me unlikely that the true seals will ever assume this type of flipper characteristic of ichthyosaurs, chiefly for the reason that all but the manus is contained within the body contour, the function of the long bones appearing to be chiefly that of a scaffold upon which are hung some of the muscles of importance to swimming, rather than an integral part of a flipper.

The part which the antibrachium plays in the economy of aquatic mammals differs considerably in various sorts. It is difficult to know in just what light it should be considered in the Phocidae. In the latter the position of the external axillary region is opposite the wrist, which is capable of extreme mobility, and hence the antibrachium cannot be considered as functioning with the manus. And yet the humerus acts as independently of the forearm as the exigencies of the position of both within the body contour will admit. Although in consideration of the functional differences the osteological similarity of the forearm, consisting of a broadening of the bones, in seals and sea-lions is really phenomenal, the two should be considered separately. In both of them the elbow joint is synovial but flexion of the forearm is apparently reduced, especially in the sea-lion.

In the seal *(Phoca hispida)* which I have dissected there is a remarkable broadening of the distal radius and proximal or olecranol part of the ulna, which does not affect the opposite ends of these bones. No useful reason for this specialization is positively known. As the anterior limb is used for neither propulsion nor as a specialized rudder, as shown by form of the manus, there could not be this stimulus for a broadened forearm, and besides, the recession of the segment within the body contour renders any such mechanical adjustment useless. Although there may be perfectly good obscure reasons for this condition there is only one apparent. The divisions of the phocid triceps muscle are truly enormous and in addition have phenomenal leverage by virtue of their attachments. Thus the longest division extends from the vertebral border of the scapula to beyond the middle of the fore arm. In action it not only extends the antibrachium but sweeps the whole arm to the rear; and the large, broad olecranon gives it just so much more leverage. The origin of the two extensores pollicis muscles from the lateral olecranon well away from the joint give to them a marked ability to supinate the manus. Upon the medial side of the bone an unusually large area is provided for the broad and powerful flexor digit. communis, with consequent power of digital flexion, and the broad ulna provides a better lever arm for the adductor action of the peculiarly developed abductor digiti quinti longus. I can see no definite muscular advantage in the broadened distal part of the radius save chiefly in providing better leverage for the extensor metacarpi pollicis, and possibly to a slight extent of the pronator teres and supinator brevis. So far as we know there is no terrestrial nor natational use for such a specialized equipment for powerful motions in several directions of the manus, and I am accordingly compelled to believe that it has other functions of exceeding specialization, which the very shortness of the external arm facilitates. Perhaps the reported use of the manus of northern phocids for scratching holes through the ice is of more importance than might appear at first glance. But this function could not be of influence in the case of those tropical genera in which the nails are well developed, so perhaps the manus is particularly useful for some such purpose as scratching about on the bottom for echinoderms or similar fare.

In the *Phoca hispida* dissected there were very deep grooves (fig. 35) upon the distal radius for the passage of the tendons of the extensores metacarpi pollicis, digitorum communis and lateralis, and a shallower one for the extensor pollicis longus. These grooves are almost as well defined in *P. fasciata* but much less so in *P. groenlandica,* while that for the extensor metacarpi pollicis is the only one at all marked in the genus *Monachus.* Neither are they a character in *Odobenus* and they are entirely absent in the Otariidae. These grooves, when they occur, are for the purpose of fixing the position of the tendons concerned. They would indicate that contraction of the respective muscles, effecting extension of the whole manus as well as abduction of the pollex, is often repeated and rather strong; but this is perhaps incidental. Their real significance is to prove that the chief action of these muscles is effected at a time when the manus is already considerably extended, or, at the very least, on a line with the antibrachial axis; and further, I believe that the fact that these grooves are sunk into the bone not at a perfect right angle but facing somewhat in the ulnar direction is an indication either that the chief extensor action of the muscles is combined with some definite abductive movement of the manus, or else that the manus is usually definitely abducted during their action. By means of these grooves and ligaments above them the tendons are prevented from pulling away from the bone. This action of the wrist will be discussed further.

In the sea-lion *(Zalophus californianus)* which I dissected there is an even more marked broadening of the proximal ulna, while the breadth of the distal radius is no greater than in the seal. Unlike the latter animal, however, there is a purely mechanical need for a broadening of the sea-lion antibrachium which the former seems now to lack. In otariids the normal posture places the external axillary angle at about the middle of the fore arm, while in phocids it is situated opposite the wrist. Hence in the former this segment is influenced by the stimulus for becoming flatter and broader so as to offer less resistance, and has been more completely so in the past. The development of the seal manus seems to indicate that this same quality of stimulus has at no time been a very important factor in its evolution. Further, in the sea-lion the distal fore arm is essentially a mechanical part of the paddle (manus), as it is not in the seal, and that the broadness of the flipper in the former pinniped has had some real influence in continuing such broadening onto the fore arm, thus affecting the radius, while the corresponding broadening of the olecranol part of the ulna should rather be attributed to muscular stress. In view of this evidence I am of the opinion that the largely similar osteological specialization of the fore arm in these two pinnipeds is likely not ascribable strictly to the same stimuli, but rather to two (or more) rather diverse influences which have effected similar conformation or convergence.

That the sea-lion has need for an antibrachium broader than the bones alone have been able to supply may be shown by the fact that upon the radial edge there is a thickened, partly fibrous development of tissue which is continuous with the pectoralis profundus; or this may have been built up as a buffer by the action of water resistance. In otariids the purpose for which the flipper is used necessitates that it normally be held more extended than the phocid finds advisable. Observation of the segments during dissection, correlated with the position of the humeral condyles in relation with the axis of the bony shaft, indicates that the brachial-antibrachial angle in static posture is about 125 degrees or less in *Phoca,* and 155 degrees in *Zalophus.*

As in *Phoca* the lateral aspect of the olecranol surface of *Zalophus* is occupied by the very broad origins of the extensores pollicis longus and metacarpi pollicis, while the triceps longus and lateralis, for extension of the fore arm and flexion of the whole limb, gain added leverage by virtue of the broad olecanon. The two pollicis extensors are extremely broad in their tendinous parts and are well fitted for supinating movements of the anterior flipper border. I believe, however, that the chief stimulus for the remarkable broadening of the proximal ulna of the Otariidae has been the origin of the palmaris longus (fused with the smaller head of the flexor carpi ulnaris). This is a very specialized muscle, not only at origin but distad, where it extends in a tendon 25 mm. broad at its narrowest point, thence widening to cover the whole wrist before splitting in two sheets, the more robust going to the anterior and the lesser to the posterior border of the flipper. This effects a definite cupping action of the palmar surface as discussed later.

In pinnipeds as well as Cetaceans there is no appreciable tendency for either the fusion of the radius and ulna, nor for the reduction of one of these bones at the expense of the other, which is as one would expect. The stimulus is for each bone to become independently broad, which is apparently facilitated by non-fusion, and ultimately after the entire abandonment of the land, for the formation of a toughly fibrous connection of all the bony elements, which gives a desirable amount of resilience. The antibrachial bones of the Sirenia, however, are anomalous among aquatic mammals of a high degree of specialization. In the dugong the radius and ulna are distinct and quite simple, without any real indication of broadening. In the manati there is some slight tendency toward flattening of the ulna and this bone is firmly fused at both extremities with the radius, although the shafts of the two bones are curved and quite wide apart. In *Hydrodamalis* there is little or no propensity for flattening but the extremities not only are firmly fused but the shafts as well (fig. 40), save for one small interval that has doubtless remained as a foramen for the passage of nerves and blood vessels. Hence one finds the unexpected condition that in this detail the Steller sea cow does not resemble its (presumably) nearest relative, *Halicore,* but the more distant *Trichechus,* and it must be inferred that in the first and last mentioned genera there is no need either for pronation or supination of the manus save what is possible through the carpus, while presumably in the dugong some little rotation of this segment is possible. Stronger extension of the antibrachium in the dugong is indicated by the better definition of the olecranon. A synovial elbow joint presumably of the normal sort is possessed by this order.

In all known cetaceans, both living and fossil, the radius and ulna are distinct and neither is appreciably reduced in comparison with the other. Almost always the radius is slightly more robust than the ulna, which might be expected because of its exposed position upon the anterior border of the arm, where it would encounter full water resistance; but at times (as in *Platanista*), the ulna may be the larger. Apparently zeuglodonts retained the synovial character of the elbow joint, and this may also have been the case with some Miocene whales, for *Eurhinodelphis,* and likely others, have articular surfaces upon the bones, although these are of lessened area and definition, so that it is more logical

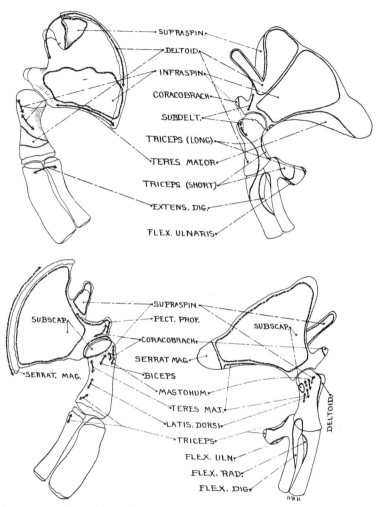

FIGURE 37. Areas of muscle attachment upon the pectoral limbs of a toothed whale (*Monodon*, to the left) and a whalebone whale (*Balaenoptera borealis*, to the right, redrawn from Schulte). Lateral view above, medial below.

to presume that the joint had lost at least a part of its original mobility. As the antibrachium is essentially a mechanical part of the paddle the tendency is for an increasing flattening of both bones, and for their simplification, following the disappearance of functional antibrachial musculature.

The only bony details that might be ascribable to present or past muscular stimulus is the olecranon. This was large and somewhat pinniped-like in zeuglodonts, and to a lesser degree in *Eurhinodelphis*. It is usually present in balaenopterids as a well defined process, which at times is of large extent. Schulte (1916) has shown that in the fetal state of at least one species this is truly of phenomenal size and as a distinct bony arm, stretching somewhat distad, may be half the length of the humerus (fig. 37). In *Eubalaena* it does not constitute a real process, although there is some projection of the ulnar head as a substitute. Usually in odontocetes there is a slight process at this point, but occasionally (as *Platanista*) in the sorts in which the antibrachium is most reduced there is no indication whatever of an olecranon.

The reason why the olecranon is of such enormous size in the fetal balaenopterid is unknown, but it is only natural that at this stage it should be larger than in the adult, in which the muscles of the lower arm are so reduced. The muscles most intimately concerned with the olecranon Schulte found to be a long and a short triceps head. He further found as indubitably present flexores ulnaris, radialis, and communis, and an extensor digitorum communis, as well as a ligamentous band which he considered as representing the biceps. This is essentially in accord with the findings of other investigators of balaenopterid musculature, except that he failed to encounter a plamaris longus, as reported by Carte and MacAlister, or a flexor sublimis as mentioned by Perrin. The details of these muscles do not concern us in the present connection. What is of interest is the fact that although they (excepting possibly the long triceps) are either entirely or virtually nonfunctional, there are a number of them clearly present in the arm of mysticetes, and this constitutes one of the chief items of evidence that this group as a whole may be less highly specialized than odontocetes.

Clearly recognizable fore arm musculature has often been reported in the Odontoceti. For instance Schulte (1918) stated that in *Kogia* there were triceps, extensor communis, flexores carpi ulnaris, digitorum radialis and digitorum ulnaris, as well as interossei. In my own dissections of *Monodon* and *Neomeris,* however, I failed to find any of these as clearly recognizable muscle remnants. In the former there was a tendinous band stretching from the humeral head to the olecranon to which I referred as a triceps; but there was nothing else distad except fibers identical with those strengthening the tissue between the digits. In *Monodon* I distinguished and tentatively named by virtue of their position a triceps, biceps relic, flexor digitorum and extensor digitorum, but the two latter were rather loosely associated fascicular bundles and I cannot even be entirely sure without histological preparation that they really contained muscle fibers. It is thus indicated that at least in a great many odontocetes the intrinsic musculature of the flipper proper is considerably more atrophied than in mysticetes, and is on the rapid road to total disappearance.

As mentioned there is a tendency in the Cetacea toward flattening of the antibrachial bones, and this is especially pronounced in those sorts in which this segment is the shortest. Like the case of the sea-lion this is entirely expected, for the antibrachium is a mechanical part of the flipper and is subjected to the same influences as those which have operated to shape the manus. Also like the sea-lion there appears to be need for an antibrachium that is externally broader than the breadth of bone can furnish, and there is disposition of tough fibrous tissue not only on the radial border, but softer tissue in more generous amount upon the ulnar border as well. As previously mentioned the tendency in most sorts of odontocetes at least would seem to be toward the reduction of the radius and ulna to the size and conformation of the larger carpal elements.

In considering the manus of aquatic mammals we come to the critical part of the fore limb. All other parts of the arm are directly influenced by it or by certain body factors; hence the manus may be expected to reflect in most perfect degree the stimuli that have operated upon the pectoral appendage. In no two diverse sorts of aquatic mammals is this detail alike, for the reasons that in no two sorts has it been used in exactly the same manner, and that diverse kinds of aquatic mammals split off from the main stems and have constituted individually separate lines, each developing its own peculiarities, since before the manus became so highly specialized. In consequence it is unsafe to venture any broad generalities except that if the manus be used as a propulsive organ or an equilibrator it can be expected eventually to assume the characteristics of a paddle or flipper in which the bony element's become flattened and simple, without movable joints. Usually there will also be apparent a stimulus for a lengthening of the manus at the expense of the more proximal brachial elements.

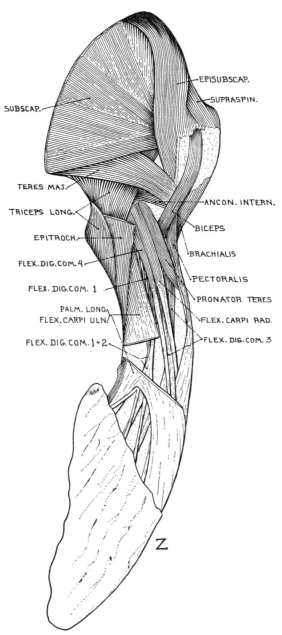

FIGURE 38. Superficial musculature of the medial aspect of the left pectoral limb of a sea-lion (*Zalophus*).

[240]

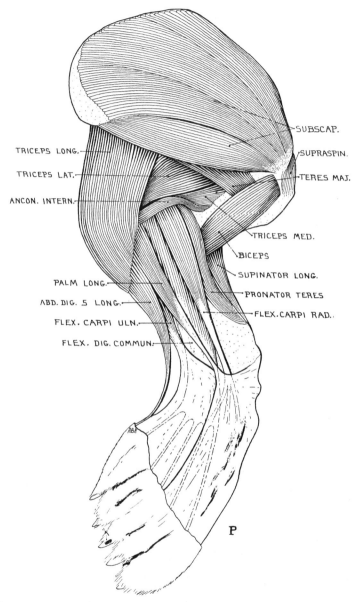

TRICEPS LONG.
TRICEPS LAT.
ANCON. INTERN.

PALM LONG.
ABD. DIG. 5 LONG.
FLEX. CARPI ULN.
FLEX. DIG. COMMUN.

SUBSCAP.
SUPRASPIN.
TERES MAJ.

TRICEPS MED.
BICEPS
SUPINATOR LONG.
PRONATOR TERES
FLEX. CARPI RAD.

P

FIGURE 39. Superficial musculature of the medial aspect of the left pectoral limb of a seal (*Phoca*).

[241]

There would also seem to be another character which the manus of the aquatic mammal may be expected ultimately to adopt. In most terrestrial mammals the bones of the metacarpus and digits consist of a shaft and but a single epiphysis. The latter is situated upon the the distal end of the shaft in the four lateral metacarpals, but at the proximal end of the first metacarpal and all the phalanges. Apparently there is a tendency in aquatic mamals for all of these bones to develop an epiphysis at *each* end. Kükenthal (1891) appeared to think that this is attributable to the retarded rate of ossification in the elements of the manus. Double epiphyses are well differentiated in distal digital bones of at least most cetaceans, and Kükenthal stated that this character is advanced in sirenians, and among pinnipeds in *Mirounga, Hydrurga* (=*Stenorhynchus*), *Otaria,* with indications in *Odobenus,* and in the pes only of *Cystophora.* But Flower (1876) said that they are double in the manus only of *Mirounga,* while Weber (1886) stated that in the pinniped manus the epiphyses are normal, being double only in the proximal elements of the pes. Kükenthal also said that there are indications of double epiphyses in *Ornithorhynchus, Hydromys, Hydrochoerus* and *Castor.*

All this is as may be. To establish the incontrovertible presence of double epiphyses is not always easy without extremely careful investigation, and without taking any especial pains in preparation I have been able to satisfy myself that this situation obtained only in the case of sundry Cetacea and a very few pinnipeds of the proper age to exhibit the character to good advantage. I might point out two likelihoods, however; that it is extremely easy to interpret cartilaginous irregularities at the end of the phalanges as early stages of epiphyseal ossification, and that if one searched for this condition among strictly terrestrial mammals with as much zeal as it has been sought in partially aquatic forms he might very well have equal success.

It is almost an invariable rule that when a mammal first takes to the water the use of its anterior limbs as natatory organs will be merely incidental. The probabilities are that these members will be used with the hind limbs in swimming "dog-fashion" but the important function of all four legs is for terrestrial progression. As the hind feet are ordinarily larger than the anterior, these will have a tendency to increase in proportional importance for propulsion. In most cases I believe that the aquatic development of the manus will be extremely slow, for it has not sufficient area to be of decided import either in steering or swimming. But bodily conformation may introduce disturbing elements. Without a tail of fair proportions or with a body of unusual width (hippopotamus) the dog-fashion method of swimming will long be followed, the important function of the fore limb then being largely as an antagonist, to neutralize the oblique, sidewise impulse furnished by the separate kick of each hind foot. In such case the manus might be expected to acquire aquatic modification almost if not quite as rapidly as the pes. But still, each sort of mammal is a law unto itself, and accordingly it is necessary that each be considered separately, with some unavoidable repetition of facts.

In the majority of insectivores and rodents the manus is totally unmodified. They appear to have no need for especial speed in the water, the food being mainly herbivorous or insectivorous, and the feet are presumably folded against the body so as to offer no resistance while swimming. They can have no use for propulsion and in the water can act only as grasping organs for securing prey. Manual activity will thus be in direct proportion to the amount of terrestrial locomotion indulged in, and as the latter decreases one would expect that a tendency for reduction of the limb would increase.

Concerning various sorts of slightly aquatic mammals the statement is frequently made that the fore feet are partially webbed, when an examination of the specimens fails to indicate the fact. Apparently some authors have been prone to interpret the membranes stretching between the *base* of the toes as aquatically-adapted webbing when as a matter of fact this is no better developed than in a great number of comparable forms with habits that are exclusively terrestrial. Undoubtedly in the desman, Australian water rats and some otters the webbing of the anterior appendage is slightly better defined than what may be considered as normal, but it is of very slight degree indeed and may be partly illusory because there is also a slight tendency for a lessened definition of the digits following decreased use of the limb. In *Chimarrogale,* however, there is more of a hairy fringe to the manus than terrestrial insectivores show, this matching a similar but far stronger development upon the hind feet.

Here a question intrudes itself that should be accorded brief mention. Irrespective of the fact that hypertrophy of the posterior limbs is often accompanied by atrophy of the anterior ones (compensational development), is there not a tendency, however slight, for the manus to adopt the same quality of equipment as that occurring upon the pes: the same quality of nails, webbing, or hairy fringes? Needless to say, *if* there be such a tendency it is often set at nought by diversity of function of the two members.

At any rate the few manual changes in those mammals mentioned above are too slight in degree either to be of much use to their possessors or to be of much significance. They may or may not be indicative of a line of future progress.

The development of the manus in the platypus is comparable in some respects to that of turtles, and in others to the sea-lion. Apparently there has been ample time since this queer beast took to the water for it to experiment fully and to adopt in marked degree those peculiarities best suited to its needs. Its body is definitely flattened and it is likely that this always has been a characteristic. Its tail probably assumed its present form not primarily as a rudder for swimming but to keep the animal on the bottom while feeding. Hence in order to overcome oblique movements which would be affected by the alternate kicking of one pair of limbs it probably, throughout a very long period, swam by diagonal kicks of all four feet. As a result it is likely that the webbing of these four members progressed for a while at a fairly uniform rate. But it is the invariable rule among mammals that when aquatic specialization has attained some considerable degree of perfection the tendency is for a single member or pair of members to take over the primary means of propulsion. If for no other reason than that the fore limbs are better situated for acting in co-operation with the tail for keeping the animal near the bottom it is likely these would receive the greater stimulation for development. This might be either increased still further, or else hindered, according to the quality of the action involved, by the presumable function of removing the bottom débris and stones in search for food and of digging burrows. At any rate they have gained the evolutional ascendancy over the hind limbs. But the future of this creature is hardly predictable. The manus may or may not eventually assume the form of a flipper, or for that matter the tail might ultimately take over the function of a primary propulsive organ.

It is seen that the manus of the platypus (fig. 42) is very highly and peculiarly modified, the characters, in fact, being entirely unique. These, however, are not satisfactorily brought out in Burrell's photographs. In a spirit specimen before me the membrane extends considerably beyond the nails, but projecting even farther are leathery thickenings of the membrane. There is but one of these to the first and fifth digits, extending in continuation of the digit. Each of the other three digits has a pair of such thickenings, however, which diverge very slightly as they extend distad. Apparently they all have a sort of hinge at the base, for the predigital part of the membrane is folded back against the palm well out of the way when the animal is not only scratching and digging, which is accomplished chiefly by means of the well developed claws, but also during terrestrial progression. The claws have undoubtedly been retained for these purposes. Wood Jones has given the length of the digits as in the order of 4, 3, 2, 5 and 1, but it is seen that their development is to all intents bilaterally symmetrical.

It is not at all easy to determine from the literature the precise actions of the platypus manus that are involved in swimming. From an examination of the skeleton it seems that the humerus must operate

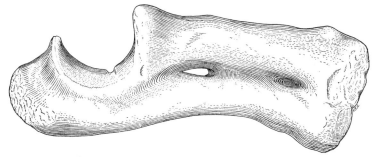

FIGURE 40. Bones of the right fore arm of the Steller sea cow (*Hydrodamalis*). from a specimen in the U. S. National Museum.

chiefly in the transverse vertical plane, and the fore arm mostly parallel to the body axis and also in the vertical plane. Hence the manus probably operates on the thrust-and-recovery method, rather than on some system which involves oblique action against the water.

In some important respects the aquatic stimuli encountered by the limbs of the capybara and the hippopotamus are similar. It is true that the former animal lacks the markedly broad beam of the latter, which makes it convenient for the hippopotamus to counteract the tendency for the alternate kicks of the hind legs to deflect its course first to one side and then to the other. But both lack a functional tail and in neither has the development of one pair of limbs gained decided ascendency over the other pair, which may be considered as somewhat unusual in mammals of their degree of aquatic predilection. Apparently, as stated, all four limbs are habitually used in diagonal alternation while swimming, but it is not meant to imply that an observer can never detect variations of action, just as a human swimmer may play about and keep afloat by several expedients.

Neither fore nor hind feet of the hippopotamus are appreciably modified in an aquatic direction, and it is readily seen that it would be extremely difficult for a foot of its type, upon which such a ponderous body is imposed, to change into a flattened propulsive organ, at least as long as terrestrial locomotion is a necessity. The fore feet of the capybara are furnished with interdigital membranes which extend practically to the tips of the toes, although these are indented, rather than straight, along their free margins. This is equally the case in the hind feet.

The fore feet of the otters are chiefly of interest in the present connection because I deem it likely that the ancestors of the Cetacea were beasts of somewhat similar conformation, swimming by the same methods. It is usually stated that the manus of the common otter is partly webbed, but this character is not more pronounced than in many terrestrial mustelids. There is no webbing at all in the African clawless otter (*Aonyx*, fig. 42), while Allen (1924) shows that it is particularly extensive and broad in the African *Lutra maculicollis* (fig. 42). It is to be expected that during swimming the fore feet of the latter animal are employed in a manner somewhat different from the case of the former. Other genera of otters are presumably variously intermediate between these two extremes.

The common otter is very nimble and so given to sportiveness while in the water that it is very difficult to determine any really important uses to which the manus may be put while in this element. The anterior limbs are often used for grasping, but hardly as a definite aid to propulsion, nor apparently for equilibration. Certainly they will eventually become more modified, but in the mean time they may be expected to exhibit some tendency toward reduction of size, as is indeed the case to a slight extent in the sea otter.

Not a great deal of significance can be said about the manus of the Sirenia. They do not use the fore limbs as a direct aid to propulsion and probably never have since the days when they swam dog-fashion, if the present evidence is trustworthy: nor apparently are the limbs used as equilibrators for elevation and depression of the body after just the same fashion that the flippers of the whale are now employed. Rather should their action be compared to that of the seal, but to a more marked extent, being thrust to the side for making sharp turns or waved about as assistants to the accomplishment of a variety of rather languid evolutions. In addition they are said to be habitually employed for bringing herbage toward the mouth, and it is frequently stated that the females use the flippers for clasping the young.

Externally the manus of the Sirenia is paddle-like though of a rather irregular and blunt form, rather than gracefully shaped like that of the sea-lion and most whales. I would not regard any of the known stimuli as particularly strong for the assumption of a fore limb of this character and it seems likely that an immense space of time has been necessary for its attainment. Small blunt nails are present apparently in all forms of the manati except the African *Trichechus inunguis,* and it is interesting to note that these are situated upon the flipper border where they would be most available for such acts as scratching, although they hardly project sufficiently to be very useful in this respect. Nails are absent in *Halicore* as was probably also the case in *Hydrodamalis.*

The synovial character of the wrist joint seems to be completely retained in the Sirenia. As these mammals are entirely helpless out of water there is not the need for abduction to 90 degrees of the wrist that the sea-lion has, but yet this joint is said to be definitely and quite surprisingly mobile, probably because of feeding needs. The pisiform bone is absent in this order. In the dugong the carpal elements are astonishingly reduced to three, one subtending the radius, a second the ulna, and a third elongated bone distad with which the first four metacarpals articulate. In the manati there are six carpal elements (Flower said seven) arranged in two transverse rows of three each. The significance of these carpal details is unknown.

The sirenian metacarpal and digital bones are somewhat flattened, especially in the manati, as is usual in aquatic mammals, the ungual phalanges are of irregular shape and particularly flat as in the sea-lion, and the pollex is definitely reduced. The fourth digit is the longest, this representing the tip of the flipper. Beddard (1900) has mentioned that in this order "hyperphalangy is also met with but to a very small extent." In most skeletons any predigital cartilaginous elements which might be present would be lost during cleaning, but in the mounted dugong in the National Museum the left manus (fig. 35) shows a small cartilaginous nodule, now shrunken, upon the tip of the third digit, this being the fourth phalangeal element. Furthermore this is of the exact character as the single premetatarsal element of the first, and the third of both the second and fifth digits, which is of significance in show-

[247]

ing how much similarity may exist between actual phalanges and a pre-digital item. This is of much interest and will be discussed later.

The manus of the seal and sea-lion could not be more utterly unlike were they totally unrelated, and they must be discussed from a different angle.

In the normal posture assumed by the seal the manus is the only part of the arm that projects beyond the body contour, the external axilla of the specimen dissected falling opposite the pisiform bone. The manus is not appreciably altered in relation to body length from the condition obtaining in such a terrestrial carnivore as the cat, but as so much of the arm is out of sight the hand appears unduly short. It is very broad, however, this being largely due to fatty tissue, skin, and hair. The digits are all completely connected by membranes, but these are narrow and do not permit spreading of the digits so as to present a broad surface to the water. In *Phoca* the first digit is the longest and heaviest, the others being successively slightly shorter, and all are furnished with heavy nails. And this is the case in most seals, although perhaps in none are the nails relatively heavier than in *P. hispida,* while a few have them slightly lighter, and Beddard (1900) has stated that in *Ommatophoca* they are quite rudimentary. In mounted *Mirounga* examined they are short and blunt as though worn down by much use. The conformation of the arm would permit of the nails being used for scratching over but a negligible area of hide, so they must have some other function, else they would almost certainly have become atrophied. This may consist either of scratching holes in the ice or prying about on the ocean floor.

As there has been no rotation of the long bones of the pinniped arm the radial border of the manus is presented directly forward, and it can accordingly be pressed flat against the body without either pronation or supination from the normal. While dissecting *Phoca* it was found that the static posture of this segment is markedly abducted with respect to the antibrachial axis, this being through an arc of more than 45 degrees. From this position adduction cannot be great, the chief inhibitor being the peculiar abductor digiti quinti longus, but the latter helps in further abduction, which is possible to an angle with the antibrachial axis of at least 90 degrees. Provision is apparently not made upon the articular surfaces of the phocid scapholunar and ulnare for extension of the manus to 90 degrees, but nevertheless such extension seems possible because of the looseness of the ligamentous connections. Of course these may not have been so loose in life, but it is likely that they were almost so. This same character allows of flexion of the manus after the integument and tissue has been removed to the excessive point where the metacarpals are actually parallel to the fore arm; so the flexion-extension arc is through 270 degrees. This would hardly be possible in life because of the confining influence of the surface tissue.

With all of the arm except the hand within the body contour and with the segments somewhat flexed at that, a relaxed position of the manus of the normal sort (i. e. on a line with the fore arm) would bring the palm almost against the body. In this position the body would accordingly prevent much flexion of this member (although if the arm be extended or rotated this would not be so and extreme flexion would then be possible), and in order that the manus might be thrust horizontally it would have to be extended to almost 90 degrees. Furthermore, such extension would seem usually to be initiated from a normally abducted position of the manus (in respect to the antibrachium). That this is not only so, but that such motions are strong and oft repeated, is indicated by the depth and somewhat lateral inclination of the extensor grooves upon the radial surface in this mammal, as already discussed.

The change in posture of the phocid manus has brought about some alteration in the carpus. Its ulnar side is weak and has a mitred appearance, because of the normal abduction of the rest of the manus. Carpale 2 is partly interposed between the scapholunar and carpale 1. Crowding has resulted in, or at least has been accompanied by, the assumption of a rather conical shape by the carpales, this being especially pronounced in the third, in which the apex is presented dorsad so that from this aspect it appears as a bony point. The ulnare (cuneiform) is reduced in size and the fifth metacarpal subtends that, rather than the lateral side of the unciform, which has the function purely of a fourth carpale. In consequence the fifth digit really appears to be more opposable than the thumb. The abducted posture of the manus places the pisiform at the side of the fifth metacarpal.

It is seen (fig. 35) that there is a gradual reduction in size of the metacarpals from the first to the third, while the fourth and fifth are abruptly smaller, and this, in connection with carpal details, places the metacarpal-phalangeal articulations on a line. The first phalanx of the pollex is more than twice as long as the first two phalanges of any of the other digits, and this causes the pollex to be the longest. It appears that the assumption of an abducted posture by the hand is partly effected not by pure adduction at the bases of *all* the digits, but only part of them. Thus there may be said to be pure abduction only of the fifth and

[249]

fourth digits, while in the others this is complicated by a flexional tendency progressively in a radial direction, so that the pollex is almost as much flexed as abducted. It is difficult to express this in words but the illustration (fig. 35) shows it quite clearly. In reality there has been partial rotation of the first digits and an increasingly lesser amount in the others, so that extension of the thumb really helps to adduct it (in relation to the antibrachial axis).

With the exception of the radial grooves or pulleys, as already noted, the extensor muscles of the phocid manus show nothing that may be deemed of much significance. The flexor muscles, however, exhibit several points of interest. The tendon of the palmaris longus broadens so as to cover the whole radial half, including the border, of the palm. The flexor carpi ulnaris tendon does the same to cover the entire palm. The peculiar abductor digiti quinti longus has already been mentioned. The origin of these three muscles from the broadened olecranol part of the ulna, in connection with the normally abducted posture of the manus, makes it likely that all three of them have more the function of effecting still more marked abduction of this member, than of flexors. Apparently the chief muscle for the accomplishment of flexion is the flexor communis, as indicated by the excessive breadth of its origin and of its common tendon, and it acts chiefly as a flexor of the manus as a whole, rather than of individual digits.

In recapitulation it may be said that the anatomical characters of the phocid manus, and what they indicate, is precisely in conformity with what one would expect did one bind down his own arm in phocid posture and attempt to move the manus similarly. Almost all mobility would be through the carpus, which would have to be adapted for an extreme amount of abduction and extension. Such mobility is useful in fiddling movements of the manus in making slight adjustments of posture while floating, just as in the case of a human swimmer, but in order to explain the degree of alteration which we find in this member of *Phoca* it is necessary to assume that it has some other important action, involving a considerable degree of strength, than any which is now surely known.

As in the seal the radial border of the manus of the sea-lion is presented directly forward so that it offers the least possible resistance to the water. The external part of the limb has the shape of a long, rather falcate paddle, thicker upon the anterior border and thinner upon the hinder edge. The external axillary angle is about at the middle of the fore arm, so that the distal half of this segment in reality constitutes a part of the flipper proper, for which it is fitted partly by the broadening of the radius and partly by subcutaneous tissue deposited upon the radial and ulnar borders.

The manus of the sea-lion differs in many respects from that of the seal. It is not carried in an abducted posture and there has been no crowding of the carpal elements. The scapholunar is large to correspond with the great width of the distal radius, and its proximal articular surface, as well as that of the ulnare and the metacarpals, extends farther onto the dorsal surface than in the seal. This is so as freely to allow extension of the manus to an angle of 90 degrees with the antibrachium, which is essential for terrestrial progression. The ulnare is not on a transverse line with the scapholunar but above it, to correspond with the more proximal position of the distal extremity of the ulna. The reason for this is obscure. The first carpale is very large to match the size of the first metacarpal, and the unciform upon which the fifth metacarpal abuts, is on a line with the three carpales. The latter have no tendency to adopt a conical form as in the seal and are not crowded. Also there is perhaps less tendency than usual for the crowding of the proximal ends of the metacarpals. These latter are on a line in the first four, and somewhat more proximad in the fifth. The first digit is much more robust than the others and is the longest. This is not accomplished almost entirely by the elongation of the first phalanx, as in the seal, but the metacarpal is also markedly lengthened. Ulnad of the pollex the other digits are progressively shorter in gradual sequence. The increase in the robustness of the pollex is to be expected, in order that the anterior border of the flipper should be strengthened, for it is this part of the manus that furnishes the chief impulse in swimming, while the posterior border plays a more passive rôle. All the digital elements are much more definitely flattened than in the seal, and this is most pronounced in the ungual phalanges, which are also of very irregular shape, this likely having followed the atrophy of the nails, for the same situation obtains in the Sirenia.

There is a considerable deposit of fibrous tissue upon the entire radial border of the flipper and even more at the ulnar border proximad of the tip of the fifth digit, this helping to give a broader surface. This, and the leathery covering of the manus greatly hinders adductive and abductive action. After this tissue has been removed it is found that in static posture the axis of the manus is abducted from that of the antibrachium by not more than 15 degrees. Thence practically no further adduction is possible, but abduction is permitted to about 90 degrees.

I would deem that in life the latter figure might be cut to one third by the confining effect of the integument. Supination and pronation, even in the partially dissected specimen, was through only some 45 degrees with respect to the humerus (this therefore including rotation of the antibrachium). Flexion of the manus is through an arc of 90 degrees, which is more than one would expect because one cannot see that such an action would be useful.

It has been seen that the effectiveness of the otariid paddle is increased by the extension of its border beyond the digital tips. Unlike the case of less modified aquatic mammals in which the dorsal and palmar integument of the digital webbing is in close contact with practically no subcutaneous tissue intervening, there is a liberal amount of

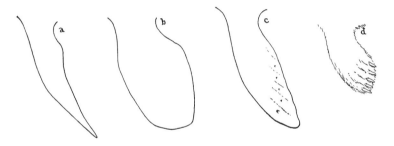

FIGURE 41. Outlines of cetacean and pinniped flippers: (*a*) *Globiocephal*ı (after Murie); (*b*) *Orcinus* (from a photograph by E. P. Walker); (*c*) sea lion (*Zalophus*); and (*d*) seal (*Phoca*).

such tissue in the manus of the sea-lion and whale, as well as the sirenians. This has the effect of padding out the interdigital depressions so that the flipper presents a plane surface both above and below.

In the Otariidae the effective length of each digit is increased by a cartilaginous rod, the extent of which can be determined in the entire animal by the distance from the respective nails to the flipper border. Presumably there has been a strong stimulus either for a flipper longer than the bony elements of the manus have been capable of supplying, or else for an elastic border.

It has been mentioned that the predigital part of the membrane of platypus is strengthened by thickenings of the integument, and it is not unlikely that these may eventually be replaced by cartilage. But these are double in the three middle digits of the platypus and single in the sea-lion, so although the two are analogous in some respects they need not be homologous. In other words the same need has been fulfilled by somewhat different structures. But the chances are that in both sorts the membranous border of the manus extended well beyond the digits before there was much stiffening needed, although in the platypus the stiffening elements have now outstripped the intervening membrane.

Presumably there must be a contrivance in the sea-lion for overcoming involuntarily the tendency for the water resistance while swimming to force the manus to extension with the fore arm through 90 degrees; in other words to prevent the manus from assuming its usual terrestrial posture. It is likely that this need is provided for by special tonus of the palmaris longus. The latter is exceedingly broad and powerful, and partly tendinous throughout a considerable portion of its length, so that it could well perform such a static function. Furthermore its palmar tendon splits into two parts, the more robust of which extends along the radial border of the hand and the other along the ulnar border. Flexion while forcing the palm against the water would accordingly have a cupping action upon the palm, the force being more pronounced along the anterior border. The fourth head of the flexor digitorum communis and the flexor carpi radialis, in addition to the extension of a part of the pectoral, are also well disposed to act upon the radial border of the flipper, so that the latter has an unusually powerful equipment for sweeping the flipper obliquely through the water after the manner that will prove most effective in propulsion.

The nails of the manus in the sea-lion should receive further brief mention. As the part of the flipper distad of the bony termination of the digits cannot be folded back against the palm the nails can no longer be used in scratching the body or in any other effective way. Accordingly there is no stimulus for their retention and they are now reduced to insignificant nodules which do not project beyond their integumentary pits.

It is frequently stated that the manus of the walrus is of a character intermediate between that of the seal and the sea-lion, but this is not quite correct. In size this is so, but in characteristics it is merely a less developed otariid flipper that is shorter and broader, but with digital cartilages and rudimentary nails also.

Just as there is considerable variation in the details of the remainder of the cetacean arm, so is there corresponding differences in the manus.

To attempt to interpret the reason for this lack of uniformity would be idle because we can have no conception of the precise manual needs of a particular sort of whale, nor what form would constitute an ideal equilibrator for its precise body form. I think one should work on the theory not that the cetacean pectoral limb has evolved from a single flipper form, but that diverse kinds of whales split off from one or more stocks long before the manus had assumed a paddle shape, so that subsequently that of each has developed independently, with many somewhat different lines of specialization according to individual needs.

The width of the cetacean manus depends upon the disposal of the phalanges. In a broad flipper the digits are spread somewhat fanwise and are relatively short. The breadth of flipper is marked, according to Flower (1876) in *Physeter, Hyperoodon, Monodon, Delphinapterus, Inia, Platanista, Orca,* and *Orcella.* In the latter genus (fig. 41) this broadness in relation to moderate length is especially marked. A very narrow flipper is usually long, just as in the case of the wing of a bird. The limb of *Globiocephala* is especially noteworthy for this feature, and accompanying, and contributing to, the condition is the fact that the border digits are much reduced in length, while the second and third are remarkably elongated. The second may have about eleven elements in the adult and as many as nineteen in the late fetal stages, and a comparable situation obtains in at least some of the other porpoises (as *Phocaena*). Leboucq, I believe, was the first to make this claim, later (1888) denied by Weber, but reaffirmed by Kükenthal. This is, of course, evidence that the flipper of such porpoises is undergoing reduction in length, but it must not be inferred that this is necessarily the case in *all* cetaceans. That great length of flipper is not invariably associated with extreme hyperphalangy is shown by the case of the humpback (*Megaptera*). This genus has the longest flipper of any living whale, the exposed part of the arm reaching a length exceeding twelve feet, although in the mounted specimen in the National Museum there are in the second digit but 6 phalangeal elements in addition to the metacarpal, while there are 5 such plalanges in the mounted *Sibbaldus.* These facts have aroused much speculation. It is commonly reported that the flippers of *Megaptera* are used for giving gargantuan and resounding love pats to the opposite sex, and it has been claimed that they are also employed for clasping during mating, as well as to help "herd" schools of fish into the cavity of the capacious mouth. This is all that we have to go on, unless we fall back on the theory that the great arm length has been purely in the nature of an overspecialization.

It should here be noted that whereas the flippers of all other whales are flat, of a shape determined by the disposition of the tips of the digits, and without marked serrations of the borders or uneven protuberances, that of the humpback has numerous sharply defined prominences of a warty appearance, as also occur upon other parts of its body, and one of these is always situated dorsad of the tip of each digit.

But few remarks can be offered regarding the great variation in the shape of the cetacean flipper. Just as might be expected the extremes in shape are not encountered in the most speedy sorts. In the latter this member is always moderately falcate and moderately pointed, with graceful proportions. We can but presume' that the excessively broad flipper of *Orcella* and the excessively narrow one of *Globiocephala* have developed in response to some particular need of the animal in question.

Following the complete abandonment of the land the cetecean carpus lost the synovial character of the joints. At present the rate of manual ossification is much retarded, probably following this loss of synovial articulation, and the cartilages representing the various carpal elements are closely packed together in a flattened mosaic, without movement save that afforded by the elastic character of the tissues. In some sorts of odontocetes complete ossification of the carpal bones appears to be attained with age.

In most sorts, however, it is even slower, and this is especially marked in mysticetes. In adults of the latter the carpal elements are mostly cartilage, while the bony centers are relatively insignificant and sunk within the hyaline substance, reaching the surface of the cartilage only with age.[3] This is in contrast to the situation in the majority of highly modified aquatic reptiles, for in these the carpal bones appear to be as closely packed as possible.

The homology of the cetacean carpal bones has often been discussed but without any very convincing success, and it is likely that the question can never be settled. Malm (1871) gave it much attention. Briefly, in the Odontoceti the pisiform, or what appears to be this bone, is occasionally represented. Besides this there are either five or six elements, or even additional ossicles which are believed to be supernumerary in character, for the number may not be uniform upon the two sides of the same animal. Usually because of their position a scaphoid, lunar

[3] It should be noted that the drawings of mysticete limbs (fig. 36) have been made from mounted specimens in which the cartilage has been replaced by some modelling substance. Hence the cartilaginous elements of the carpus are not individually defined.

and ulnare may be identified, but at times (see *Platanista,* fig. 36) the conditions are more uncertain. The second row of bones cannot be named with any feeling of confidence. Some of them are surely compound but their constituent elements can not be determined. In *Globiocephala* (fig. 36) the fifth metacarpal may be fused with the ulnare, thus in effect connecting this digit with the ulna, and in *Platanista* (fig. 36) the same end is attained by the direct conjunction of the metacarpal with the ulna. Variations of this character might be enumerated at great length but without much profit.

In the Mysticeti the carpus may have as many as nine elements (in *Eubalaena,* according to Holder, 1883) including a presumable pisiform extension, five in *Sibbaldus,* or even as few as three in *Balaena,* according to Flower (1873), but the latter is not clear as to whether this includes the pisiform, which is present.

I do not see that the undoubted trend of the cetacean carpus can be determined. Is it in the direction of a bony mosaic, like it has been in marine reptiles, or toward a completely cartilaginous carpus, as the condition in Mysticeti might indicate? The decreased rate of ossification even in odontocetes might bear out the latter possibility, but for all that we know the two groups may ultimately attain opposite goals in this respect.

In the Cetacea the metacarpals of those digits having considerable length are absolutely indistinguishable from the proximal phalanges. They all have precisely the same function, or rather practical lack of function, and hence the two elements may be considered to comprise a homogeneous complex.

There are invariably five digits in the Cetecea except in the case of the rorquals, which have four. As might be expected the development of the border digits is largely dependent upon the conformation of the flipper as a whole. In those sorts having a pollex this is always short, but the fifth digit may be either practically as long as any of the others (*Platanista*) or extremely rudimentary in a flipper of such extreme falcate shape as that of *Globiocephala*. In those sorts with narrow flippers the digits are situated close together, while if these members are broad the digits are disposed somewhat fanwise. Invariably, I believe, the second one of the digits present is either the longest, or practically as long as any other. (*Platanista*). Unlike the case of the Otariidae the digit upon the anterior border of the flipper is (as a rule) not markedly more robust than any other, although Flower and Lydekker (1891) have stated that in *Sotalia* the two outer digits are heavier than

[256]

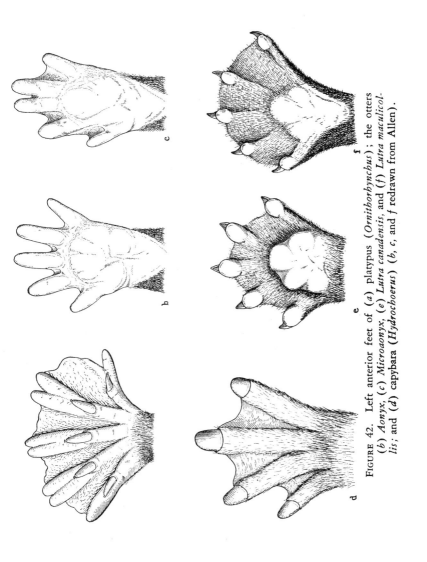

FIGURE 42. Left anterior feet of (*a*) platypus (*Ornithorhynchus*); the otters (*b*) *Aonyx*, (*c*) *Microaonyx*, (*e*) *Lutra canadensis*, and (*f*) *Lutra maculicollis*; and (*d*) capybara (*Hydrochoerus*) (*b*, *c*, and *f* redrawn from Allen).

the other three.   It can accordingly be deduced that if the flipper of an aquatic mammal be employed as an oar the digit upon its anterior border will likely be more robust than the others, while if it be used purely in equilibration there will probably be no digit markedly heavier than the others.

As in the sea-lion and sirenians the subcutaneous tissue between the cetacean digits has been built up so that the flipper surfaces are plane. This interdigital tissue is very tough, and at least in adults of some sorts (*Neomeris*) is strengthened by a network of fibers which have a tendency to converge distad from the bases of the digits, suggesting the manner in which the heads of the femur and humerus are ordinarily strengthened osteologically.

As there is such variation in the width of the cetacean manus it is only to be expected that there would occur a corresponding stimulus for digital differentiation, and there might conceivably be a tendency for digital reduction in one group of whales and the opposite tendency in another.   Consequently, in spite of the fact that there are but four digits in the rorquals, it was not inconsistent with the possibilities for Kükenthal to claim that at least the broad-handed beluga (*Delphinapterus*) exhibits a trend toward the polydactylous condition of the ichthyosaurs (in which there occur as many as nine digits) and fish, because he discovered in the white whale that the two phalangeal elements (but not the metacarpal) of the pollex were double, one pair being situated beside the other.   This is a possibility that can not be lightly discarded, but as the only beluga in the National Museum with articulated manus has a pollex entirely normal, and no one else besides Kükenthal has ever encountered such a doubling in this or any other cetacean, I must presume that his material was pathologic.

The same investigator (1893) argued with his customary vigor that it is not the pollex which is absent in the rorquals but the third digit. Attempting to carry this argument still further he made the remarkable statement that in *Balaena mysticetus* the metarcarpus of digit 1 is not really a part of the pollex but represents a prepollex, in spite of the fact that in *Eubalaena* the same detail is provided with two phalanges.   His reason for this course of action was that he had "occasionally" discovered in *Sibbaldus musculus* a structure which he interpreted as constituting atrophied phalangeal segments lying between digits three and four, and entirely disassociated from the carpus.   Such a situation would be quite astonishing.   In the first place it would be extremely unlikely indeed that the more distal part of a digit could persist as a remnant in the connective tissue of the flipper while its metacarpus and all carpal sign of its original situation had disappeared without leaving the faintest trace. In the second place supernumerary digits, of greater or lesser completeness, not infrequently occur upon the manus of divers sorts of mammals, even to the point where these are perfectly functional, and this does not necessitate the implication that derivation was from an ancestor which normally had six or seven digits.

In the Cetacea the rate of ossification of the digital centers is very slow and one seldom encounters a specimen of sufficient age to show the bony parts closely approximated each to its neighbor. More frequently in the articulated but otherwise cleaned manus there is a considerable hiatus between each element, this space being occupied by cartilage that has shrunk as it dried. Each phalanx is flattened, truncated at the ends, and somewhat constricted in the middle. The proximal bone (metacarpus) of each digit is the longest and most robust, and there is a progressive decrease in size distad. Almost invariably these ossicles are longer than broad, but Flower and Lydekker (1891) stated that except for the proximal phalanges of digits two and three, the reverse is the case in *Orcella*. The shorter digits may have only two or three phalanges, but the longer ones are furnished with a greater number, there occasionally being as many as eleven (at least) elements in the adult, and a considerably greater number may occur in the young (*Globiocephala, Phocaena*). This is the condition known as hyperphalangy, which occurs in no mammal other than the Cetacea, although it is characteristic of certain of the large, extinct, aquatic reptiles. Before considering the possible explanations for the manner in which this has been brought about it will be well to examine the conditions now obtaining in cetacean digits.

Flower (1876) has stated that in the Odontoceti the phalanges are often connected together by imperfect synovial joints. So far as I know no other investigator has independently made this claim and although it cannot be denied unqualifiedly, I would consider it as extremely unlikely that this is ever the case. There is not the faintest sign of a synovial cavity in the young odontocetes which I have had the opportunity of examining and it is not likely that such would develop only in adult life. In a cetacean fetus of say one third term size, it is seen that each digit is composed of normal hyaline cartilage with a small center of ossification at its middle. Investing each cartilage is what appears to be perichondrium, and this covers the ends of the elements as well. There is not the least sign of an irregularity, cavity, or articular cartilage, and the latter is undoubtedly never differentiated at all in these mammals, so that there is no process of changing over of a synovial joint into a syndesmotic or fibrous one, but rather a later, direct alteration of the interphalangeal perichondrium or periosteum into fibrous tissue as there develops need for interphalangeal strengthening. In effect, then, the phalangeal articulations of the Cetacea are in the nature of synarthroidal (rather than synchondrosial) joints or sutures, differing from most sutures as we know them only in that they occur between cartilages rather than bones, at least until old age.

The subject of hyperphalangy in the cetacea has been productive of a perfect furore of argument and the most violent partisanship for one theory or another. The theses that have been advanced in the endeavor to explain this condition may be said to number five, and to these may be added a sixth, which hardly seems promising, and a seventh consisting of the theory to which I incline.

(1) Leboucq (1889) considered that the Cetacea are descended directly from swamp-inhabiting creatures on the order of amphibians, rather than from a terrestrial mammalian stock; hence that hyperphalangy has been directly inherited from this ancestor. Presumably, then, he would consider that this condition can be explained in the same way that Howes has argued (in 3).

(2) Steinmann (1912) advanced the belief that the Cetacea are derived directly from the ichthyosaurs rather than from land mammals. It was therefore necessary that he consider that hyperphalangy has been inherited from these reptiles, which does not explain its origin.

(3) Howes and Davies (1888) considered that hyperphalangy has been attained by means of the independent ossification of intercalary syndesmoses.

(4) Kükenthal (1889) was persuaded that this condition was brought about through the initial separation and subsequent independent development of the phalangeal epiphyses, and a number of others have favored this view.

(5) Weber, Bauer and others have held that hyperphalangy was initiated through the secondary division of a predigital strand of cartilage, presumably on the order of those present in the Otariidae.

(6) There may be added to this group the rather unlikely theory that supernumerary digital elements have occurred through the fundamental division of the phalangeal anlage.

(7) A further premise, and one which to me seems to hold the most promise, is that hyperphalangy was initiated by the addition of

one simple cartilaginous element after another, according to need, upon the tip of the normal digit. I claim no great originality for this. It has been implied, at least partially, in the past, but I know of no one who has really championed it as the method by which hyperphalangy has been developed.

(*1* and *2*) I believe that the countless anatomical details of the Cetacea that are either fundamentally or precisely similar to those of terrestrial mammals constitute overwhelming evidence that the order is directly derived either from a strictly mammalian stock, or at least from the same group of mammal-like reptiles from which the land mammals took origin. I hasten to affirm that the latter statement is made purely in the line of argument, and that there is not as yet the slightest reason for believing it to be so. Furthermore, if one believed that the Cetacea were derived directly from amphibia, or "swamp-inhabiting creatures", then anatomical details render it obligatory also to embrace the hypothesis that all mammals had a similar direct ancestry, rather than from terrestrial, mammal-like reptiles. The possibility that the Cetacea sprang from the ichthyosaurs is utterly untenable, in which opinion every anatomist will agree. Before progressing to the next question, however, it is in order to examine more thoroughly the quality of the hyperphalangy that occurs in the manus of aquatic reptiles. This, of course, varied in direct degree to the completeness of aquatic adaptation, and relative to other considerations. Thus the flipper of the highly aquatic *Geosaurus* (fig. 34) had but one more phalanx than the normal mammal, but the single carpal element upon its anterior border, the first metacarpus and first phalanx of the pollex, are enormously more robust than the other comparable ossicles, which probably indicates that the flipper was long used as an efficient oar before it became so reduced in size. In *Elasmo-saurus* (a plesiosaur) hyperphalangy was about as far advanced as in those whales which show this to best advantage (although other elements of the arm differed considerably). Some ichthyosaurs had as many as twenty-six phalanges in the third digit, and in these reptiles the bony elements of the entire arm were essentially similar, save for the slightly larger size of the humerus, radius and ulna, and were all closely packed in a bony mosaic. Williston (1925) mentioned that this condition could not be due to ossification and separation of epiphyses, for these reptiles had none of the latter. Whatever was the mode of their development, it is not necessary to infer that this was the same as in the Cetacea. Both groups encountered stimuli for an increase of phalangeal elements and both have exhibited this convergence, but it is by no means certain that the structures are entirely homologous. It seems odd that the turtles have apparently been unable to attain the condition of hyperphalangy, although of course it is entirely possible that the stimuli which they have encountered were of a different sort. Be that as it may, in spite of the fact that the manus of the most specialized turtles have the phalanges very much elongated, no additional elements have been acquired. Presumably their flippers are entirely satisfactory as swimming organs except for the fact that the length of the phalanges renders the bones more liable to breakage, and signs of the previous fracture of the manus is frequently to be seen in skeletons. In such turtles as the leatherback the fact that the first digit is not much the heaviest, in spite of the flipper being used as an oar, is without great significance, for it is not elongated like the middle three.

(3) Perhaps the majority of present-day cetologists incline to the theory of hyperphalangy by intercalary syndesmoses. Howes and Davies advanced this thesis as applying to the Cetacea after having studied the small accessory phalangeal element in several amphibians. Their contention was that the interosseous spaces are filled with fibrous tissue which takes the place of synovial joints. This structure might appear to be a derivative of the phalangeal investing tunic, but rather is it a differentiation of the mass from which the phalanges themselves are derived, although intimately related to the sheath; so that the phalanges and syndesmoses are, together with the investing sheath, differentiations of a continuous common blastema. Thickenings of the fibrous interosseous tissue might then become phalangeal elements. Thus is indicated a possible intercalary origin, from articular syndesmoses, of the supernumerary phalanges of the Cetacea. This may have been associated with loss of the ungues in a manner similar to that in which elongation, by regular segmentation, of the cartilaginous rays in the paired fins of the Batoidei (skates and rays) would appear to have been connected with the disappearance of the horny fin rays.

I have no criticism to make of the above facts and theories in so far as they concern amphibia, but they should not have been applied to the Cetacea. In the first place, as these intercalary bones are derived from fibro-cartilage they are in effect nothing but sesamoids and although occasionally these may reach a considerable size, there is no reason for believing that in mammals they can ever develop to the *exact size and proportion* of a phalanx derived from hyaline cartilage, nor that such could ever truly become interpolated as a part of the digital series. Even if they could do so in amphibians it is no proof that this would be the case in mammals. Furthermore, in spite of the fact that hyperyhalangy is much more marked in some whales than others, and even in some digits of a particular individual than in other digits, all phalanges are perfectly and uniformly graduated in decreasing size distad from the carpus. If the supernumerary phalanges had intercalary origin of this sort it is, to my mind, entirely inconceivable that all of these added elements could now invariably occur in perfect and complete development, being indistinguishable from the true phalanges. Rather would there frequently be encountered a number of elements throughout one particular phalangeal series, at least in some whales, which were markedly smaller and different in shape from the true phalanges. Finally, what appears to render the theory incontrovertibly untenable insofar as concerns Cetacea is the virtual certainty that were it a fact there would be embryological evidence to support it. In the first place there would, during embryonic development of the arm, likely be a stage by no means short during which the intercalary elements would be relatively much smaller than the phalangeal ones. In the second—and this is the most important point of all—had any of the phalanges ever had origin by intercalary syndesmosis, *the cartilage in which they are preformed would have to be fibro-cartilage, while that of the true phalanges would be hyaline cartilage:* but all of them are uniformly hyaline.

(4) Kükenthal argued with his customary ingenuity that hyperphalangy in the Cetacea is attained by means of the separation and subsequent development of the phalangeal epiphyses. According to this theory, then, the retardation of the ossification of the phalangeal shafts allowed greater individuality of the epiphyses, and these, lacking the stimulus for conservativeness usually imposed by synovial joints, were permitted to increase in size, following a trend toward longer digits, until they had attained the exact conformation of the phalangeal shafts. The facts that this process could account for but eleven or twelve phalanges, and that the proximal ones still show double epiphyses did not deter the sponsor of this theory in the slightest; nay, it but stimulated him to claim that after this had once taken place, it occurred all over again a second time, or a third if necessary, giving birth to as many elements as needed. Unlike Zeus, however, these epiphyses could not spring forth full panoplied, as perfect phalanges, and if they had such origin there would surely be encountered various stages in their development, so that in a single digit of at least some whales there would not be a uniform diminution in all the phalanges, but some would be abruptly smaller than their neighbors, while some phalangeal shafts would have epiphyses

and some would not. This should invariably be more marked in the embryo, but no indications of it have ever been discovered.

(5) Weber, Bauer and others have subscribed to the belief that hyperphalangy has been accomplished by the segmentation of a predigital strand of cartilage. It has been mentioned that in *Ornithorhynchus* the part of the webbing of the manus that projects beyond the nails is slightly stiffened by linear thickenings of the skin, running in continuation of the digits. These may be partly or wholly subcutaneous in situation, however. Of course skin and cartilage is very different in structure, but it is well known that where there is real need for the stiffening of a membraneous extension of the skin in mammals this will be accomplished, where practicable, by rods of bone or cartilages (as in flying squirrels and bats). Hence it is reasonable to suppose that the predigital cartilages of the sea-lions had their inception in just this manner. And where there is well formed cartilage, ossification may, without difficulty, eventually take place. The developmental process involved in the initiation of predigital cartilages is, of course unknown, but it cannot reasonably be doubted that this was intimately correlated with the fact that there was no longer the stimulus furnished by repeated contact of the tips of the true digits with a hard surface. There were other factors concerned, however, as indicated by the accomplishment of digital extension in bats and marine turtles by means of elongation of the bones. There seems to be nothing particularly remarkable in this process. Rather would the incomprehensibility lie in the question of how the cartilaginous strands could become segmented so as to develop separate centers of ossification for the formation of phalangeal elements precisely like those of the true digits. One would reasonably expect that the ossicles situated distad of the latter would exhibit some abrupt transition in character, and even if this modification originally took place so far in the past that this transition was not apparent in the adult, it would almost certainly show to a striking degree in some embryonic stage.

(6) Unlike the case of the axial skeleton the bones of the appendages develop from somatic, mesenchymal condensations. Occasionally, from unknown causes, some of these anlage may experience reduplication. This is known because sometimes one hears of a man, cat or pig with six or more well-formed fingers, or one phalanx more or less than normal upon each digit. This, apparently, is different from the process known as twinning, in which the tip of the finger is split, and even the second phalanx may be double. A sixth finger should not be looked

upon as a splitting into two parts of any particular digit, but as an initial laying-down of six instead of the usual five digital elements.

Why could not the hyperphalangy of the Cetacea be attributable to a similar supernumerary duplication of the phalangeal anlage? This, according to my notion, constitutes a theory that is plausible in more respects than that of either intercalary syndesmosis or epiphyseal differentiation, but nevertheless there are particulars which render it unsound. Duplication of the individual phalangeal anlage would not be sufficient to account for it. There would have to be reduplication not once but several times, in the case of the digits of those Cetacea in which hyperphalangy is most pronounced, of the entire digital complex. And this would have to occur not side by side, as occasionally encountered among Mammalia, but in tandem, so that an entire new digit with three phalanges would be superimposed upon the end of the original digit. But this has never been reported in mammals. If, however, we presume for the sake of argument that such an occurrence be possible, the hypothesis can be entertained only until the embryological evidence can be examined. If it were fact then the embryological development of each phalanx would be obliged to progress at approximately the same rate as of all other phalanges. The entire digital complex would have been derived from the three original phalanges, and the ultimate cetacean phalanx would develop at the same rate as the third. But this is not the case. The differentiation of each phalanx is a trifle slower than that of the one next proximad, which effectively disproves the possibility that the supernumerary elements could have been derived by this sort of reduplication.

(7) Beddard (1910) indicated an opinion that conditions in the Sirenia are more pertinent to hyperphalangy than those in the Otariidae, but he therefrom progressed to a consideration of the Amphibia, which is an entirely different question.

Both Flower and Lydekker (1876, 1891) have stated that there are never more than the usual number of phalanges in the Sirenia, but sometimes in an articulated manus (at least of the dugong) that has been properly cleaned there is to be seen a small, cartilaginous button upon the tip of the ungual phalanx (fig. 35), as mentioned by Beddard. This I consider to constitute the initiation of hyperphalangy.

The fact that the digits of the terrestrial Mammalia exhibit a quite remarkable conservatism in that they never develop phalanges in excess of the normal complement is probably partly attributable to the conservative influence supplied by the almost invariable frequency with which the tips are applied to some hard surface—usually the ground. If such habits be entirely relinquished it is only natural that some development should follow which one would never expect to encounter in the normal mammal. This should be further stimulated, or complicated, by the tendency for the disappearance of the nails, a decreased rate of ossification which seems to be characteristic of the mammalian flipper, trend toward the alteration in the character of the manual joints, and an unknown number of other factors. Certainly the reduction of the nails has been followed by an irregularity in the form of the ungual phalanges not only in the Sirenia, but the Otariidae as well. It does not seem at all remarkable to me that this should be followed in the former group by the addition of a cartilaginous, predigital nodule. The condition that has brought it into existence will doubtless stimulate it to further differentiation, accompanying what may be termed the simplification of the true phalanges: so there is doubtless a tendency for a convergence in the characteristics of the ungual phalanx and the predigital element. What can happen once can happen again, and a second predigital nodule could be added in course of time, as the first increased in size, and this would continue to occur as long as there was an activating stimulus. As a matter of course there would eventually appear a center of ossification in each nodule as it attained sufficient size. Presumably the start of this process occurred at a sufficiently remote period so that the transition in form between the true ungual phalanx and the first accessory element is now almost insensibly gradual. The growth of the distal digital elements is therefore very similar to the successive distal caudal elements, as they become defined after birth in a rodent or other mammal with unusually long tail.

But there can be nothing to this theory unless it be upheld by embryological evidence. This should consist of a condition in which the digital bones are not all laid down at once, but after the embryological differentiation of the first four phalanges (including metacarpus), there should be a gradual addition, one after the other, of digital cartilaginous segments, in each of which ossification is slightly less pronounced than in the segment next proximad. And as this process progresses the flipper length should gradually grow distad from the tip of the original limb bud. This, apparently, is exactly what happens, and this theory to account for hyperphalangy is the only one yet advanced to which the known embryological evidence supplies any confirmation.

Presumably the difference between this sort of addition of cartilaginous nodules, one by one, to the digits, and the predigital cartilage of otariids, is rather basic. In the former case the cartilages are added as fast as there is room for them between the original digit tip and the slowly extending flipper border. In the sea-lion it appears probable that the flipper border extended considerably beyond the phalanges before there was any indication of cartilaginous stiffening. This suggests that comparatively recently, geologically speaking, the latter animals were in the habit of folding back the predigital part of the flipper in precisely the same manner as they now fold back the predigital part of the pes.

Leboucq (1889) claimed to have found a thickening of the cetacean epidermis above the digits well back from the flipper border. This he interpreted as the remnant of a nail, and Kükenthal held the same view. I believe this is no longer considered as probable, if for no other reason than that after the disappearance of a true nail any thickening of the epidermis in the same situation would speedily be obliterated. Assuming that the vestige of nails still persists, however, there has been considerable argument over the question of where such rudiments would be situated—whether upon the flipper border of farther back. The answer to this hypothetical question depends entirely upon what one believes to have been the origin of hyperphalangy. If one subscribe to the belief that this had intercalary origin of some sort then he must also believe that the true ungual phalanx, and therefore any rudiment of nail, is situated upon the flipper border. If he entertain the conviction that accessory phalanges were added distad of the original digit, then must he argue that any possible nail rudiment must be situated above the fourth phalangeal element distad of the carpus.

Chapter Eleven

The Pelvic Limb

In some respects the tendencies followed by the hind limbs of aquatic mammals are similar to those shown by the anterior appendages, and in others they are very different. In at least one detail the stimuli for the aquatic specialization of the posterior limbs are less complicated than of the anterior pair for the reason that the latter are often, or at least for a long time, of great service in the securing and consumption of food, while the hind limbs never have this function. Thus there are three primary stimuli acting upon these members of an aquatic mammal; the influence introduced by the mode of terrestrial locomotion, aquatic locomotion, and equilibration, but unless specialization be very far advanced the two latter are often very intimately combined. Terrestrial locomotion introduces a variety of factors, for when an animal first takes to the water its hind limbs have been fashioned not only for a particular mode of walking, but as a support, and differ according as the weight is heavy or light, the foot plantigrade or digitigrade, et cetera. Hence when a mammal first swims it already has an equipment that may predispose it to certain definite motions of the limbs to accomplish swimming, and prevent or handicap it in indulging in other kinds of actions. Naturally the difficulties which a shrew must overcome in evolving a paddle from its feet are very different indeed from those which an elephant would be obliged to surmount. For these reasons it is entirely impracticable to unravel the separate factors and be sure in all cases that we know just what we are talking about. But average trends may be evaluated and probabilities advanced.

It seems certain that in practically all of the terrestrial ancestors of living aquatic mammals strictly quadrupedal locomotion was employed, and that none of them was a leaper, or a flyer, or any other sort of animal in which one pair of limbs was used chiefly in progression. Hence, during the first inefficient attempts at swimming all four limbs were kicked with as nearly equal force as the muscular equipment of the animal would allow. In most mammals the hind legs are not only appreciably more powerful than the fore limbs, but the hind feet have a larger area. Hence from the very start these members had inherent ascendancy for potential development, which, coupled with the fact that the most efficient method for propelling a body through the water is from the rear, introduced the extreme likelihood that they would at some stage of aquatic specialization be of more importance for swimming than the fore limbs, *unless* the conditions were complicated by other factors.

In general the function of the hind limbs of a mammal that is exclusively aquatic may be of three sorts: that of propulsion, of steering, or of no use at all. Naturally, partial terrestrial dependence will introduce complex difficulties that may vary in their influence from 100 per cent to zero, according to the individual case. In a mammal of the shape of a hippopotamus and without a respectable tail, it would be expected that always the development of all four limbs will be at a fairly even rate, neither pair at the expense of the other, as in the case of some of the aquatic reptiles of somewhat similar body conformation. This is so for the reason that the mud-turtle type of swimming is employed, as already discussed. To some extent this is also so in the case of the capybara, and is not so essentially dependent upon the fact that in both animals there is great dependence upon the land.

Lack of ascendency, or at least delayed ascendency, in the development of the hind limbs may also be attributable to the fact that the tail was already of such a form that it was enabled to take over the function of chief swimming apparatus before the hind feet could become highly modified for this purpose, as seems to have been the case in the river otters and the insectivore otter (*Potomogale*).

If the body be not unduly broad (hippopotamus) or if the tail be not already large and powerful (river otter), the hind feet may be expected usually to gain ascendency over the fore, and, for a time at least, to be the principal means of swimming. This statement needs qualification, however. The anomolous exception is the sea-lion, and we are utterly incapable of stating whether the development of this order is because of some individual idiosyncracy in its make-up, or whether such a tailless mammal as the capybara might be expected to follow the same course, if it should ever become pelagic, rather than that of the seal. I judge, however, that the sea-lion is an exception, chiefly for the reason that the most efficient means of propulsion is from the rear.

In most sorts of mammals that become aquatic the tail is of goodly length but of small diameter at base, and in such there will always be ascendental development of the hind feet. This takes the form of enlargement of these members, and either a growth of stiff, bristly hairs

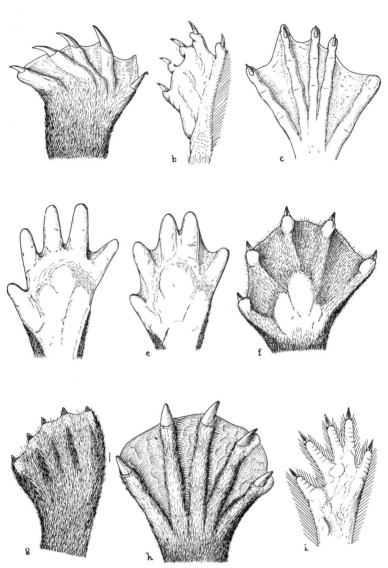

FIGURE 43. Left hind feet of (*a*) platypus (*Ornithorhynchus*); (*b*) the insectivore *Galemys*; (*c*) the marsupial *Chironectes*; the otters (*d*) *Aonyx*, (*e*) *Microaonyx*, and (*f*) *Lutra maculicollis*; (*g*) sea otter (*Enhydra*); (*h*) beaver (*Castor*); and (*i*) muskrat (*Ondatra*): (*d, e,* and *f* redrawn from Allen).

along the borders or webbing between the toes, or a combination of the two, and this specialization increases, theoretically, until the time when the tail shall have become sufficiently modified so that it is capable of furnishing greater propulsive power than the feet are capable of producing, when the latter will rapidly fall into disuse and tend to disappear.

If a mammal use its hind feet for swimming by any such method as that employed by the seal, with lateral motions in a sole-to-sole position, the action will be entirely symmetrical and the tendency will be for the feet to become symmetrical, with fifth and first toes of equal length and longer than the third, giving the foot a shape comparable to the tail of a fish. Any oar-like use of the hind feet will tend to produce, or rather to accentuate, an asymmetrical foot, with either the first or the fifth toe the longest, according to the precise method in which the foot is used. Usually this manner of swimming consists in alternate strokes of the feet, these reaching either down or out so as to act upon relatively undisturbed water. There is no mechanical reason why an aquatic mammal should not swim by placing the broad part of the feet horizontally upon either side of the tail and oscillating the entire hinder end in the vertical plane, but no mammal is surely known to do this, although it is not unlikely that the sea otter may employ this motion. It is also not improbable that asymmetrical development of the feet, comparable to that in the sea otter, might follow the phocid method of swimming provided there were also a tail of moderate length; but then the feet would not be the sole means of propulsion, for the tail would help.

The potentialities of the mammalian foot for furnishing alternating propulsive strokes of the character that is employed by most water birds seems to be limited, and the two can hardly be compared. If a jerboa (*Dipus*), with its elongated, fused metatarsals now took to the water it might eventually evolve a propelling foot very similar to that of a duck or grebe, but so far as the evidence goes no aquatic mammal ever had a terrestrial ancestor with pedal equipment of this type. Nor is it probable that any highly aquatic mammal has evolved from the cursorial ungulate type with fused metatarsals, and there is probably no stimulus connected with a life in the water that would produce such fusion.

In an aquatic bird such as a duck the propelling mechanism gains proper leverage by being upon the end of a tarsus of considerable length. The latter is relatively narrow and offers but slight resistance during recovery, while at the same time the foot collapses almost like a closed umbrella. Undoubtedly the important features, except for the webbing, of this equipment were evolved before the avian ancestry became aquatic. But in mammals the only terrestrial types roughly approximating this condition are saltatorial sorts of a desert habitat which would never be expected to become aquatic. It seems apparent that every mammal which has ever become highly aquatic originally had a shank but slightly narrower than the foot, and there seems to be no stimulus furnished by an aquatic existence for narrowing the shank. If the foot and shank increase in breadth together, as is frequently the case, so that a greater surface may be presented to the water, the shank would then be productive of too much retardation through recovery during swimming and the result would be impossibly inefficient. If the shank should become shortened or withdrawn into the body contour, as is now the case in the Pinnipedia, not only would even a greater amount of resistance during recovery take place, but the propelling feet would be so close to the body that they would lack requisite leverage and the animal would fiddle along at a slow pace. Hence, if an aquatic mammal is going to swim at speed by means of the hind feet it will have to evolve some other method of using them than by the alternate thrust-and-recovery of the duck. And this has been accomplished by the seals in the way in which they place the feet sole to sole and oscillate them from side to side. It may also be so after a different fashion in the sea otter. This method of swimming is the most effective of any that uses the hind limbs as the sole method of propulsion, but it could not be highly developed in a mammal with a long tail for the reason that the latter would quickly outstrip the feet as a propeller. Just how and why the feet of the seals adopted this style of function while those of the sea-lion did not is a question about which one can only speculate.

If a mammal that is highly aquatic swim chiefly by means of the tail the hind feet will not play any important part as equilibrators for the reasons that they they are too near the propulsive organ to be efficient, and as but one pair of equilibrators in the same plane seems necessary in mammals, the fore feet, being better situated for the purpose, will assuredly take over this function. The only mammal swimming chiefly by the fore feet and with a tail suitable for equilibration is the platypus. It is probable that its hind feet were at one time a greater aid to active propulsion than they are now. Whether these members are now really necessary for assisting the tail to function in equilibration is unknown. But it is requisite that a flat-bodied animal such as the platypus, or marine turtle, swimming by partly or chiefly vertical movements of the anterior limbs, should have an equilibrator or rudder for vertical steering at the hinder end. In the mammal the tail is of such form that it should be able to accomplish this unaided, while in the turtle, without an effective tail, the hind feet are specialized for this purpose.

Attention may here be called to conditions in some of the extinct aquatic reptiles. In no highly modified reptile of this sort do the hind limbs seem to have disappeared completely, although in the most specialized sorts, as *Ichthyosaurus,* they are considerably smaller than the anterior ones. In those reptiles in which the transition of the tail into a swimming organ was only partial, however, the hind limbs were practically as large as the forward pair. In some of the more primitive ichthyosaurs, for instance, such as *Cymbospondylus,* the body was more anguilliform, there was certainly no broadly spreading caudal fins, and all four paddles were about equally well developed. These details are suggestive that not only were the inherent variational capabilities of the reptiles different from mammals, as already claimed, but that most of these large aquatic reptiles were descended from broad-beamed ancestors, which obliged swimming after the mud-turtle method, somewhat like the case of the hippopotamus, for long ages. But the ichthyosaur of the more familiar type hardly has need for two sets of equilibrators in the same plane and the posterior feet were evidently in course of elimination when these creatures became extinct.

Perhaps there are some who will question my conviction that in air breathing mammals, and presumably in reptiles, but one pair of equilibrators is necessary in the same plane, for they may recall that in many fish there are all sorts of accessory fins in a variety of situations. But this seems to be an entirely different matter. The equilibration of small fish is a much more delicate matter, and fish have precision apparatus for remaining in one exact position in spite of currents and other disturbing influences. In larger forms of different structure fins must be much more substantial (i.e. thicker), and furthermore, an air breathing vertebrate cannot spend an indefinite amount of time drowsing in one spot or threading its way delicately in and out of submarine growth deep beneath the surface. Even in the sharks, as in some other large fish, there is the same tendency as exhibited in cetaceans and ichthyosaurs for the elimination of accessory fins and the development of a single pair of equilibrating paddles, in addition to a dorsal fin. The tendency in large aquatic forms thus appears to be for the acquisition of three equilibrating devices spaced at more or less regular intervals.

[273]

If this be so, as seems probable, it is only to be expected that when the tails of more primitive cetecea and sirenians had become sufficiently specialized so that they were capable of sustaining higher speed than the hind limbs alone were capable of attaining, the latter would be folded back against the body and, playing a decreasingly important part in the economy of the completely marine mammal, would finally suffer atrophy and disappear.

For the purpose of the present study the posterior limbs of aquatic mammals will be discussed under several headings as follows:

1. Hind feet the chief or exclusive swimming organs
2. Hind feet used chiefly as equilibrators
3. Hind feet absent or definitely subordinate in swimming

In this simple classification there are naturally several forms placed under one heading that might with almost equal propriety be put under another.

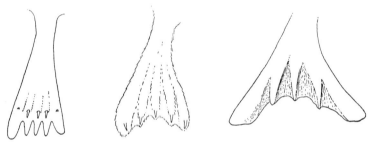

FIGURE 44. Dorsal outlines of the feet of a sea-lion (*Zalophus*), seal (*Phoca*), and elephant seal (*Mirounga*).

1. Under this first heading are included (*a*) those forms in which the fore feet are used but very little or not at all in swimming but which have a tail of sufficient length to be of marked potential value in natation, although the latter is not as yet sufficiently specialized to be the exclusive agent for propulsion; (*b*) all aquatic mammals which employ all four feet in swimming and whose tails are of insignificant size; and (*c*) such forms as swim exclusively by means of the hind feet and in which the tail is too small to be of use in this function.

(*a*) Under this heading should be placed the single aquatic marsupial *Chironectes,* all Insectivora except the insectivore otters *Limnogale* and *Potomogale,* and all aquatic rodents except the capybara.

[274]

Enlargement of the hind feet frequently results from a life in the water but this character may develop slowly and one must be careful always to compare forms in which approximately the same method of swimming is employed. It would be entirely without significance to compare in this respect the feet of hippopotamus, seal, otter and beaver, because the stimulus for foot development is different in each. In comparisons it must also be remembered that various terrestrial activities tend to increase the size of the hind feet. Thus the percentage of the hind foot length to head and body length may approach 30 per cent in arboreal squirrels. Probably for the reason that it is slightly given to arboreal activity this percentage in the wood rat *(Neotoma)* may be greater than 20. But in several terrestrial shrews of different genera the above percentage varies around 16.5, in various terrestrial rodents, such as the common brown rat (*Rattus norvegicus*) it is about 17.5, although in the same species it may vary from 16 to 19. In various other non-aquatic shrews, however, such as *Sorex pacificus,* the foot may be from 19 to 23 per cent of the head and body, but this is probably because of some definite stimulus of which we do not know. On the whole it seems safe to say that this percentage in most terrestrial insectivores of the shrew type, and of rodents, will usually be found to be less than 20. Even in a rodent with definitely aquatic propensities it may fall below this figure (as little as 17.5 in *Arvicola*). Selecting a series of aquatic forms that chance to be available it is found that in the slightly aquatic genus *Dasymys,* without other modification of the feet, this percentage is 19 to 21, in *Neomys foidens* about 21.5, in *Atophyrax (Sorex bendirii)* 22.5 *Ichthyomys* (one specimen) 23, *Hydromys* 22.5 to 25, *Galemys* 26 to 30, *Ondatra* 38.5, and *Castor* 38 to 40. These figures are only approximately correct as they may be based on but a single specimen and the collector's measurements may not be accurate. They do show, however, that in such small mammals as swim chiefly by means of the hind feet there is a tendency for the elongation of these members, in degree according to aquatic modifications in other respects. Such elongation may be barely appreciable until after the toes have become webbed or fringed, and may finally result in a foot that is twice as long as in a more generalized, terrestrial genus. There is frequently an appreciable lengthening of the toes in relation to the foot proper, and a more marked spreading of these members so as to present a greater surface (by means of webbing or fringing) to the water.

Another modification which is often encountered in aquatic rodents and insectivores is, according to the spirit specimens that have chanced to be available for examination, the alteration of the foot posture in the direction of pronation, or tendency for the elevation of the outer border of the pes. The aquatic shrews *Neosorex* and *Atophyrax*, the genus *Galemys,* and the muskrat (*Ondatra*) all show this pronation to an extent of about 30 to 45 degrees. This is accomplished not through the ankle joint, or even, apparently, in the tarsus, but through the metatarsus and digits, thus affecting mainly the part of the foot which furnishes the chief propulsive power, and is very noticeable in the alcoholically hardened specimens. Just what degree of further pronation, or of supination, from this posture is possible in life I do not know. Nor is it possible to determine the precise reason for this modification. All that can be said at present is that it is an adaptation by means of which these animals can employ the feet in a more effective manner for the purpose of swimming, for the action of the feet against the water depends entirely upon the position in which the legs are held. If the axis of the femur be maintained in the vertical plane then the feet would also be kicked in this plane, and the pronated posture of these members would tend to force the hinder part of the body mediad at each stroke, thus operating to neutralize the propensity which the anterior end would otherwise have of swinging toward the opposite side after each vigorous kick. Or if the femur and pes be held largely in the horizontal plane during swimming, then the pronated foot posture would tend to force the rear of the animal in a downward direction, which would have to be overcome by some antagonistic body posture or action.

The pronated foot posture of these mammals, and of *Ornithorhynchus* as well, results in a slight rotation of some of the toes so that when the digits are slightly flexed the nails of the more lateral toes have a tendency to point in a lateral direction.

In feet of small aquatic mammals a comparatively early modification is the acquisition of webbing between the toes or a fringe of stiff, bristly hairs upon the sides of the foot and toes. The latter is more often encountered, but such fringes would be of little or no help in the case of a body of large size and in such instances they are never encountered. It seems that it is easier for a small foot to develop fringes than webbing but it is utterly impossible to state the reason for this difference. Apparently it is just a case of two mammals being able to respond differently to the same stimulus, as seems to be the case so frequently in other details.

There is often a great difference in the precise character of the bristle fringes. They are present in varying definition and form in *Desmana, Galemys, Neosorex, Atophyrax, Neomys, Chimarrogale, Crossogale, Nec-*

[276]

togale, Ondatra, Ichthyomys, Rheomys and *Anotomys*. They may occur
as a single fringe upon the outer border of the foot (*Desmana*, and
Galemys fig. 43, b), upon both borders, upon the toes only, or the
entire foot, including toes, may be heavily fringed (*Ondatra*, fig. 43, i).
Webbing may be either absent or present. Needless to say this fringing
either supplements webbing or else is a make-shift to take its place.
I say make-shift for I can hardly believe that it can be as efficient as
webbing, although it undoubtedly helps in propelling a small body
through the water.

In many sorts of terrestrial mammals the bases of the toes are joined
by rudimentary webs, and it is only natural that as these are of such
importance in the economy of an aquatic mammal they should increase
in area. Evidently they are easily developed by all classes of vertebrates,
but why they should be lacking in such an essentially aquatic animal as
the muskrat remains a puzzle.

The form and relative proportions of the toes of the above sorts of
aquatic mammals differ. In some the plan of these members is essentially normal. In some there is lengthening of the hallux so that the
foot is practically symmetrical (*Chironectes* and *Castor*), and this is
undoubtedly the most suitable shape for such mammals as swim by motions that involve actions resulting in equal water force being applied
to both borders of the feet. But while it is true that most rodents and
insectivores are not as yet sufficiently aquatic for there to have been very
profound changes in the pedal equipment, it must not be assumed
that an asymmetrical foot falls short of the animal's requirements, for it
may employ some method of swimming that involves asymmetrical foot
action.

Nails of the normal sort are present in the aquatic mammals mentioned and where differences occur these are undoubtedly attributable to
some habit that has nothing to do with the aquatic life. In *Ornithorhynchus* the nail of the first digit is straighter and more spike-like than
the others. In *Chironectes* the hallux nail is not claw-like as are those
of the other digits but is of the same general shape as the nails in man
and is little more than a callosity (fig. 43, c).

The pes of the beaver (*Castor*) appears to be a more efficient paddle
than that of any existing rodent. It is large, broad and fully webbed,
and is operated by powerful muscles and heavy limb bones. Parsons
(1894) found that its calcaneal tendon is very readily separated into its
component parts. In this animal the nails of the first and second pedal
digits are slightly more slender than those of the others (fig. 43, h).

[277]

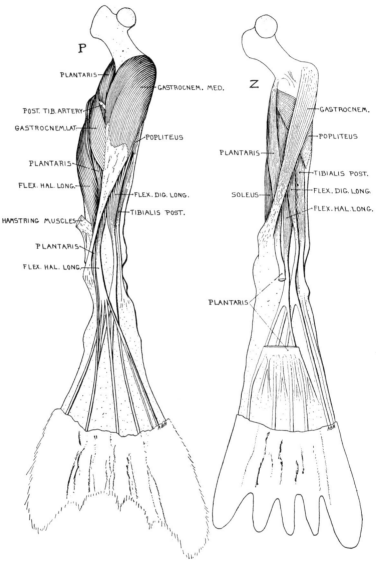

FIGURE 45. Posterior view of the leg musculature of the seal (*Phoca*, left), and sea-lion (*Zalophus*, right).

That of the former is subtended by a soft pad, and of the latter by a similar pad in addition to a serrated, horny growth which Bailey (1923) affirmed functions like a fine-toothed comb in dressing the fur.

An additional modification in these as well as all other aquatic mammals is the tendency for the elimination of the tubercles or prominences upon the soles of the feet.

Such slightly aquatic mammals as the marsh and swamp rabbits and water rat (*Arvicola*), as well as the genus *Neofiber*, need no further mention in this chapter, for there has been no change in the feet. The feet of the coypu (*Myocastor*) are moderately webbed. Those of the Australian water rats, *Crossomys, Hydromys* and *Parahydromys* vary from somewhat webbed in the first to unwebbed in the second.

The aquatic stimuli brought to bear upon the hind feet of all the mammals discussed in the present connection (not *Ornithorhynchus*) seem largely the same or essentially similar. All have tails that are at least moderately long and I believe that without exception all swim by alternate kicks of the hind feet while the fore feet are rarely utilized to any important extent during natation. At one time (Howell, 1924) I reported catching a glimpse of a water shrew (*Neosorex*) swimming by kicking the hind feet in unison after the manner of a frog, but I am now of the opinion that I was mistaken, either in identifying the creature so hastily seen or else in my observation, for no one else has ever reported similar action in a small mammal.

As already partially discussed in a previous chapter, small mammals of this sort quickly abandon any swimming motions of the fore limbs, chiefly, it seems, for the reason that the area of their hands is so much less than that of the feet that not only could they furnish very little propulsive power, but for the same reason they could not act efficiently in neutralizing the asymmetrical action imparted by the hind legs acting alone. For this latter purpose there is an efficient equilibrator already at hand in the tail, which furnishes as much stabilization in the water as does the tail of a kite in the air. Presumably the precise action upon the water of the hind feet kicking in alternation varies in different mammals, but it is a point extremely difficult of determination. At any rate there is one uniform result in that as the feet are kicked the effort swings the hinder end of the body from side to side, thus imparting a sinuous motion to the tail. This movement should be involuntary at first, as far as concerns the tail proper, but as the muscles concerned develop with use it is capable of being employed as a definite help to locomotion. Evidence shows that as a result the tail in some manner is modified so that its lateral aspects become broader than the vertical ones, either by the acquisition of a hairy fringe above or below or a flattening of the member in the transverse plane; and of course with such a specialization the efficiency of the tail as a swimming organ is thereby increased. This is the rule, but there are some exceptions. Presumably *Chironectes, Myocastor* and the Australian water rats are sufficiently aquatic to have developed a flattened tail, but the latter is perfectly terete. Whether they actually swim by some slight variation in the movements of the hind feet that does not introduce a stimulus for tail flattening, or whether the caudal appendage merely has unusually inherent resistance to such change but will later accomplish it, is unknown.

The speed which the hind feet operating upon the kick-and-recovery principle are capable of imparting to the body is very definitely limited, as previously discussed. In mammals with a tail of respectable length there will almost certainly come a time (save possibly in the case of the beaver) when the tail and its musculature will have become sufficiently specialized so that it can propel the body at a faster pace than can the hind feet alone. The latter will then quickly fall into disuse unless they be employed for other important purposes. It is not unlikely that the tails of *Desmana* and *Ondatra* are now almost sufficiently specialized to accomplish more speedy swimming than the feet, but for a very long time at least the latter will continue a necessity to both animals for terrestrial activities.

Most of these small mammals discussed are not sufficiently modified for there to have occurred marked change in any part of the leg above the ankle. Dobson (1882), however, has remarked that in *Desmana* the femur is but little more than half the length of the tibia, and the superficial biceps and semitendinosus have broadened considerably (fig. 48), both of these alterations thus being in the direction followed by the seal. The gluteal muscles are also unusually robust, as in so many other aquatic mammals. In all aquatic mammals having an enlarged hind foot one will almost certainly find that the shank and thigh muscles are correspondingly strong.

(*b*) Next for discussion is the case of such aquatic mammals as employ all four feet in swimming but have a short tail, involving the polar bear, capybara, aquatic rabbits, hippopotamus and tapir. All have great dependence upon the land and in all the tail is insufficient in size to act as a rudder or stabilizer, and in all but the rabbit the hind feet are not much larger than the fore feet. The latter may thus function effectively to overcome the asymmetrical, wabbling motion that the kicking of the hind feet alone would otherwise impart to the virtually tail-less body, and the animal accordingly trots through the water. The aquatic rabbits may possibly constitute an exception to this rule. If already the hind feet are not held sufficiently close together during pro-pulsion so that no stabilizing action of the fore feet is necessary to over-come any lateral deviation that would otherwise tend to occur, this might possibly be an eventual adaptation.

The tendency in these mammals, except the rabbit with its negligible aquatic specialization, is for all four feet to develop equally, limited naturally by inherent functional difference, as, for instance, in the moder-ate webbing of the capybara. But the dependence upon the land is still too great for there to have been much change.

This four-square method of propulsion is far from efficient and really high speed can never be attained without some adaptation pro-viding for an oblique action against the water by all four feet, such, conceivably, as that now employed by the walrus. For the same reasons as in the walrus it is now utterly impossible to predict whether the an-terior pair will gain the functional, and hence evolutional, velocity, as in the sea-lion, or the posterior pair, as in the seal, if indeed they do not have constitutional limitations that would prevent them from fol-lowing either course to a satisfactory conclusion.

(c)   Under this heading (as of page 274) there should be included *Enhydra* (which will be left for the last), the Phocidae, and the Odo-benidae partially. The latter need be given scant attention for in both conformation and function the hind feet of the walrus is fairly inter-mediate between the seals and the sea-lions. In shape these members partake of the characters of both. They may be placed flat upon the ground but the definition of the posterior part of the astragalus indicates the probability that they are developing the limitations in this respect of the Phocidae. The hind feet are said to assist swimming by lateral oscillations like the seals, or they may be used purely for equilibration like these members in the sea-lions. Which pair of limbs will ultimately gain the ascendency is a matter for pure conjecture.

In the seals (Phocidae) the external features of the part of the hind limb above the foot are very similar to those of the sea-lion, and for the reason that the anatomical differences are more significant when treated comparatively, this part of the phocid limb will be discussed in connection with the otariid. Just why the seal adopted, or could adopt, its present method of swimming it is extremely difficult to say. Its foot is really much less remarkable than that of the sea-lion. Externally the

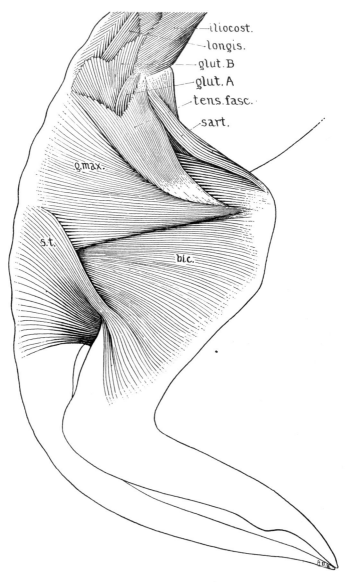

iliocost.
longis.
glut. B
glut. A
tens. fasc.
sart.

g. max.

s. t.

blc.

FIGURE 46. Lateral view of the right hind limb musculature of the sea otter (*Enhydra*), from a specimen in the U. S. National Museum.

noteworthy features consist of the hairy sole, or at least that it is as hairy as the dorsal part of the foot; the retention in all (apparently) genera but *Mirounga* of sharp nails of fair size when there seems to be no use to which they could be put, the fact that the first and fifth digits are longer than the others, while the third is the shortest, giving to the pes when expanded a lunate shape comparable to the tail of a fish or whale; and the expected fact that the webbing between the toes is complete.

The foot of the seal has retained its proportional length in relation to body length, as judged by such a terrestrial carnivore as the cat, and therefore has not suffered the marked shrinkage experienced by the shank and, especially, the thigh. In a partly dissected specimen of *Phoca hispida* the tendency of the feet was to adopt a position with soles not strictly parallel but with the upper (outer) borders slightly converging, but in the live animal this seems to be overcome by muscle or other tissue tension so that the soles are perfectly parallel. From a trailing position of the feet, movement through the ankle joint proper is apparently only about 30 degrees, but greater freedom is allowed by means of the articulations of the astragalus and calcaneum with the centrale and cuboid, this amounting to about 65 degrees. As a result the tarsus may be flexed approximately at a right angle and by this essentially hand-like action one foot is enabled to follow through and aid by its medial motion the lateral movement of the opposite foot while executing a swimming impulse. This tarsal joint, as it may be called, also enables the animal to progress very slowly by a manner first called to my attention by Breder (MS) and since observed personally. With feet palm to palm these members are flexed through the tarsal joints, the tips of the digits remaining together. The feet are then abruptly extended, imparting a gentle though definite forward impulse to the entire body. This cannot be strong enough, however, to constitute a useful accomplishment.

No great significance can be attached to details of the phocid tarsal bones, other than that already mentioned and the extension of the astragalus, discussed later. It is somewhat narrow, however, and thus sidewise (horizontal to the sole) motion is facilitated through looseness of the tarsal articulations. This tarsal narrowing causes a crowded, somewhat overlapping condition of the proximal metatarsi. Spreading of the toes, in this animal at least, seems to be facilitated by an arrangement of the joints whereby this is accomplished by a partially oblique extension of the first digit and a slight flexion of the fifth, this being comparable to conditions in the muskrat, desman etc. The hallux is elongated and considerably more robust than the other digits, while the fifth, although equally elongated, chiefly by lengthening of the first phalanx, is only very slightly more robust than the three middle toes.

No mammal swimming by means of the hind limbs alone could attain by any thrust-and-recovery method the speed of which the seal is capable, but would be obliged to develop some method whereby the pedes act obliquely upon the water. Mechanically there appear to be only three ways in which such a mammal as the seal could attain this end. One is by employing the hind limbs in the same fashion that the sea-lion uses its anterior flippers, adducting sharply in the transverse plane. This would probably be an impossible specialization. Not only does it seem utterly out of the question for the adductor muscles of the hind limb to evolve into anything comparable in efficiency with the enormous pectorals of the sea-lion, but any such development would almost certainly place the greatest body diameter in the pelvic region, and this would introduce very profound complications in the mechanics of its swimming. The second possibility might be for the feet to be held in the horizontal plane, soles up, with their inner borders touching, and oscillated in the vertical plane, as suggested for the sea otter. The third course left to it would be to take the direction which it has actually followed, and to swim by oscillations from side to side of not only the adpressed hind feet but the entire posterior end of the body.

Just how the animal was enabled to do this is a somewhat puzzling question. It may, however, have depended on some determinant which in itself was very simple. Thus any habits which did not tend to develop the back muscles in swimming and did tend to throw the greater work upon the fore limbs would be apt to predispose the sea-lion ancestor toward its present course of evolution. On the other hand some mode of swimming when all four feet were being used for the purpose, tending to develop the back muscles, as they evidently are now being developed in the tailed muskrat, possibly coupled with some slight anatomical detail which one might judge of very minor importance, should have been sufficient to start the seal on the road to its present state.

I have searched the literature and written to everyone likely of whom I could learn, including Alaskans who have spent years about the Aleutian Islands, in an endeavor to ascertain precisely how the sea otter swims. No two accounts have been the same, for probably none of these informants has himself been sufficiently close to a sea otter moving through clear water at speed to be sure of the motions, or having observed, was satisfied with determining that there was movement of the feet. Consequently we must content ourselves for the time being with a judgment of the swimming method according to anatomical evidence.

A brief examination of the external features of the sea otter will show that the fore feet are of small area but that the hind feet are much enlarged, with the fifth toe the longest and the others successively shorter, while the sole is densely haired. The remainder of the hind limb is unusually short. The tail is of moderate length, extending to the rear about the same distance as the tips of the toes when the limbs are trailed, and is said to be slightly expanded in a lateral direction.

The anterior feet of *Enhydra* are too small to be of any practical use for swimming, and for the same reason the tail is not capable of acting alone. The hind feet are, however, sufficiently modified so that we can be sure that they furnish the chief propulsive force. Precisely how fast the animal can swim I have no idea, but according to accounts several men in a light skiff have all they can do on a long pull to catch up with one. It seems that they may be able to travel as fast as ten miles an hour, possibly with higher speed for a short distance. One can be sure that they are at least reasonably speedy swimmers, too, because otherwise a pelagic mammal of this size would quickly be exterminated by sharks and killer whales, even though kelp beds are their favorite habitat.

If this probability be conceded, then they cannot swim by direct thrust-and-recovery movements in alternation by the hind feet, for, as previously described, even moderately high speed cannot be attained thereby. On the contrary some motion must be employed involving oblique action of the feet against the water, comparable to the swimming actions of the seal. Can the sea otter swim in the same way as the seal, with lateral motions of the feet held palm to palm? Indications are that it does not do so for the following reasons. In the seal the feet are shaped like the tail of a fish with lunate posterior border, and for the same reasons. Both first and fifth digits are elongated to accomplish this. If the sea otter swam in a similar fashion it is sufficiently specialized for it to show definite indications of the same development. This, however, might be complicated by the presence in this animal of a tail of moderate length. The tail of the seal is too short to be of any consequence, but it is possible that in the sea otter this member, of the same length as the feet, might in a mechanical sense take the place of the uppermost toe held against it in swimming posture. But the *fifth* toe is the longest and in order to act with the tail so that the posterior border of the swim-

ming organ (feet and tail) would present the requisite lunate form
the feet would have to be held back to back, instead of palm to palm,
and this is either impossible or would at least require changes in posture
which would be conspicuous externally. Of course there may be some
idiosyncrasy about the swimming of this animal for which a non-lunate
propulsive organ, and one with a longer upper than lower border would
be advantageous; and this must be entertained as a possibility, but it
is hardly likely.

If the above arguments be correct, it seems that the only way left for
the sea otter to swim efficiently with its existing equipment is to place

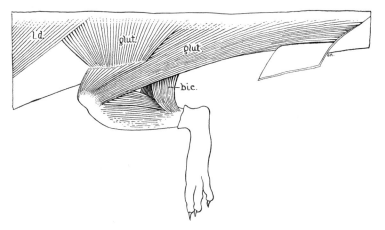

FIGURE 47. Lateral view of the left hind limb musculature of the insectivore
otter *Potomogale*, redrawn from Dobson.

the feet horizontally to the rear, one upon either side of the tail, sole
up, and oscillate them in the vertical plane on the same principle that
the whale uses its flukes. It is true that no other mammal is known
ever to have employed this method of swimming and accordingly there
are doubtless those who will claim that it is illogical, but on the contrary it appears to be a more logical method of swimming than that
now employed by the seal. The external details of the sea otter support
this presumption. The long fifth toes would form the outer borders
of the swimming organ, as they should, and the tail, of almost if not
quite precisely the length of the extended feet, would comprise a
strengthening, central element. The whole would accordingly present

in a satisfactory degree the lunate rear border theoretically desirable in such a propulsive apparatus.

Fortunately it is possible to present some myological evidence in this discussion of the sea otter. There is available from the National Museum the skinned carcass of one of these animals that was brought from Copper Island by Dr. Stejneger in 1897. This was evidently dried out at least once, for the muscles are leathery and the nerves brittle. Nevertheless it is entirely suitable for investigating those muscles which might be expected to shed some light on the method of swimming employed. Accordingly the more significant muscles of the hind limb are described and figured (fig. 46). Nothing further is attempted for I have no wish to encroach upon the chosen field of E. R. Hall in investigating the comparative anatomy of the Mustelidae.

Reference to the illustration shows that the muscle usually termed caudofemoralis is large and powerful. It arises from the dorsal fascia, superficial except caudally where origin is deep to the semitendinosus. Insertion is upon the distal half of the femur and to the knee. At least in the majority of cases this muscle is merely the chief part, usually the whole, of the true gluteus maximus, for it is served, in the case of the sea otter as in others, by the inferior gluteal nerve. And what is so often called the gluteus maximus, where the term caudofemoralis is also employed, is proven by the innervation to be the gluteus medius.

The tensor fasciae latae arises by tough aponeurosis from the ilium, and mediad from over the gluteus medius to an aponeurotic insertion upon the femur for a considerabe distance distad of the greater trochanter. It seemed to be innervated by a nerve thread coming from deep between the main gluteal mass and the quadriceps. This was undoubtedly a branch of the superior gluteal nerve, as usual, but was so brittle that it broke and I was unable later to pick it up again.

As is so often the case the main gluteal mass was separable only with difficulty, the whole being fused toward insertion. It was so deep (23 mm.) and hardened that the only way the innervation could be determined was to pick the muscle to pieces, fiber by fiber. Three parts of origin could be detected. Gluteus A arose thinly from the tip of the ilium. The portion B arose from the dorsal fascia both craniad and mediad of the ilium. Both A and B increased rapidly in thickness and soon fused with the deeper and more robust part C, which arose from the entire lateral aspect of the ilium, the whole inserting very strongly upon the greater trochanter. The arrangement of the fibers was slightly multipenniform and the mass constituted a muscle of great strength.

The deeper and most of the superficial parts were indubitably served by the superior gluteal nerve, but whether or not the caudal border of B may have received twigs from the inferior nerve cannot be stated with any certainty. If so then this part would have gluteus maximus affinity.

The superficial biceps femoris was highly modified much in the manner that it is in the Phocidae (fig. 23). It arose from the tuber ischii and spread fanwise to a fascial insertion over the shank from the knee almost to the instep. The deep biceps was almost an exact counterpart of this muscle in both seal and sea-lion. It was very slender, arising from the deep dorsal fascia alone and perhaps a trifle anterior to the acetabulum, and passed to the vicinity of the outer malleolus of the ankle.

It is frequently difficult to establish the homology of the semitendinosus and semimembranosus for the reason that they often shift about and the innervation is not particularly distinctive. Hence homology can seldom be indubitably established and one must name them as best he can, judging by generalities. In the sea otter the apparent semitendinosus is phenomenally developed, to a greater extent than in any mammal known to the writer. Origin is very broadly by fascia, the muscle increases in thickness and most of it passes to the medial side of the leg, where it inserts broadly along the shank, distad to its middle and partly under cover of the gracilis. Oddly enough, however, the fibers along the cranial border do not pass mediad of the shank with the remainder, but to the lateral aspect, inserting as a caudal extension of the superficial biceps. Accompanying the hypertrophy of the semitendinosus is atrophy of the semimembranosus, which is a weak slip of a muscle, some ten millimeters broad, beneath the other.

There is a peculiar development of the sartorius. This arises from the tip of the ilium, with origin extending broadly onto the ventral border of the bone. As a broad, thin muscle it inserts for 40 mm. onto the knee and a bit distad. In continuation still farther distad over the proximal part of the medial shank is the insertion, for another 40 mm., of the gracilis.

The lumbar part of the erector spinae was very robust, the powerful iliocostal part causing a flaring outward of the ilium, to which it is attached. The hypaxial lumbar muscles were, however, definitely weak and no larger than one would expect in any terrestrial mammal of equal size.

In considering the above muscles it cannot be claimed that any particular swimming method is clearly indicated. Because the interaction of the muscles in any strong motion is so intricate it is not safe to state that the muscular equipment is not fitted for the same swimming method as employed by the seal. On the other hand, there are considerable differences, which contribute evidence to the probability that the sea otter swims by vertical oscillations of the expanded feet. In such actions the muscles of the back are involved to elevate the pelvis. The erector spinae is certainly fitted for this. The hypaxial muscles are *not* fitted for depression, but the strong rectus abdominis can accomplish it, with a more favorable lever arm. The lesser gluteal complex is much

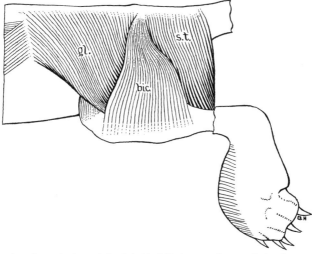

FIGURE 48. Lateral view of the left hind limb musculature of the desman (*Desmana*), redrawn from Dobson.

specialized for a purpose not particularly clear, for its action depends upon the habitual posture of the femur and the way in which other muscles co-operate. The gluteus maximus is powerful for further extension of the femur after it is partly extended, and the superficial biceps, semitendinosus and gracilis form an extraordinarily efficient complex for elevation of the shank when the feet are extended to the rear, thus accomplishing upward thrusts of these members. On the whole the muscular details do not detract anything from the above theory of the way in which this animal swims, but, on the contrary, contribute considerable coroborative evidence.

The partial binding down of the hind limbs effects a static posture with these members more trailing than would otherwise be the case, and for the same reason the terrestrial posture must be with the back more arched, so that the sacrum is more elevated toward the perpendicular in a manner suggestive of the sea-lion's posture. The evidence clearly shows that the feet *are* placed in plantigrate position with ease, although I have seen it stated in a letter that this is impossible.

In examining a skeleton of *Enhydra* in comparison with one of *Lutra* it is seen that although the proportions of the ilium in regard to post-acetabular dimension is the same, the entire innominate is somewhat heavier, to accommodate stronger musculature, and the ilium flares considerably in a lateral direction, for a stronger iliocostal attachment. Taylor (1914) has stated that the pelvis is more nearly parallel to the vertebral column also and mentions that there is no evidence of a ligamentum teres, as there is in *Lutra*. He also showed by drawings reduced to the same size that the thigh and shank bones of the sea otter are the heavier and that their relative proportions are not appreciably altered. The greater trochanter is slightly broader, because of the specialization of the lesser gluteal complex, and more seal-like. The metatarsals and phalanges are slightly flattened, and manipulation of the hardened spirit specimen seems to show that there is unusual freedom of movement in the articulation of the astragalus and calcaneum with the centrale and cuboid—a functional tarsal joint as in the seal, making the foot more handlike to facilitate oscillating movements when the pes is held on a line with the shank.

Why the sea otter should have developed a swimming method essentially different from that seemingly in course of elaboration by the river otters is not entirely clear, but the fundamental reason which appears mose likely is that in the former animal the tail has always been too short and light to be used for propulsion alone. Presumably the hind feet will continue their specialization and become still more efficient, while the tail may either continue to function as an accessory, central stiffener for the swimming organ formed by the hind feet, as it now appears to do, or become shorter as the feet have less need for a support of this kind. Conceivably, however, the tail may ultimately increase in power and take over the function of primary propeller.

2. Aquatic mammals which use the hind feet chiefly for equilibration include *Ornithorhynchus*, the Otariidae or sea-lions and fur seals, and the Odobenidae or walruses partly. The latter need no further consideration here save to mention once more that their feet partake of

the character of those of the sea-lion, with cartilaginous extensions of the digits, and in form somewhat resemble those of the seal. Apparently they are rather intermediate in their uses as well.

The hind feet of the platypus are much less modified than the anterior ones, and in a somewhat different way. In all probability all four feet were formerly used for propulsion, on the mud-turtle principle, until the tail had become flattened and by more frequent use the fore feet were enabled to gain developmental ascendancy. Now the hind feet apparently are but rarely used to assist propulsion, but rather is their action comparable to that in the marine turtles—mainly for vertical and to a lesser extent for horizontal steering, in the former direction aiding the tail in depressing, or in elevating the body. They are fully webbed and articulation of the four lateral toes is such that these digits flex in a direction more toward the outer border of the foot and away from the hallux than in a strictly palmar direction. A variation in this flexional peculiarity has been noted in the case of *Galemys, Sorex, Phoca* and *Ondatra,* the advantage evidently being either for attaining a slightly different angle of the membranes with respect to the articulation of the ankle, or more probably so as more easily to secure the maximum expansion of the membranes.

For the purpose of more readily evaluating the differences between the pelvic limbs of seals and sea-lions these two groups of pinnipeds will be discussed together under the present heading. But first it should again be mentioned that seals swim by oscillations from side to side of the hind feet, placed palm to palm, and of the entire hinder end of the body, and that the hind feet are useless in terrestrial progression for the reason that the sole cannot be placed flat on the ground. The sea-lion, however, can place the sole upon the ground, but only by curving the sacral vertebrae in a position perpendicular to the surface. In the water the feet are trailed, usually sole to sole much as in the seal, but they seem never to be used in the slightest degree as an aid to purposeful propulsion. When playing or lolling about in the water the feet may be flexed in all manner of positions, assisting to alter the posture of the body, and I have watched a young fur seal progress very slowly by repeatedly rolling the entire body over and over, the hind feet thus acting to some extent in a cork-screw manner as a propeller; but such inconsequential actions need not be considered.

In scrutinizing the influences to which the pelvic limbs of the Pinnipedia have been subject one must begin with the spinal musculature. In the Otariidae the latter is not appreciably modified and need not be considered. In the Phocidae the spinal muscles, both apaxial and hypaxial, are enormously developed and play a more important part in swinging the feet from side to side than do the intrinsic muscles of the limbs. The chief effect which this development has had upon the pelvis of this group is the fact that the ilium has been projected sharply to the side (figure 49) so that it may accommodate a powerful attachment of the iliocostalis lumborum. In other respects the innominates of these two groups exhibit many resemblances, chiefly brought about by the fact that in both the legs are always maintained in a trailing position. In both, also, a significant result of the latter posture has been a binding down of the shank so that the "crotch" consisting of the tissue joining the limb to the vertebral column, falls opposite the heel. As a consequence there is little independent movement of the limb bones, which means that flexion and extension of the thigh is very definitely circumscribed, and that although rotation and abduction may be strong, they can be only through a short distance. This has resulted in reduction in the length of the femur, and of the ilium, or pre-acetabular part of the innominate, which accommodates the lesser gluteal muscles. The latter is but 16 per cent of the entire innominate length in a species of *Phoca,* and 32 per cent in *Zalophus* (an otariid), while in such a terrestrial carnivore as the cat this percentage is 59. In compensation the post-acetabular measurement, or pubo-ischial portion, is 74 per cent of the innominate length in *Phoca,* 55 in *Zalophus,* and 33 in a cat. Apparently the stimulus for the increase of this posterior part of the pelvis has been the increase in importance of the function of the muscles attached thereto as adductors of the shank. Especially is this true in the case of the seal, and parts of the ischium and of the pubis have been projected in a dorsal and a ventral direction respectively, which furnishes greater leverage chiefly to the superficial division of the biceps femoris and the gracilis (fig. 49). Ryder (1885) has claimed that the pinniped pelvis is degenerate but this statement indicates an incorrect viewpoint. It is permissible to view the ilial part as having suffered degeneracy in *length,* and the sea-lion innominate is definitely less robust in some particulars, but the posterior part has experienced relative hypertrophy and the whole is little or no shorter in relation to body length than in the cat (21 per cent in *Zalophus,* 26 in *Phoca,* and 25 in a cat). Neither is the sacropelvic connection weaker than in many terrestrial mammals.

Accompanying and intimately involved with the shortening of the pinniped ilium is shortening of the femur. In *Zalophus* this segment is now 11 per cent of the body length, 12 in *Phoca* and 35 per cent in a

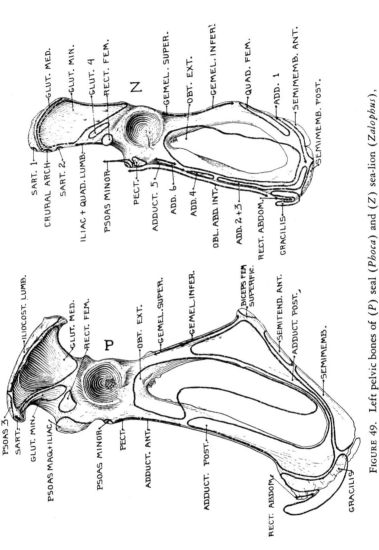

FIGURE 49. Left pelvic bones of (*P*) seal (*Phoca*) and (*Z*) sea-lion (*Zalophus*), showing areas of muscle attachment.

cat.[4] There is a complex of interacting conditions involved in this situation, some constituting cause and some effect. As the heel is bound down to the body a long femur is not desirable unless really needed for other purposes, and this is not the case, for the binding down of the leg prevents great movement of the thigh from serving any useful purpose. Furthermore, a short femur can be held abducted without projecting from the side to an undesirable extent. Especially is this true of flexion, and atrophy of the flexor muscles accompanies the shortening of the ilium and the femur. With the flexors serving little useful purpose there can be but slight need for muscles accomplishing only extension. Accordingly those extensors which usually insert near the knee have shifted their insertions to other situations and because of the increase in post-acetabular length of the innominate, the origins of the adductores, and of the obturator externus, have shifted so as more effectively to accomplish what extension of the femur is needful, in addition to the regular adduction and rotation. Because the actions involved in swimming by the seal comprise relatively simplified but constantly repeated motions there is apparently need only for simplified adductors, and these are reduced to two, whose position in respect to the external obturator is so intimate that the three have previously been mistaken for one single obturator. Probably for the reason that the sea-lion has need for very intricate, though slight, movements of adjustment for steering, the abductores now occur in six slips, and the sartorius is double. As a result of the above conditions the arc of movement of the pinniped femur in those specimens dissected seemed to be only through about 25 or 30 degrees.

In this order the femur is relatively very broad transversely and narrow in the sagittal plane. The breadth may be attributable to the fact that good leverage is needed through a greater trochanter which extends well away from the head, and a broad distal end for articulation with the shank. Its thinness is permitted on the one hand by the fact that little mechanical strength in flexion and extension is needed, and promoted because a flat femur can accomplish a sharper angle of flexion with the shank. The latter point appears to be of lesser consideration in the seals.

[4] I regret to say that in the last paragraph of page 124, Howell, 1929, an error was made in stating the percentage of femoral length in relation to body length, and the figures 22 and 29 as given are really those for the tibia. They should read 11 and 12 respectively, as was stated on page 36.

It will be noted (fig. 52) that the sea-lion has a lesser trochanter defining insertion of the psoas magnus and iliacus (so-called), while in the seal these muscles have fascial insertion over the knee and there is no lesser trochanter. In this illustration it may also be noted that while in the sea-lion the inner condyle of the femur extends but a trifle more distad than the outer, in the seal this discrepancy is very marked indeed. The result of this is that when the femur is held at a right angle to the body axis the position of the condyles will cause greater supination of the pes, so that the soles may be maintained parallel with greater ease. If the femur be extended beyond a right angle, then the condylar conformation will cause the feet to be held more elevated.

A peculiarity which affects function in the pinniped knee is the length of the tibial collateral ligament. The distal attachment of this is not at approximately one-tenth the distance from the knee to the ankle, as in the cat, but at a point about one-quarter or two-fifths this distance. This allows for considerable rotation through the knee, with the fibular ligament as the pivot. In most other mammals such rotaton of the shank can be accomplished through the hip joint, but because of the position of the short femur in the Pinnipedia such action is impossible in this order without adductive-abductive movements of the thigh.

A further accompaniment of the above conditions of the femur and its more intimate musculature is that the function of those muscles which normally act as extensors of the thigh through an attachment in the vicinity of the knee joint are enabled to alter their function. Their insertions have either migrated wholly or in part distad, and accordingly they now operate to bind down the shank and accomplish its adduction or elevation, according to posture. There is considerable variation in the form of these muscles as between the seal and the sea-lion (figs. 22 and 23), but in both animals the gracilis, semitendinosus and semimembranosus act especially in this capacity and their insertions not only do not approach the knee but are extended onto the heel by fascia in the case of the sea-lion, and in the seal by a more tendinous structure passing to the plantar fascia. In the former animal the biceps acts with particular efficacy as a binder, but not to the same extent in the seal. The conformation of these muscles has naturally affected the position of the crotch, or posterior integumental border between the leg and body, or in this case the tail, so that in both animals this lies approximately at the heel and all of the hind limb except the pes is contained within the body contour.

The pinniped shank is considerably shortened (length of tibia being 22 per cent of the body length in *Zalophus,* 29 in *Phoca,* and 36 per cent in a cat), and because it is so bound down by muscles and other tissue its possible movement, judging by the specimens dissected, seems to be through an arc of only about 15 degrees. In this order the fibular head really is not concerned with the knee joint and its conformation has been affected by the posture habitually assumed. In the *Phoca* studied the angle which the femur formed with the shank was a trifle more than 90 degrees, while in the *Zalophus* it was considerably less—apparently about 45. Accordingly in the former animal the fibular head slopes moderately, and in the latter quite steeply, enabling consequent sharper bending of the knee without mechanical interference. But there is some slight generic and even specific variation in this detail.

Upon examining the distal ends of the shank bones (in fig. 13 of Howell, 1929) it will be found that there are no deep grooves in *Zalophus,* but several in *Phoca.* Upon the lateral aspect of the latter there is a deep one for the tibialis anticus and another for the peroneus longus; upon the medial aspect there are three, for the posterior tibialis, flexor digitorum longus, and peroneus digiti quinti. As in the case of the radius these grooves have doubtless resulted from the constantly repeated oscillations of the feet, made during swimming.

The significance of the absence of the soleus in *Phoca* and of the lateral gastrocnemius in *Zalophus* is unknown. In both, the plantaris does not join the tendo calcaneus but passes mediad thereto. In the *Phoca,* but not the *Zalophus,* the tendon of the strong flexor hallucis longus passes over a posterior extension of the astragalus, as already described, and it is the tension of this tendon alone which prevents the foot from assuming an angle of 90 degrees with the shank. But this does not necessarily mean that if it were not for the tension of this muscle the seal could walk like a sea-lion. To do this it would have to bend the vertebral column so that the sacrum is vertical to the ground, and it might find such a posture inconvenient, if not actually impossible of assumption.

In bony details the foot of the sea-lion is of especial interest in the present connection in having the first and fifth toes approximately equal in length and longer than the middle three, as in the seal, and the bony elements are definitely flattened, especially the hallux. The question of epiphyses has already been discussed in chapter ten, for the situation in this respect in the manus applies equally, with minor variations, to the pes.

In two specimens of sea-lion examined, one fresh and one preserved, the relaxed posture of the foot was latero-craniad at an angle of about 45 degrees with the body axis. In life, however, an angle of approximately 90 degrees may frequently be assumed, and in young fur seals observed the latter was the invariable posture. This facilitates the placing of the feet sole to sole when in the water. In the Otariidae the

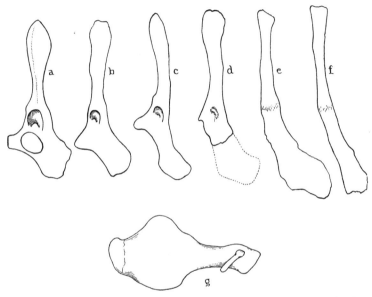

FIGURE 50. Pelvic bones of sirenians. Above is illustrated progressive development according to the fossil evidence: (*a*) *Eotherium* (Middle Eocene); (*b*) *Eosiren* (Upper Eocene); (*c*) *Halitherium* (Middle Oligocene); (*d*) *Metaxytherium* (Middle Miocene); (*e*) *Halicore dugong* (recent); (*f*) *Halicore tabernaculi* (recent); and (*g*) *Trichechus latirostris* (the manati) (redrawn from Abel).

interdigital membranes are capable of practically no expansion, and the foot is correspondingly narrow as compared with that of a seal. It is relatively very long, however, because of the predigital cartilages, as in the manus. But in the foot, these can be folded back against the palm so that the body may be scratched with the slender nails of the three middle digits. The nails of the first and fifth digits, are, however, rudimentary like those of the manus.

In the sea-lion manus the terminal border of the foot is practically straight, but in the pes it is deeply serrated, because the predigital cartilages extend considerably beyond the interdigital membrane borders. The length of foot including cartilages, in the fur seal *(Callorhinus)* appears to be relatively greater than in the sea-lion *(Zalophus)*. Also, judging by but two specimens measured, this length is proportionately greater in young of the latter genus than in an adult, which may indicate that a longer rudder was needed in the past, with somewhat less perfected swimming apparatus, than is now requisite.

The development of the pes in the Otariidae is somewhat puzzling in certain respects, chiefly because of its length in respect to breadth, the former augmented by the peculiar predigital cartilages. It may really be said to be more highly specialized than either the manus or the pes of the Phocidae. At one time I was inclined to entertain the possibility that at a not very distant geologic date it might even have been of some definite, active aid in locomotion, but this I have abandoned. Evidently the stimulus for a long neck that is highly mobile is a strong one, and this mobility necessitates a highly specialized and highly efficient rudder fully capable of compensating for abrupt movements of the head, say in snatching at prey, that would otherwise tend to deflect the animal from its desired course. Such a rudder should project well to the rear, and does, but its "hinge" should be at the body junction, and is. Hence the leg has been shortened and is contained within the body covering while the foot has been lengthened by predigital cartilages, either because these fulfill some need which longer digital bones could not accomplish or because they were more easily developed than longer phalanges.

3. Under the third heading mentioned on page 000 will be discussed those aquatic mammals in which the hind feet are absent or definitely subordinate in swimming or steering. Under this will be included the mink for the sole reason that there is nowhere else to place it without making a separate heading for its reception. It swims chiefly by all four feet, as far as I can learn, but it should not be placed with most other mammals employing this method of propulsion, for its tendencies are different. It is clearly what may be called an incipient otter which may be expected to develop its swimming abilities along precisely the same lines as has the latter, and accordingly it will be given no further attention.

For discussion under this heading there are the aquatic Tenrecidae, *Limnogale,* and *Potomogale,* the Luttrinae or river otters, the three recent families of the Sirenia, and the Cetacea.

I am unable to discuss *Limnogale* with any profit for I have never seen a specimen and have been able to find nothing of significance in the literature save that the toes are webbed and the distal part only of its powerful tail is laterally compressed. *Potomogale,* the insectivore otter (fig. 4) is peculiar in that there is no present evidence that the hind feet were ever used for efficient aquatic propulsion, for the legs are short and the toes unwebbed. In the case of other aquatic insectivores, as well as rodents, the feet are used for swimming pending the modification of an adequate tail, which is theoretically the more efficient propulsive organ, and during this time the former are confidently expected to become quite notably altered. Hence it is requisite to assume that even from the first this insectivore had a phenomenally powerful tail, robust at base, and consequently somewhat different from this member in most representatives of the order; and that these characters, coupled with an elongated body, enabled it to swim quite well by lateral oscillations with little or no help from the pedes.

The only alteration in the pes of this animal is that there is a thin, cutaneous extension of the lateral border. This is believed to be for the purpose of enabling the feet to present fewer inequalities when they are folded back against the base of the tail during swimming. As a matter of fact it is doubtful if anyone knows just how the feet are then held and these lateral extensions of the foot may have some entirely different function. Dobson (1882) has shown that in this genus there is a very remarkable specialization of the hip musculature. Evidently the femur is fixed by its flexor muscles, and with this bone thus acting as an origin the caudofemoralis or gluteus maximus extends far backward onto the tail (fig. 47), thus helping in the lateral oscillations of this member. There has been no definite reduction in the length of the femur as in the desman, and Dobson stated that although in general form the pelvis resembles that of *Centetes* it is narrower and without a true symphysis, interpubic connection being ligamentous.

The river otters are of particular interest in the present connection because I regard it as most likely that the terrestrial ancestors of the Cetacea were beasts of very similar body form, namely, with long, sinuous trunk, short legs, and powerful tail, especially at base. The otter is such a playful creature that it is often difficult during observation to be sure which actions are concomitant to greatest efficiency in swimming, and which are attributable to exuberance of spirits. In the common genus *Lutra* the hind limbs are frequently kicked in a sort of galloping action, seldom entirely rhythmically but at odd intervals. I have gained the impression that this use is not invariably so much for the purpose of speeding progression as it is for maintaining proper posture in the water. Frequently when the greatest speed is attained the limbs are not used at all. It is highly probable that at one time in the past the tail of the otter was a less efficient swimming organ than is now the case, and undoubtedly the limbs then constituted the chief propulsive apparatus, as now in the mink. But in all generalized mustelids the body is long and sinuous and the legs short. Hence, body motions, such as arching of the back, are important in terrestrial locomotion and it is natural that a similar use was made of the body and tail when the animal first took to the water. Because of this and the shortness of the limbs it is not improbable that the latter did not respond so readily to the stimulus for aquatic modification of the extremities, or else that this stimulus was not as strong as in the case of most other mammals, so that the tail more readily gained the ascendancy.

The hind feet of otters (fig. 43) vary greatly from an unwebbed condition to one in which the interdigital membranes are broad and full. But as nothing is known about the finer points of the swimming methods of these less familiar sorts the subject cannot be pursued further.

One would hardly expect much change in the otter as long as it follows its present mode of life. If it should increase in size and take to coastal waters then its tail should broaden into a more efficient propeller and its hind limbs tend to become reduced as it may become more independent of the land. Eventually its pectoral limbs should assume the form of equilibrating paddles and it is likely that superficially at least it should finally become very much the same sort of beast as is the cetacean.

The subject of the sirenian pelvic element is one which I approach with considerable hesitancy. Flower (1876) has said, the pelvis is rudimentary, composed in the dugong of two slender bones on each side, placed end to end, commonly ankylosed together. He considered that the upper, attached to the vertebral column by a ligament, represents the ilium and the lower the ischium, or ischium and pubis combined. He also stated that although there was a vestigeal femur in *Halitherium* this bone is not represented in any living member of the order. But other observers, possibly working with different species, have reported otherwise. The innominate of the dugong is usually figured as a single, slender bone, made up of two terminal elements, situated with the long axis vertical, connected by ligaments dorsad with the transverse process of one vertebra, and ventrad to its fellow of the opposite side. Abel

(1906) has illustrated some of the steps in its evolution by the fossil evidence (fig. 50) whereby it is indicated that the ilium has become more slender but has not been shortened, the pubis has suffered suppression, and the ischium slight elongation and alteration in shape.

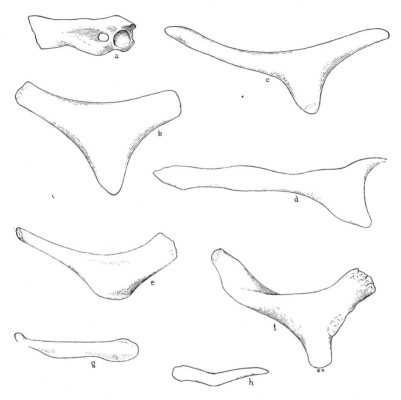

FIGURE 51. Pelvic bones of Cetacea: (*a*) the zeuglodont *Basilosaurus;* (*b*) *Sibbaldus;* (*c* and *d*) *Balaenoptera physalus;* (*e*) *Megaptera;* (*f*) *Physeter;* (*g*) *Mesoplodon;* and (*h*) *Lagenorhynchus*: (*b, d,* and *e* are redrawn from Augustin, and *c, g, f,* and *h* from Abel). These are supposed to be in right latero-ventral aspect.

Abel's figure of this region in a manati from Surinam (fig. 50) indicates that in this family there has been a fundamentally different procedure. He claimed that a vestigeal femur is present, attached near the anterior end of the pelvis; hence that the ilium is greatly reduced. The pubis has probably been eliminated and the ischium shortened and contorted. This situation is broadly similar to what is indicated in a zeuglodont, as discussed shortly, but very different indeed from what has taken place in the dugong. Clearly the reduction in the hind limb of these two families of sirenians has been upon entirely different principles, likely involving fundamental difference in the uses to which the hind limbs were previously put, and it seems unsafe to hazard any opinion in regard to just what particulars of the disappearing hind limbs may have been involved.

In other respects the modications of the Sirenia are much more puzzling than in the Cetacea and it is correspondingly difficult to reconstruct the probable steps that were followed. In the case of the whales the salient characteristics are such as to point clearly, to me at least, to numerous hypotheses that have a high degree of probability. This is not the case in the Sirenia. It is difficult for me to envision any sort of primitive mammal of proboscidean affinity taking to the water in the sluggish manner which one would naturally connect with the sirenian prototype and developing the tail which we now see in connection with the total disappearance of the hind limbs. Evidently the latter situation followed abandonment of any use of these members, but more I do not feel like saying at the present time, for further statement would be pure, unsupported speculation.

Indubitably many groups of the Cetacea have been distinct from one another for a great length of geologic time, presumably since the period when they had external hind limbs which were quite well formed. As the latter tended toward more and more complete disappearance there naturally occurred differences in precise details. Hence it is entirely according to expectation that in examining the anatomy of this order one finds hardly any two distinct sorts with the details of the pelvic musculature entirely the same. In fact the differences are so great that it is often difficult to correlate the pelvic muscles of two sorts of whales, and utterly impossible yet to homologize them with muscles of terrestrial mammals for the reason that the nerves are so greatly altered, and in the present state of our knowledge it is unwise even to attempt to do so.

The cetacean innominate is now so different from the generalized condition that very little can be proven about the homology of its parts, nor does the fossil material, except in the Archaeoceti, help us to any extent. Usually one area of the bone is thicker and broader than the rest, and this, presumably, marks the previous situation of the hip joint, but there may be exceptions. In some cases existing projections from this area may be processes recently and secondarily developed for some specialized purpose, and in others an anterior and a posterior process may actually be remnants of both ilium and ischium, as suggested by the skeleton of a young sperm whale in the National Museum. In this there are two small pelvic bones on each side, one before the other, which one would expect to have fused at a later age. Again, in those whales in which the pelvis is broader at the anterior end there has probably been virtual elimination of the ilium as there seems to have been in *Basilosaurus*.

In shape the cetacean pelvis varies greatly. In what seems to be its simplest condition it takes the form of a short, practically straight, slightly flattened bone (in many porpoises). Or it may be bent somewhat after the fashion of a boomerang, but with one end broader than the other *(Megaptera)*, or quite broadly triangular *(Sibbaldus)*, or narrow at one end and abruptly expanded at the other (some individuals of *Balaenoptera, Monodon*, etc.). In a large rorqual it will not exceed a length of 18 inches, and in a small porpoise may measure about one inch. At least those which are more elongate occur practically parallel with the body axis and almost directly above the genito-excretory orifice. It has been forced ventrad of its usual situation, evidently by the hypertrophy of the hypaxial musculature and has no direct (bony) connection with the vertebrae. In its simplest form it is doubtful whether the cetacean pelvis should really be looked upon as constituting a disappearing remnant of the hind limb, but rather that the hind limb has *already* disappeared and what remains is merely a necessary anchor for the perineal musculature, thus, possibly as greatly reduced as it will ever become. As these muscles are the more powerful in the male the pelvic bones average larger in this sex than in the female.

There has been considerable controversy over the question of whether the cetacean pelvis represents all, or but one, of the elements of the normal innominate. Flower (1876) stated that this bone ossifies from one center and probably represents the ischium. Struthers was of substantially the same mind, while Delage (1885) argued that all three bones are represented. Whether the bone actually develops from three centers of ossification seems to me a purely academic question of not much import in the present connection. The chief thing is that there has been marked reduction, almost if not quite to the point of elimination, of certain elements of the normal pelvis, and relative increase in the importance of at least one other. The important point at issue is whether the present conditions in this regard can be interpreted.

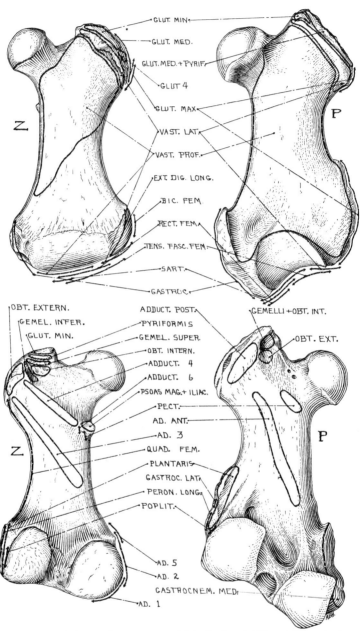

FIGURE 52. Left femur of (Z) sea lion (*Zalophus*) and (P) seal (*Phoca*), showing areas of muscle attachment: anterior aspect above and posterior below.

[304]

I think the evidence is clear that there has been reason for retention of a part of the ischium, for perineal attachments, and it may reasonably be conceded that a considerable part of the present pelvis—say at least half—is of such derivation. *If* another pelvic element has been retained in appreciable amount it probably consists of the ilium, for the muscles usually attached to the pubis have disappeared or relinquished pelvic connection.

This brings us back to conditions in the Sirenia. It is incontrovertible, I think, that the plan of pelvic reduction in the Trichechidae has been on a basically different principle from that in the Halicoridae. If this can be so in the Sirenia it can apply with equal logic to the Cetacea, and accordingly it must be postulated that in the whales the pelvis *may* have suffered reduction on at least two, and possibly more, basically different schemes, each perhaps varying in minor details according to the group concerned. Of those conditions involved in a possible retention in definite degree of the ilium nothing can be told, but considerable can be inferred about that plan by which the ilium may have been eliminated, just as in the manati, by analogy to the pinnipeds and zeuglodonts, as follows.

The pelvis of the zeuglodont *Basilosaurus* (fig. 51) was considerably less advanced in specialization toward simplification than in any living whale. This is a roughly quadrangular bone with an indubitable acetabular depression upon its extreme anterior part, for attachment of the rudimentary femur, and posterior to this, a foramen which is obviously a relic of an obturator foramen. One can hardly escape the conclusion that this situation represents an extreme atrophy of the ilium and the entire disappearance of all but possibly that part forming the anterior margin of the acetabulum. Evidently the zeuglodont ancestor had begun to trail the hind limbs and there ensued a reduction in the length of the ilium, just as is now to be seen in the pinnipeds, and further atrophy of the hip muscles arising from this bone and relinquishment of its iliocostalis attachment would naturally be followed, in the course of time, by the entire suppression of the pre-acetabular part of the pelvis. Accompanying the reduction of the muscles attached to the ilium would be a similar atrophy of those of the hip and thigh that were attached to parts of the ischium and pubis. There would naturally follow atrophy of these parts of the innominate bone, but there would not be the same relative degree of atrophy of the ischial portion which afforded attachment to the perineal muscles, and these would increase in proportionate importance. As a matter of course the form of the complex would change in an unpredictable manner. And, this, apparently is just what has transpired in the case of this zeuglodont.

A change which was fundamentally of this sort has doubtless occurred in *some* of the Cetacea, and in these there must be virtual if not complete elimination of the original ilium, as well as the pubis, but in other sorts there may be an entirely different situation, which the existing pelvis is too greatly altered to indicate with any clarity.

According to Kellogg (1928) no zeuglodont is yet known having more of a hind limb than a rudimentary femur. In no member of the Odontoceti except the cachalot *(Physeter)* has any sort of a femoral rudiment been found, but vestiges, in varying degree, of the pelvic limb have been reported from most if not all species of Mysticeti examined, although the frequency which which it has been stated as entirely lacking in certain specimens leads one to suspect that this may be an individually variable character in some sorts. At least in *Balaenoptera physalus* it apparently is never more than very rudimentary indeed, and Abel (1908) illustrated this detail in a number of individuals, mostly collected from the literature. Struthers (1887-88) reported that the humpback *(Megaptera)* has a cartilaginous femur inclosed in fibrous tissue, and indicated that he considered that the pubis constituted the greater part of the pelvis. Eschricht also reported a femur in this genus. Flower found that in the blue whale *(Sibbaldus)* it occurred as a bony nodule, which Beddard stated was attached to the pelvis by one anterior and two posterior ligaments in which there were a few muscle fibers. It is in the balaenid whales, however, that the cetacean limb is least reduced, and several authors have reported both a bony femur and a cartilaginous tibial head present in the Greenland or bowhead whale *(Balaena mysticetus)*. The former was stated by Beddard to be from 4 to 9 inches in length. According to Struthers (1880-81) the femur is flexed forward and the tibial rudiment is horizontal, while there is a synovial bursa between the femoral head and the pelvis, and another between the femur and tibia. He (1893) has also stated that there are three muscle slips from the femur to the pelvis.

That the loss of the external hind limbs in the Cetacea occurred at a rather remote time is indicated by embryologic conditions. Guldberg and Nansen (1894) found that in a 17 mm. fetus of the porpoise *Phocaena* the pelvic buds were one-third mm. in length, while in one of 7 mm. these were three-quarters mm. long (the fore limbs being one and one-half mm.) and shaped like an oval leaf. Histologically they consisted of undifferentiated mesoderm without sharp separation from the epidermis, and were entirely comparable, except in size, with the hind limb buds of other mammals. Apparently their lesser definition in the 17 mm. *Phocaena,* and still less in a 27 mm. *Lagenorhynchus,* indicates that a reduction in absolute size begins as the external cetacean characters (chiefly the relative increase in tail size) are initiated. In another *Phocaena* fetus, of 18 mm., the limb was represented merely by a papilla, which is the penultimate step in external disappearance.

Before summarizing the discussion of the cetacean pelvic limb it may be well to consider briefly the case of some of the extinct aquatic reptiles. In almost all the larger sorts known the pes was practically as large as the manus, even though adaptation had progressed to the point where hyperphalangy was already far advanced. It has previously been argued herein that this was probably attributable largely to a difference in the methods of swimming employed by these reptiles compared with the case in mammals, partly because of difference in bodily conformation as well as equipment and inherent tendencies. It was only when aquatic adaptation had become very far advanced indeed, to the point comparable with that now attained by the Cetacea, that the hind limbs had begun really to shrink in size, as illustrated by *Ichthyosaurus.* In bodily form these reptiles seem to have been as highly modified for an active aquatic life as cetaceans and it is not likely that the former had need for retaining the pedes as balancers. Hence it appears necessary to believe either that the external hind limbs of aquatic reptiles were much slower to disappear, because employed for a much longer time for active swimming in a four-limbed manner, or else that the Cetacea were phenomenally amenable to influences which resulted in elimination of the pedes, for which there is not the slightest evidence.

To complete this chapter it is only necessary to offer a brief summary. If a mammal be well modified for an aquatic life but without present evidence that its hind limbs have ever been used to an important extent for either swimming or steering, these may remain relatively unaltered in most respects, or become peculiarly specialized in an unpredictable manner (as in the gluteus maximus of *Potomogale*). If these members are used as the chief method of swimming or steering and are highly altered accordingly, one may expect with a considerable degree of confidence that there will be a shortening of the femur, frequently accompanied by shortening of the ilium. If the latter does not occur the lesser gluteal mass should be very robust, and if it does, the gluteus maximus may be very powerful. The crotch will tend to migrate toward the rear and there will very often be a broadening of the superficial division of the biceps femoris, usually to cover most of the shank. It may, or more often will not, be correspondingly broad at origin. There will frequently occur a great increase in the power of the semitendinosus, possibly accompanied by a similar development of the gracilis, and semimembranosus, the insertions of all three extending farther distad toward the heel.

If the animal be large and the feet need not be used for extensive terrestrial progression the pes will usually be webbed, and symmetrical providing it be used in a symmetrical fashion, in which case it may be

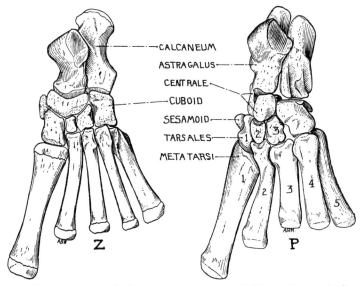

FIGURE 53. Dorsal view of left tarsus and metatarsus of (Z) sea-lion (*Zalophus*) and (P) seal (*Phoca*).

expected that the first and fifth toes be of equal length and longer than the middle three. If the animal be of small size the pes may be either webbed or fringed, is less apt to be symmetrical, and the tendency will most frequently be for the gradual enlargement of the member. No matter in which active manner the Cetacea once employed the hind limbs for swimming the present result could have been attained providing they were equipped with a relatively long and rather heavy tail and that swimming methods entailed curvature of the backbone (and tail) in the vertical plane. It is not probable that an actively predaceous mammal of the sort that it is likely the whale has always been would evolve a tail expanded in the horizontal plane for the purpose of keeping it near the bottom, as seems to be the case in the platypus, but rather that it developed along the same lines that the tail of the otter is now following. The speed of development of such a tail should be materially assisted by the presence of short, and relatively weak, rather than long and powerful, pedes, and such small feet would, presumably, disappear more rapidly than would those which had become large and highly specialized. Hence there seems to be considerable evidence that the Cetacea could most readily and quickly have evolved from some sort of terrestrial ancestor having many of the chief bodily characteristics now occurring in the common river otter.

Chapter Twelve

Other Soft Parts and Physiological Features

THE WRITER has not concerned himself, save very incidentally, with any part of the physiology of aquatic mammals other than of the muscular system, nor with the internal organs. Nevertheless, the present contribution would be very incomplete without including some consideration of these features, because some of the most interesting problems concerning the Cetacea are involved with them. Accordingly it is aimed to offer brief mention of what seem to be the most important of these questions, and to point out some of the possibilities, without any attempt at an exhaustive consideration. These points have frequently been discussed with Doctors E. K. Marshall and G. B. Wislocki of the Johns Hopkins Medical School, who are working on some of them, and to whom I am accordingly obligated.

There is a veritable host of interesting problems concerned with the physiology of the more highly modified aquatic mammals, especially the Cetacea, which are particularly difficult to solve largely for the reason that it is not easy to experiment on the live animal, and because they involve alterations in quality from those encountered in ourselves, so that a very real obstacle is our inability to compare conditions with what we know obtains in the case of terrestrial mammals. Perhaps the most important, or at least spectacular, of these questions is concerned with the ability of some whales to descend to depths in excess of a mile, and to remain thereabouts for more than an hour. How do they withstand the pressure? How do they take down sufficient oxygen? More important still, how do they get rid of the carbon dioxide in the blood? If we hold our breath for a minute we are in distress, not because we need more oxygen at once, but because there is imperative need to dispose of the accumulation of carbon dioxide.

Another point, of particular interest to students of the human brain, is that although the porpoise apparently has less need for intelligence than almost any other living mammal, and little if any more than had the ichthyosaurs, the convolutions of its brain are more marked than in man, thus indicating the probability that this character is not as significant of intelligence as many now believe. Other details of the cetacean brain are its large cerebellum, and the fact that it is larger in transverse than in sagittal measurement, the latter feature undoubtedly being a reflection of the peculiar skull development.

FATTY TISSUES

It has already been remarked that aquatic mammals are particularly prone to develop fatty tissue of one sort or another. An extensive blubber layer irrefutably fulfills a physiological need in aiding the retention of body heat by those sorts which inhabit cold waters. This does not apply with equal force to tropical and subtropical sirenians and cetaceans. Apparently they could make other provision for withstanding the very moderate temperatures to which they are subjected, and yet in spite of their abundant fat the former are particularly susceptible to any slightly unusual chilling of their aquatic habitat. It thus seems not unlikely that the blubber layer may have other uses than purely that of insulation.

Seals, at least, often have a great deal of intestinal fat, and the Odontoceti usually have accessory fatty equipment in the form of a frontal adipose cushion of a specialized quality of fat, and other deposits about the angle of the jaw. Of greater significance still is the enormous and unwieldly spermaceti organ of the cachalot, and what I have previously referred to as a sort of circulatory system for oil. This, apparently, is present to a very slight degree in the porpoise *Neomeris,* but has not been reported in other cetaceans. In a fetal narwhal *(Monodon),* however, which I dissected, the subcutaneous tissue, and much of the intermuscular tissue, was given the appearance of a rubber sponge by an amazingly intricate and extensive system of oil ducts, intercommunicating with each other and with larger subcutaneous oil reservoirs. Even some of the larger muscles had small oil sinuses throughout their substance. About the neck at least these oil ducts were very numerous, of very small diameter, and apparently had intimate relation with the retia of the vascular system. How extensive the flow of oil through these ducts from one part of the body to another may be in the live animal is of course unknown, but while dissecting the back, if one pressed upon the dorsal region oil welled up and out in copious amount. In life muscular pressure could attain the same result, and there would accordingly be an irregular, and possibly somewhat limited, flow of oil from one area to an adjoining one.

This specialization of the narwhal cannot be fortuitous any more than can the spermaceti organ of the cachalot. Any such unique and strongly marked character must have been developed to fulfill some definite physiological need. What this function can be is entirely unknown, but it it only logical to presume that as specialized fat órgans such as are being discussed occur only in the Cetacea they have something to do with the ability of this order either to hold the breath for a phenomenal length of time, or to assist in withstanding great pressures—an explanation which seems unlikely.

In what way could this oil system of the narwhal assist the animal to hold its breath? The only possibility that occurs to me is that there might be sufficiently close connection between capillaries of the oil system and vascular capillaries for there to be possible interchange of carbon dioxide by diffusion. The process then would be for the oil to take over a part of the carbon dioxide as this accumulated in the blood during prolonged submergence. When at the surface once more some eight or ten breaths might then enable the oil to give up its excess of carbon dioxide by the same process reversed. And it is well known that after a lengthy submergence a whale must spend an approximately equal period of time at the surface, frequently inflating the lungs, before it can again repeat its dive. The above is a possibility, I say, but it should be emphasized that it is not *probable* and it is offered merely as the only explanation that presents itself.

If this detail of the narwhal has been developed to facilitate long submergence one would naturally suppose that this animal can remain below the surface for a longer period than can the majority of porpoises, but it is not known whether this is actually the case. It appears only reasonable, however, that cetaceans of the order of the narwhal and beluga which occur where they must often be hemmed in by extensive ice fields should have unusual ability for lengthy submergence, and it will be of interest to ascertain whether the latter has also developed this oil system.

The sperm whale or cachalot is believed to have the ability to submerge for a longer period than any other cetacean, and it is the only one which has developed a spermaceti organ of phenomenal size. I am unable to believe that these two circumstances have no connection, but there is no clue to the manner in which they may be dependent. The fact that the fatty equipment of the narwhal and cachalot are so very different may be without much significance, for they were undoubtedly independently developed and could fulfill the same need. The difficulty lies in the statement that the spermaceti organ is a "closed system," and no mention is made of any marked vascular networks.

Through the efforts of H. C. Raven and H. I. Wordell I secured two samples of sperm oil and E. K. Marshall kindly determined for me their power for absorption of carbon dioxide. One hundred cubic centimeters of each dissolved respectively 92.8 and 94.6 cc. of carbon dioxide. This was left standing for 72 hours and the samples then refused to absorb more carbon dioxide, indicating that none of the original gas taken up had suffered change. One hundred cc. of water, alcohol, and petroleum will absorb respectively 54.6, 230, and 82 cc. of carbon dioxide; so it is seen that these experiments do not show that sperm oil is remarkably solvent of this gas. It should be pointed out, however, that the question is by no means settled. The samples used were of old sperm oil which had been altered by the action of the free fatty acids which it contains in considerable amount. The experiment should be repeated with *fresh* material, and under pressures increasing to one ton to the square inch.

In abandoning this topic it should again be emphasized that we know nothing whatever about any physiological process, other than for the conservation of heat, which the fatty tissues of aquatic mammals may have. All that is now possible is to advance tentative and unsupported hypothesis as a basis from which to work.

DIGESTIVE SYSTEM

The digestive system of pinnipeds and cetaceans at least is called upon to furnish an unusual quantity of raw material for the production of both blood and fat; but it is difficult to know which features constitute phylogenetic inheritances from terrestrial ancestors and which may have been developed by an aquatic life. The stomach of the Cetacea, both Odontoceti and Mysticeti, is very complicated and with numerous divisions. It is usually stated that most forms have four of these with a fifth dilation of the duodenum, but ziphioid whales may have "9, 10, even 13 or 14" divisions, according to Beddard, quoting other authors, and in the latter group the structure suggests that the usual first division may be missing. In this character of complexity the cetacean stomach is comparable to that of ruminants, but there are fundamental differences and their resemblances are probably attributable to convergence rather than to relationship. Apparently conditions may be interpreted as follows.

The typical porpoise stomach consists of a relatively simple division which appears to be a dilation of the esophagus. It communicates by a passage in its upper wall with the second division, whose walls are heavily plicated and of great thickness. It is here that the gastric juices are secreted, and apparently these are regurgitated into the first division. Indigestible matter, like cuttlefish beaks, is probably vomited forth, from time to time, but the gastric juices are doubtless capable of dissolving fish bones. At any rate the stomach contents must be thoroughly liquefied before leaving the second division for its exit is minute and of little greater diameter than that of a stout probe. The third and subsequent division appears to be nothing but a specialization of the pylorus, contorted and with several constrictions.

The cetacean intestinal tract is not usually remarkable and is of moderate length. Owen (1868) quoted Hunter to the effect that in a *Balaenoptera acutirostrata* 17 feet long the total intestinal length was 93-3/4 feet. But on the other hand the same author stated that in *Hyperoodon* the whole intestinal canal was sacculated in a surprizing manner.

The stomach of the dugong is remarkable for being very thick and muscular, and for having two caecal appendages or diverticula. Owen gave the intestinal length of a half-grown individual as 27 feet. The caecum of this animal is single, but it is bifid in the manati.

The stomach of the Pinnipedia is not particularly noteworthy but the intestinal tract is of phenomenal length. Engle (1926) has recorded a sea-lion *(Eumetopias)* with intestinal length of 264 feet, which is over twice as long as has been reported for a fur seal (107 feet). Stones of considerable size and of an aggregate weight of several pounds have been reported as found in the stomach of sea-lions. It has been suggested that this is for the purpose of weighting down the animal, but I would regard this added weight as insignificant to a body weighing several hundred pounds. More attention should be paid to the time of year at which stones are or are not present, and if both sexes are so equipped. If they should prove to occur only in males during the mating season then it might be inferred that they are for the purpose of preventing undue atrophy of the stomach, through functioning as a sort of gastric "chewing gum," during the many weeks that this sex is without food while guarding the harem.

URO-GENITAL SYSTEM

Potentially there is much to be done on the urinary system of the marine mammals. A fundamental question is how do these creatures get the water which they must drink. If they obtain it by drinking sea water, then how do they eliminate the excess of saline matter, which they would be obliged to accomplish by some specialized means? Actually, however, there are practically no details of this system so far known that are surely attributable to an aquatic life. It is usually stated that the cetacean kidney is lobulated, but this is a misleading term. That of most if not all large terrestrial mammals is lobulated, but in the Cetacea it takes a different form, each kidney appearing to be an aggregation of a multitude of small kidneys closely packed and contained within a single envelope. The significance of this condition is unknown.

Although of doubtful significance aquatically, there are many interesting features of the gonads of highly adapted aquatic mammals, some of which may here be accorded brief mention. Engle (1926) has reported on some of the equipment of the males as follows:

	Seminal vesicles	Prostate	Bulbo-urethral gland
Odobenus	none	present	none
Eumetopias	none	present	none
Phoca	none	present	none
Halicore	1 pair	one	none
Phocaena	none	paired	none
Dephinapterus ..	none	present	none
Sibbaldus	none	present	none

As previously mentioned there is a tendency coupled with the aquatic life to eliminate the scrotum together with other bodily protuberances. Accordingly, among Pinnipedia, Sirenia, and Cetacea, the Otariidae and Odobenidae are the only ones which retain this feature. In the Cetacea the testes are not truly abdominal, but are situated in "a pouch near the inguinal ring." As Meek (1918) has stated, in the porpoise the penis is greatly modified, the copulatory part being differentiated from the rest and separated by a joint which allows a wide range of movement. This jointed character is not present, or at least apparent, in Mysticeti. The cavernous body of the penis proper is single in the Cetacea and double in Sirenia. In both sexes of the Cetacea the uro-genital and anal orifices are situated within a common sulcus bordered by a pair of labia.

RESPIRATORY SYSTEM

The respiratory system and its appurtenances is perhaps the most critical factor in the cetacean equipment, for every activity of an aquatic mammal is governed by the necessity of renewing as often as necessary the supply of air. This need is often antagonistic to many of its other activities, as it very seldom is in the case of a terrestrial mammal.

During his investigations of the lungs of the porpoise *Tursiops*, G.

[315]

B. Wislocki (1929) found that the cartilaginous armature is such as to give unusual strength and incompressibility. In the smaller bronchioles possessing a diameter of less than 0.5 mm. there are muscular sphincter valves. Moreover in the respiratory bronchioles there is a complete lining of flattened respiratory epithelium instead of a partial one as in other known mammals. There is a tremendous development of elastic tissue, and double capillary beds of interalveolar septa instead of single ones, as in terrestrial mammals. The respiratory bronchioles do not possess sacculations and alveolar ducts are lacking.

The above sphincters probably close at the end of inspiration and ordinarily remain so until expiration begins, preventing the gradual collapse of the air spaces as outside pressure increases, thus acting in antagonism to the elastic tissue of the lungs. The individual sphincters are weak, but there are probably several million of them, and each imprisons but a minute amount of air, so that all of them working together are doubtless capable of preventing the escape of air even into the trachea under any pressure to which the animal cares to subject itself. Thus external and internal pressure can be equalized, as it must be in any animal which experiences an external pressure of a ton to the square inch. No thorax could *resist* this stress without collapsing: and this is probably the chief physiological adjustment necessary for deep diving. Others are necessary, to prevent such things as bleeding at the eyes, and likely some alteration in the action of the heart, but it seems that these should be of a more minor character. The epitheleal equipment presents an increased surface area subserving the function of respiration, and the elastic tissue forms a powerful mechanism for emptying the lungs in a minimum of time.

Pick (1907) has stated that the lungs of the dugong agree with those of the Cetacea in having a cartilaginous armature (which, however, is partially calcified) extending to the smallest bronchioles, a tremendous quantity of elastic tissue, and in the size of the air-sacs and thickness of the septa; but the distribution of muscular tissue differs.

Schulze (1906) made an estimate of the number of air cells and their respiratory surface in the sloth, cat, man and porpoise *(Phoceana communis,* which reaches a length of about five and a half feet), as follows:

	Air cells	Respiratory surface
Sloth	6,250,000	5 square meters
Cat	400,000,000	20 square meters
Man	150,000,000	30 square meters
Porpoise	437,000,000	43 square meters

He considered that the lung capacity of the porpoise was about the same as in man, but it is likely that in a cetacean of strictly comparable size it is somewhat greater. At least all investigators are apparently in agreement that aquatic mammals have a greater lung capacity than terrestrial ones of the same size.

The above facts indicate that not only can the porpoise retain air in the lungs entirely independent of the muscles of the thorax or diaphragm, but the oxygen can be more completely utilized by the lungs themselves. Not only this, but the blood of cetaceans and pinnipeds at least is richer in hemoglobin than usual, and hence can store more oxygen. In addition the intrinsic musculature of the lungs and the character of the diaphragm enable them to empty the lungs more completely, and the pressure at considerable depths also facilitates the absorption of all oxygen.

The physiology of normal respiration is too involved to present here and may be obtained from any good text book. Certain points must be considered briefly, however. The respiratory apparatus of man is not particularly efficient. Thus, about 20 per cent of the air which we breathe is composed of oxygen, but in normal, unlabored breathing we utilize only about a quarter of this amount. If nitrogen be added so that the oxygen be reduced to 10 per cent the respiratory rate will be slightly raised, and pulse accelerated, but there is no definite discomfort felt. When the oxygen content is gradually reduced to 6 or 7 per cent, however, consciousness is lost, often without much of a premonitory symptom. Thus, in spite of rapid respiration, we are incapable of making much use of one-third of the oxygen that is furnished us. It is likely that whales can use almost all of this.

The total lung capacity of the average man is supposed to be about 4,700 cc. During quiet respiration only about 500 cc. is taken in, and in spite of our best efforts we are unable to empty the lungs of their final 1,000 cc. It is likely that the whale almost completely fills the lungs at each inspiration, and disposes of practically all the residual air at each expiration.

In man the expired air usually contains about 4 per cent of carbon dioxide. When the CO_2 content of the inspiration is raised to even 1 or 2 per cent increased breathing results to a marked degree, and at a content of 10 per cent there is great distress experienced and the face becomes congested, while the respiration is multiplied quantitatively several fold. It has been said that the breath cannot be held voluntarily after the carbon dioxide content of the lungs has passed 7.5 per cent at which time the oxygen in the lungs has been reduced to a point of from 9 to 11 per cent. Whether whales differ in their tolerance of free carbon dioxide in the lungs is as yet unknown, but it is assumed that they must be much more tolerant, which would be accomplished at least partially by alteration in the sensitivity in this respect of the respiration center of the brain.

But there is another aspect from which this gas must be considered. In dogs the venous blood has a CO_2 content of about 45 cc. per 100 cc. of blood. Not more than from 2 to 2.5 cc. of this can be held in physical solution, while the remainder must be in chemical combination. Carbon dioxide can enter the corpuscles and react with alkalis combined with the hemoglobin to form a bicarbonate which in the lungs breaks down again to liberate CO_2. Not only should the efficiency of this process be quantitatively facilitated in marine mammals, with their reported increased hemoglobin content, but there may be qualitative augmentation as well. Pressure may be of critical import in this connection, as it might well be if it should prove that the oil of whales can actually take up some of the carbon dioxide from the blood. I regard it as certain that any such physiological processes cannot operate with an equal degree of efficiency at surface-water pressure and at the pressure of a ton to the square inch. For this reason the possibility does not appear fantastic that the cachalot may not be able to hold its breath for an inordinate length of time—say fifteen minutes—when at the surface, while the increased pressure at the depth of one mile might so alter its physiological processes that it would have no difficulty in remaining below for an hour. There is no evidence whatever in support of such a theory, but it must be taken into consideration in planning future experiments.

It is not always easy to determine the length of time during which a mammal may hold its breath. Thus I have taken every opportunity to observe seals and sea-lions but have never yet seen one submerge for much more than a couple of minutes. And yet it seems certain that at least the boreal seals, which spend much of their time beneath the ice, must very greatly exceed this, say to the extent of at least ten and very possibly twenty minutes.

A man may hold his breath without undue discomfort for one minute—I have just done so. The reader may exceed this, and if he first violently and completely empty his lungs several times he may be able to last as long as two and a half minutes. But this is a very moderate accomplishment. Beebe (1926) has stated that a sloth breathes about once per second, and yet he has seen one recover from 30 minutes of complete submergence, while he had heard of one doing so after 45 minutes. Burrell (1927) found that the adult platypus can remain under water for the duration of 5 to 10 minutes, while the new born young can survive a submergence of three and a half hours. Parker (1922) reported that a large manati at rest was in the habit of staying down for a period varying from 7 minutes to 16 minutes and 20 seconds, then arising to the surface for as much as three minutes and taking several breaths. A younger animal arose oftener, and fishermen told him that when these beasts are being hunted they may stay below for half an hour.

An experienced whaler whom I consider trustworthy has assured me that when it "sounds" a humpback will ordinarily stay down from 8 to 12 minutes, with observed maximum of 23 minutes; finback, 8 to 12, with 28 as a maximum; bowhead, 12 to 15, with 30 minutes as a maximum; and cachalot from 30 to 60 minutes with an observed maximum of one hour and 45 minutes. Andrews (1916) has quoted an instance where a blue whale sounded for 50 minutes, reappeared to spout 20 times and then disappeared for another 40 minutes. Certainly there is abundant evidence that the cachalot can easily submerge in excess of an hour, but records of over two hours should be viewed with suspicion. Certain other of the larger Odontoceti (as *Hyperoodon*) are notable for the speed with which they seek great depths when harpooned, and for the length of time that they can stay there. The Mysticeti are evidently less gifted in this respect, and it is doubtful whether this ability is as well developed in porpoises with littoral predilections.

That lengthy submergence is not fatal to the sloth and young platypus is doubtless chiefly attributable to their low rate of metabolism, and armadillos and ant-eaters should be expected to exhibit the same accomplishment. This has also been advanced not infrequently as the possible explanation of why whales are also able to suspend breathing for so long. There is no evidence to show that in any marine mammal the rate of metabolism is low, and several particulars indicate that it is high, among which are details of the lungs, blood and the temperature. In *Turstops* the latter is about 36 degrees Centigrade. So in the Cetacea I am inclined to believe for the present that what may be designated as the *normal* rate of metabolism is not low. But some fish are known to be able to contrive, by some obscure means, to lower their rate of metabolism very quickly when it is advantageous for them to do so. The same may almost be said of mammals which hibernate. This state, apparently, is almost entirely determined by temperature in the case of reptiles, but in mammals this is only partially so, and there seem to be other factors involved. With them metabolism almost ceases, and although the process is not well understood, it must be accomplished in a relatively simple manner, for some species habitually hibernate while other closely related species, in other climates, never do so. Furthermore, some rodents regularly aestivate during the hottest part of the year, when food is scarcest.

In view of the facts so far available it seems that in any study of the whale's ability to withstand lengthy submergence, account will have to be taken of the possibility that this order, or at least many of its representatives, may have some apparatus whereby its rate of metabolism is lowered temporarily in a more or less voluntary fashion. So little is yet known of this general subject that there is no way of predicting the probability of this being the case. It may actually be either quite high or zero; but it should be considered, nevertheless.

No mention has yet been made of the fact that during deep submergence by a human diver the pressure saturates the blood with nitrogen, and unless decompression be very gradual, nitrogen bubbles will form in the vascular system, afflicting him with what is known as the "bends" and often causing death. Whales often ascend with rapidity from great depths and at first thought it might seem that they would be obliged to have some provision for overcoming such a disagreeable situation. It must be remembered, however, that a human diver gets a new supply of nitrogen at every breath, while a whale has only the initial supply of this gas which he has taken down with him. Very likely this is not sufficient in amount for saturation of the blood to the point where the latter would give off bubbles when the animal again reaches the surface.

Very little is surely known about the depths to which marine mammals habitually descend. Almost nothing is known in this regard about pinnipeds. Reports are sufficiently frequent for us to believe that the cachalot often feeds near the bottom where the depth is in excess of a mile. How much deeper it can go is unknown. It is also believed that the other toothed whales of larger size have this ability to a pronounced degree. I have been told by a whaler of a finback which was harpooned and at once sounded vertically to the depth of 275 fathoms, where its neck was broken by impact with the bottom, and the carcass was then hauled straight up to the surface. Other reports exceed this to some degree. But it is probable that no mysticete can submerge as deeply as some favored odontocetes, and it does not seem probable that the majority of littoral porpoises are so phenomenally gifted in this respect.

Of particularly vital import in the function of respiration by marine mammals is the diaphragm, and it is only to be expected that there should occur some change in its details. In a pronograde mammal, as discussed by Jones (1913), the fixity of the fore limbs allows the muscles passing therefrom to the thorax to assist in breathing by raising the ribs, while in orthogrades this function has become obsolete. It should be still further eliminated in those marine mammals which seldom or never use the fore limbs for pressing against a hard surface. An increase in the muscular character of the diaphragm of aquatic mammals indicates that this is probably so.

Evidently in all aquatic mammals that are very highly modified for such a life the diaphragm is more sloping, or tending to assume a position more nearly parallel with the body axis, than in terrestrial mammals, and this is said to be so even in the otter. Certainly it is a character of all pinnipeds, cetaceans and sirenians, in the latter order reaching its greatest alteration. It is indicated that this character increases with ontogenetic development, for Beddard (1900) has stated that in an adult porpoise the ventral and dorsal extent of the thoracic cavity showed a proportion of 1 to 2.25, while in a young individual this was as 1 to 1.75. I cannot state this proportion in the sirenians but Murie's descriptions and figures show that in the manati the diaphragm extends from the much reduced sternum below, quite to the last thoracic vertebra. As this mammal has but two lumbar vertebrae this means that the diaphragm almost reaches the posterior end of the abdominal cavity and is as nearly parallel with the body axis as it could well be.

Müller (1898) believed that the Mysticeti breathe more with the thorax and less with the diaphragm, for the latter is less muscular in this group than in the Odontoceti, but cetacean conditions are difficult to interpret. In the Sirenia it seems that the rigidity of the costal articulations would largely inhibit much mobility of the thorax while breathing. The reduction of the sternum and the costal cartilages, however, does point to the probability that these features do facilitate mobility of the diaphragm, and this must be of extraordinary efficiency.

It seems that the pronounced slope of the diaphragm in marine mammals may have been assisted by an increase in the lung capacity, and an advantage thereby gained is that the levitation supplied by the inflated lungs is shifted by just so much toward the center of the body. In such a mammal as man the lungs tend to raise the anterior portion of the body, while there is a corresponding depression of the posterior end, in the water. This is not as yet entirely overcome in the pinnipeds, but in sirenians and cetaceans the natural position by flotation is horizontal, largely permitted by the alteration of the diaphragm and lungs.

VASCULAR SYSTEM

There are many points of interest connected with the vascular system of certain aquatic mammals, but it is as yet uncertain just what application these may have to life in the water. It is known that pinnipeds and cetaceans are abundantly supplied with blood. The accounts of sealers frequently make reference to this fact. Whether this character is more pronounced in the Pinnipedia is unknown, but it may well be so, for several authors, notably Murie, have shown that the seal, sea-lion and walrus are equipped with a capacious dilation of the vena cava in juxtaposition with the liver, absent in the Cetacea. I have verified this particular for the seal and sea-lion, and have found that in a young individual of the latter it was notably less marked, which seems to indicate that it is an ontological development. The capacity of this hepatic sinus is quite astounding and, as Murie has remarked, it would seem to occupy as much space when expanded as the liver itself. It cannot be doubted that it functions as a blood reservoir.

The cetacean heart is noteworthy in many respects as enumerated in the literature. Apparently it is larger than in a terrestrial mammal of equal mass, for G. B. Wislocki tells me that the heart of a *Tursiops*, weighing perhaps 600 pounds, is larger than that of an ox. Owen noted that the heart of a large whale may be more than a yard broad and not much less in length. The axis of the heart has also shifted, with apex more dorsal than in other mammals. The papillary muscles are usually said to be enormous and the organ of such conformation as to indicate great potential power.

Kükenthal (1922), quoting various sources, has remarked the strong enlargement of the spinal meningeal arteries within the vertebral canals of the Cetacea. He considered that by this provision the blood supply of the brain escapes the effect of great pressure when the animal dives deeply. This attitude is scarcely justified, however, for it is hardly possible that during long submergence the pressure of the blood could be much less in one part of the body than another. But *something* of importance seems indicated, for Kükenthal stated that in baleen whales the transverse foramina of the cervical vertebrae become reduced, accompanying elimination of the internal carotid and vertebral artery, in most toothed whales these foramina are still more reduced, and in *Physeter* and *Hyperoodon* they are entirely lacking, this detail thus apparently being dependent upon the possible depths to which whales descend. In Sirenia the transverse foramina are also rudimentary.

The most spectacular detail of the vascular system of aquatic mammals is the extent to which retia mirabilia, or networks of vascular anastomoses, occur. They are found extensively in the Cetacea, Sirenia (depicted with especial advantage in Murie's figures), in the Phocidae, and to a lesser degree possibly in the Otariidae and Odobenidae. But this is not in itself an aquatic adaptation, for extensive retia have also been reported in monotremes, some marsupials, some lemurs, ant-eaters, sloths, armadillos, some rodents, the Manidae, and others. Whatever the function may be we are justified in assuming that retia are not a secondary adaptation in aquatic mammals, but rather that such as exhibit this character have lacked the stimulus for fusion of the retia into larger blood vessels and that at least those of a diffuse pattern have been retained from the primitive ancestral condition, or rather that the embryonic condition is retained throughout life, as pointed out by von Baer (1835).

It is not easy to determine the extent of retia without specially injected material. They were not striking in the sea-lion which I dissected, while they were, over certain areas, in the seal. They may occur in the arterial or venous system or both. Retia may be gathered in single, sheathed bundles, which also convey lymphatic trunks, as in sloths and armadillos, or a diffuse pattern as in monotremes and sirenians. They may occur in different areas dependent upon the type of mammal considered. Many authors have noted the intracranial retia of the internal carotids at the base of the skull in ruminants, located at the sides and back of the sella turcica. These are said to be better developed in grazers than browsers, and least so in the giraffe. Some vascular clusters seem clearly to function as reservoirs for blood, as in the case of the psoadic plexus of Cetacea. This takes the form of numerous, transverse, separated blood vessels posterior to the kidneys.

Vrolik believed retia to be connected with aboreal habits, and Carlisle that they were correlated with slow movements, but the circumstance that they are present in some agile, nonarboreal mammals refutes these hypotheses. Hunter and Cuvier assigned to those of the arterial system the function of storing oxygenated blood, and Wilson, the storing of carbonized blood by the venous ones, Turner and Milne-Edwards of retarding the flow, and Murie was of the opinion that they facilitate interchange of substances with the lymphatic system. Hyrtl (1854) advanced the hypothesis that the diffuse pattern of retia is associated with animals doing heavy muscular work but of an agile character, as with the burrowing armadillo, while the cluster pattern is characteristic of mammals in which movement is slow and posture prolonged, such as the sloth. Wislocki (1928) has discussed this question at considerable length and the reader desirous of further information is referred to his paper. With the attention that the subject of retia is now attracting it is to be hoped that we may soon know more about their useful functions. At present, however, there is no strong evidence indicative of what these may be. There is no evidence whatever that a retial condition of the blood vessels is useful for the storage of reserve blood. In itself it would not assist in retarding the flow, nor is it likely that there is interchange of substances with the lymphatic system within the retial bundles, where these occur. In fact it seems that very little can yet be said except that possibly a diffuse type of retia might largely overcome any interruption of the blood flow that muscular or other pressure is capable of producing.

MAMMAE

Before terminating the present contribution brief consideration should be accorded the subject of the mammary equipment of aquatic mammals. The way in which these are used is usually of slight consequence to a terrestrial mammal, for except in particular cases as in some ungulates, the young may nurse from any position. It is of critical import, however, that an aquatic mammal shall be able before it forsakes the land entirely to contrive a method whereby the young may suckle with reasonable comfort while in the water. If this cannot be accomplished the creature must either retain its connection with the land, with consequent implication of reduced ability to develop the highest aquatic modifications, or else become extinct.

Probably in no mammal has the aquatic life caused any definite alteration in the position of the nipples, a possible exception being in the Sirenia. In Cetacea and Pinnipedia there is one pair of mammae situated inguinally. In Sirenia there is a pair of axillary ones, but they now occur practically upon the posterior border of the flippers. The coypu (Myocastor) has two pairs that are situated almost upon the back, one pair being just behind the shoulders and the other near the haunches. They are largely similar in the capybara but rather less elevated, as is also the case in some terrestrial species with octodont affinity. The coypu and capybara are said frequently to swim about with the young perched on the back, and apparently the latter are able to nurse from this position.

Sirenians are reported to nurse while the parent maintains an upright posture with the young clasped by the flippers, although I do not see how the latter could reach the nipple from this position.

There are numerous reports on the mammary equipment of the Cetacea. The gland itself is apparently of the usual histological character but there are numerous and rather large galactophorous sinuses which open into a lacteal duct or reservoir of generous proportions. Engle (1927) found that in a female humpback 44 feet long the mammary glands, in full function, had a length of about 6 feet and the greatest diameter of 18 inches. The lacteal capacity of a large whale must surely be astonishing and a great many gallons. The nipple is retracted within a slit-like orifice on either side of the genito-excretory labia.

There is a great deal of uncertainty regarding the way in which the cetacean mammillae are employed. The great storage capacity should indicate that the milk is removed at a rapid rate. The presence superficially of a part (at times disconnected from the main sheet) of the panniculus carnosus muscle makes it not unlikely that this may assist in the voluntary and forcible ejection of the fluid. And the situation of the nipples introduces a further element of uncertainty. I have been told of a female finback in full milk whose nipples protruded to the extent of one foot when she was inflated with air, and there are a very few published accounts of protruding nipples. The condition is so seldom encountered, however, that I regard it as doubtful whether it ever occurs after death save in the event that an animal in full milk has been inflated to usual degree by introduced air or by gases of decomposition, and possibly the sudden death of the calf and consequent engorgement of the mammary glands might have the same effect. On the other hand the securing under water of milk by the young would be so difficult without temporary protrusion of the nipple that I regard it as highly probable that this takes place, and this could easily be accomplished when necessary by contraction of the mammillary smooth muscle.

There has been much speculation regarding the cetacean position for nursing. Obviously if the dam rolls over on her side so that the blow-hole of the nursing young is above the surface the female's blow-hole is submerged. If hers is elevated then the young is completely below the surface; but Scammon (1874) illustrated the act in this position. Perhaps either posture is employed.

Literature Cited

ABEL, O. 1906. Die Milchmolaren der Sirenen. Neues Jahrb. f. Miner., Geol. u. Paläeont., vol. 2.

―――――. 1908. Die Morphologie der Hüftbeinrudimente der Cetaceen. Denkschr. k. Akad. Wiss. math.-nat. Kl., Wien, vol. 81 (1907), pp. 139-195.

―――――. 1924. Die Eroberungszüge der Wirbeltiere in die Meere der Vorzeit. G. Fischer, Jena, pp. 1-121.

ALLEN, H. 1888. Materials for a memoir on animal locomotion. Extracted from Report on Muybridge work at the Univ. of Pa., pp. 35-104.

ALLEN, J. A. 1924. Carnivora collected by the American Museum Congo Expedition. Bull. Amer. Mus. Nat. His., vol. 47, pp. 73-281.

ANDREWS, R. C. 1909. Observations on the habits of the finback and humpback whales of the eastern North Pacific. Bull. Amer. Mus. Nat. Hist., vol. 26, pp. 213-226.

―――――. 1916. Whale hunting with gun and camera. D. Appleton & Co., N.Y., pp. 1-333.

ANTHONY, R. 1903. Introduction a l'etude experimentale de la morphogenie. Bull. Soc. Anthr. Paris, ser. 5, vol. 4, no. 2, pp. 119-145.

―――――. 1926. Les affinites des Cetaces. Ann. Inst. Oceanog., vol. 3, pp. 93-134.

BAER, K. E. VON. 1835. Ueber das Gefässystem des Braunfisches. Nova. Acta Acad. Caes. Leop. Carol. naturae curiosorum, vol. 17, p. 395.

BAILEY, VERNON. 1923. The combing claws of the beaver. Journ. Mamm., vol. 4, pp. 77-79.

BEDDARD, F. E. 1900. A book of whales. John Murray, London, pp. 1-320.

BEEBE, W. 1926. The three-toed sloth. Zoologica, vol. 7, no. 1, pp. 1-67.

BENHAM, W. B. 1901. On the larynx of certain whales. Proc. Zool Soc. London, pp. 278-300.

BREDER, C. M., JR. 1926. The locomotion of fishes. Zoologica, vol. 4, no. 5, pp. 159-297.

BURRELL, HARRY. 1927. The platypus. Australia, pp. 1-227.

CABRERA, A. 1925. Genera mammalium. Insectivora., Mus. Nac. Cien. Nat., Madrid, pp. 1-232.

CHAPMAN, H. C. 1875. Observations on the structure of the manatee. Proc. Acad. Nat. Sci., Phila., pp. 452-462.

DELAGE, Y. 1885. Histoire du Balaenoptera musculus echone sur la plage de Langrune. Arch. Zool. Expér. Mem., vol. 1, pp. 1-152.

DENKER, A. 1902. Zur Anatomie des Gehörorgans der Cetacea. Anat. Hefte, vol. 19, pp. 423-445.

DOBSON, G. E. 1882. A monograph of the Insectivora, systematic and anatomical. London, pp. 1-172.

LITERATURE

DUBOIS, E. 1924. On the brain quantity of specialized genera of mammals. Koninkl. Akad. v. Wetensch. Amsterdam, vol. 27, nos. 5 and 6, pp. 1-8.

DUVERNOY, G. L. 1822. Recherches anatomique sur les organes du movement du phoque commun, Phoca vitulina, L., Mem. Mus. Hist. Nat. Paris, vol. 9, pp. 49-70, 165-189.

ENGLE, E. T. 1926. The intestinal length in Steller's sea-lion. Journ. Mamm., vol. 7, pp. 28-30.

————. 1926. The copulation plug and the accessory genital glands of mammals. Journ. Mamm., vol. 7, pp. 119-126.

————. 1927. Notes on sexual cycle of the Pacific Cetacea of the genera Megaptera and Balaenoptera. Journ. Mamm., vol. 8, pp. 48-54.

FLOWER, W. H. 1869. On the value of the characters of the base of the cranium in the classification of the order Carnivora, and on the systematic position of Bassaris and other disputed forms. Proc. Zool. Soc. London, pp. 4-37.

————. 1876. An introduction to the osteology of the Mammalia. Macmillan & Co., London, pp. 1-344.

———— and R. LYDEKKER. 1891. An introduction to the study of mammals, living and extinct., pp. 1-763.

GARROD, A. H. 1877. Notes on the manatee (Manatus americanus) recently living in the Society's garden. Trans. Zool. Soc. London, vol. 10, pp. 137-145.

GIDLEY, J. W. 1913. A recently mounted zeuglodon skeleton in the United States National Museum. Proc. U. S. Nat. Mus., vol. 44, pp. 649-654.

GRAY, A. A. 1905. Anatomical notes upon the membranous labyrinth of man and of the seal. Journ. Anat. Physiol., vol. 39, pp. 349-361.

GULDBERG, G. and F. NANSEN. 1894. On the structure and development of the whale. Bergens Mus., vol. 5, pt. 1, pp. 1-70.

HAY, O. P. 1928. Pleistocene man in Europe and in America. New York Herald-Tribune, July 1, pp. 1-8 (reprinted Wash., D.C., July 30, 1928, with slight changes).

HINTON, M. A. C. 1925. Report on papers left by the late Major G. E. H. Barrett-Hamilton relating to the whales of South Georgia. Crown Agents for the Colonies, London, pp. 1-209.

HOLDER, J. B. 1883. The Atlantic Right Whales: a contribution, etc. Bull. Amer. Mus. Nat. Hist., vol. 1, no. 4, pp. 99-137.

HOME, E. 1796. A description of the anatomy of the sea-otter. Phil. Trans. Roy. Soc. London, pp. 385-394.

HOWELL, A. BRAZIER. 1924. The mammals of Mammoth, Mono County, California. Journ. Mamm., vol. 5, pp. 25-36.

————. 1926. Anatomy of the wood rat. Monog. no. 1, Amer. Soc. Mamm., pp. 1-225.

————. 1927. Contribution to the anatomy of the Chinese finless porpoise Neomeris phocaenoides. Proc. U. S. Nat. Mus., vol. 70, art. 13, pp. 1-43.

—————. 1929. Contribution to the comparative anatomy of the eared and earless seals (genera Zalophus and Phoca). Proc. U. S. Nat. Mus., vol. 73, art. 15, Jan. 26 (1928), pp. 1-142.

HOWELL, W. H. 1921. A text-book of physiology. Saunders Co., ed. 8, pp. 1-1053.

HOWES, G. B. and A. M. DAVIES. 1888. Observations upon the morphology and genesis of supernumerary phalanges, with special reference to those of the Amphibia. Proc. Zool. Soc. London, pp. 495-511.

HYRTL, J. 1853. Das arterielle Gefässystem der Edentaten. Denkschr. d. k. Akad. d. Wissensch. Wien, vol. 6, p. 21.

JACKSON, H. H. T. 1928. A taxonomic review of the American longtailed shrews (genera Sorex and Microsorex). North Amer. Fauna, No. 51, pp. 1-238.

JENNINGS, H. S. 1924. Heredity and environment. Scientific Monthly, vol. 19, pp. 225-238.

JONES, F. WOOD. 1913. The functional history of the coelom and the diaphragm. Journ. Anat. (London), vol. 47, pp. 283-318.

—————. 1923. Mammals of South Australia. Adelaide, pp. 1-458.

KELLOGG, R. 1925. Structure of the flipper of a Pliocene pinniped from San Diego County, California. No. 5. Additions to the Tertiary history of the pelagic mammals of the Pacific Coast of North America. Contr. Palaeont., Carnegie Inst. Wash., Publ. 348, pp. 97-116.

—————. 1928. The history of whales—their adaptation to life in the water. Quart. Rev. Biol., vol. 3, pp. 29-76, 174-208.

KERNAN, JOHN D., JR., and H. VON W. SCHULTE. 1918. Memoranda upon the anatomy of the respiratory tract, foregut, and thoracic viscera of a foetal Kogia breviceps. Bull. Amer. Mus. Nat. Hist., vol. 38, pp. 231-267.

KNAUFF, F. 1905. Ueber die Anatomie der Beckenregion beim Braunfisch (Phocaena communis Less). Jenaischen Zeits. f. Naturwis., pp. 253-318.

KÜKENTHAL, W. 1889. Die Hand der Cetaceen. Denkschr. d. med. naturwiss. Gesell., Jena, vol. 3, pt. 2, pp. 23-69.

—————. 1891. Ueber die Anpassung von Saugethieren an das Leben im Wasser. Zool. Jahrb., Abtheilung System. Georg. Biol. Thiere, pp. 373-399.

—————. 1891. On the adaptation of mammals to aquatic life. Ann. Mag. Nat. Hist., ser. 6, vol. 7, pp. 153-179.

—————. 1893. Vergleichend-anatomische und entwicklungsgeschichtliche Untersuchungen an Walthieren. Denkschr. med.-naturwiss. Gesell., Jena, vol. 3, pt. 2, pp. 224-338.

LEBOUCQ, H. 1887. La nageoire pectorale des cetaces au point de vue phylogenique. Anat. Anz., vol. 2, pp. 202-208.

—————. 1889. Recherches sur la morphologie de la main chez les mammiferes marins. Archiv. Biol., vol. 9, pp. 571-648.

LE DANOIS, E. 1910. Recherches sur l'anatomie de la tete de Kogia breviceps (Blainv.). Arch. Zool. Exp. Gen., vol. 6, p. 149.

LILLIE, D. G. 1910. Observations on the anatomy and general biology of some members of the larger Cetacea. Proc. Zool. Soc., London, pp. 769-792.

MALM, A. W. 1871. Hvaldjur i sveriges museer. Kongl. Svenska Vetenskaps-Akad. Handl., vol. 9, pp. 1-104.

MEEK, A. 1918. The reproductive organs of the Cetacea. Journ. Anat., vol. 52, pp. 186-210.

MILLER, G. S., JR. 1923. The telescoping of the cetacean skull. Smiths. Misc. Colls., vol. 76, no. 5, pp. 1-62.

——————. 1929. The gums of the porpoise Phocoenoides dalli (True). Proc. U. S. Nat. Mus., vol. 74, Art. 26, pp. 1-4.

MIVART, ST. G. 1885. Notes on the Pinnipedia. Proc. Zool. Soc. London, pp. 484-501.

MÜLLER, O. 1898. Untersuchungen uber die Veränderungen welche die Respirationsorgane der Säugethiere durch die Anpassung an das Leben im Wasser erlitten haben. Jena. Zeitschr. f. Naturwis., vol. 25, pp. 95-230.

MURIE, J. 1870. Researches upon the anatomy of the Pinnipedia. Part 1. On the walrus (Trichechus rosmarus Linn.). Trans. Zool. Soc. London, vol. 7, pp. 411-464.

——————. 1870. Note on the anatomy of the walrus. Proc. Zool. Soc. London, pp. 544-545.

——————. 1872. Researches upon the anatomy of the Pinnipedia. Part 2. Descriptive anatomy of the sea-lion (Otaria jubata). Trans. Zool. Soc. London, vol. 7, pp. 527-596.

——————. 1872. On the form and structure of the manatee (Manatus americanus). Trans. Zool. Soc. London, vol. 8, pp. 127-202.

——————. 1873. On the organization of the caaing whale, Globiocephalus melas. Trans. Zool. Soc. London, vol. 8, pp. 235-301.

——————. 1874. Researches upon the anatomy of the Pinnipedia. Part 3. Descriptive anatomy of the sea-lion (Otaria jubata). Trans. Zool. Soc. London, vol. 8, pp. 501-582.

——————. 1885. Further observations on the manatee. Trans. Zool. Soc. London, vol. 11, pp. 19-48.

NICHOLS, J. T. 1915. On one or two common structural adaptations in fishes. Copeia, vol. 20, pp. 19-21.

OSBURN, R. C. 1906. Adaptive modifications of the limb skeleton in aquatic reptiles and mammals. Ann. Acad. Sci. N.Y., vol. 16, pp. 447-479.

——————. 1903. Aquatic adaptations. No. 1. Adaptations to aquatic, arboreal, fossorial and cursorial habits in mammals. Amer. Nat., vol. 37, pp. 651-665.

OWEN, R. 1866, 1868. On the anatomy of the vertebrates. vols. 2 and 3. Longmans Green & Co., London, pp. 1-592, 1-915.

PARKER, G. H. 1922. The breathing of the Florida manatee (Trichechus latirostris). Journ. Mamm., vol. 3, pp. 127-135.

PARSONS, F. G. 1894. On the morphology of the tendo-achilis. Journ. Anat. Physiol., vol. 28, pp. 414-418.

PICK, F. K. 1907. Zur feineren Anatomie der Lunge von Halicore dugong. Arch. f. Naturgesch. 73 Jahrb., vol. 1, p. 245.

POCOCK, R. I. 1921. On the external characters and classification of the Lutrinæ (Otters). Proc. Zool. Soc. London, pp. 535-546.

——————. 1921. On the external characters and classifications of the Mustelidae. Proc. Zool. Soc. London, pp. 803-837.

POUCHET, G. and H. BEAUREGARD. 1885. Note sur l'organe des spermaceti. Compt. Rendu Soc. Biol. Paris, vol. 37.

PUTTER, O. 1902. Die Augen der Wassersäugetiere. Zool. Jahrb., Abth., f. Anat. Jena, vol. 17, pp. 99-402.

ROUX, W. 1883. Beiträge zur Morphologie der functionellen Anpassung. I. Structur eines hochdifferenzirten bindegewebigen Organes. Arch. Anat. Physiol., pp. 76-162.

ROWLEY, J. 1929. Life history of the sea-lions on the California coast. Journ. Mamm., vol. 10, pp. 1-36.

RYDER, J. A. 1877. On the laws of digital reduction. Amer. Nat., vol. 11, pp. 603-607.

——————. 1885. On the development of the Cetacea, together with a consideration of the probable homologies of the flukes of cetaceans and sirenians. Rept. U. S. Comm. of Fisheries, pp. 427-485.

SCAMMON, C. M. 1874. The marine mammals of the north-western coast of North America. San Francisco, pp. 1-319.

SCHULTE, H. VON W. 1916. Anatomy of a foetus Balaenoptera borealis. In Monogs. of Pacific Cetacea. 2. The Sei whale (Balaenoptera borealis Lesson). Mem. Amer. Mus. Nat. Hist., new ser., vol. 1, pt. 6, pp. 389-502.

—————— and M. DEF. SMITH. 1918. The external characters, skeletal muscles, and peripheral nerves of Kogia breviceps (Blainville). Bull. Amer. Mus. Nat. Hist., vol. 38, pp. 7-72.

SCHULZE, F. E. 1906. Beiträge zur Anatomie der Säugetier lungen. Sitzber. Kgl. preuss. Akad. d. Wissenschaft. Berlin, pp. 225-243.

SONNTAG, C. F. 1922, 1923. The comparative anatomy of the tongues of the Mammalia. Proc. Zool. Soc. London, pp. 639-647 (Cetacea, Sirenia), pp. 145-151 (Pinnipedia).

STEINMANN. 1912. Ueber die Ursache der Asymmetrie der Wale. Anat. Anz., vol. 41, pp. 45-54.

STRUTHERS, J. 1880-81. On the bones, articulations, and muscles of the rudimentary hind-limb of the Greenland Right-Whale (Balaena mysticetus). Journ. Anat. Physiol., vol. 15, pp. 141-176, 301-321.

——————. 1887-88. On some points in the anatomy of a megaptera longimanna. Journ. Anat. Physiol., vols. 22, 23.

——————. 1893. On the rudimentary hind-limb of a great fin-whale (Balaenoptera musculus) in comparison with those of the humpback-whale and the Greenland Right-Whale. Journ. Anat. Physiol., vol. 27, pp. 291-335.

LITERATURE

TAVERNE, L. 1926. A propos de l'orientation differente de la nageoire caudale chez les Cetaces et chez les poissons. Bull. Mus. Nation. Hist. Nat. (Paris), p. 260.

TAYLOR, W. P. 1914. The problem of aquatic adaptation in the Carnivora, as illustrated in the osteology and evolution of the sea-otter. Univ. Calif. Publ. Geol., vol. 7, no. 25, pp. 265-295.

THOMAS, A. and R. LYDEKKER. 1897. On the number of grinding teeth possessed by the manati. Proc. Zool. Soc. London, pp. 595-600.

THOMPSON, A. 1869. On the difference in the mode of ossification of the first and other metacarpal and metatarsal bones. Journ. Anat. Physiol., vol. 3, p. 131.

THOMPSON, D'ARCY W. 1917. On growth and form. Cambridge Univ. Press, pp. 1-793.

TODD, T. W. 1922. Numerical significance in the thoracicolumbar vertebrae of the Mammalia. Anat. Rec., vol. 24, pp. 261-286.

TOWNSEND, C. H., 1914. The porpoise in captivity. Zoologica, vol. 1, no. 16, pp. 289-299.

TRUTAT, E. 1891. Essai sur l'histoire naturelle du desman des Pyrenees. Toulouse, pp. 1-107.

TURNER, WM. 1883. Cervical ribs, and the so-called bicipital ribs in man, in relation to corresponding structures in the Cetacea. Journ. Anat. Physiol., vol. 17, pp. 384-400.

——————. 1888. Report on the seals collected during the voyage of the H.M.S. *Challenger* in the years 1873-1876. Reports on the Sci. Results of the Voyage of the H.M.S. *Challenger*. Zoology, vol. 26, pp. 1-240.

VON HUENE, F. 1922. Die Ichthyosaurier des Lias und ihre Zusammenhänge, 4. Jahresversammlung der Palaeontologischen Gesell. im Aug. zu Tübingen, Verlag von Gebrüder Borntraeger, Berlin, pp. 1-114.

WARREN, E. R. 1927. The beaver, its work and its ways. Amer. Soc. Mamm. Monog. no. 2, pp. 1-177.

WEBER, M. 1886. Studien über Saügethiere. Jena, pp. 1-252.

——————. 1927-28. Die Säugetiere. Jena, pp. 1-444, 1-898.

WILDER, B. G. 1875. On a foetal manatee and cetacean, with remarks upon the affinities and ancestry of the Sirenia. Journ. Sci. & Arts, vol. 10, 10 pp.

WILDER, H. H. 1923. History of the human body. Henry Holt, pp. 1-623.

WILLISTON, S. W. 1902. On certain homoplastic characters in aquatic, air-breathing vertebrates. Kans. Univ. Sci. Bull., vol. 1, pp. 259-266.

——————. 1914. Water reptiles of the past and present. Chicago Univ. Press, 61 pp.

WILSON, H. S. 1880. The rete mirabile of the narwhal. Journ. Anat. Physiol., vol. 14, pp. 377-398.

WINGE, HERLUF. 1921. A review of the interrelationships of the Cetacea. (Translation by G. S. Miller, Jr.) Smiths, Misc. Colls., vol. 72, no. 8, pp. 1-97.

[331]

WISLOCKI, G. B., 1928. Observations on the gross and microscopic anatomy of the sloths (Bradypus griseus griseus Gray and Choloepus hoffmanni Peters). Journ. Morph. Physiol., vol. 46, pp. 317-397.

————. 1929. On the structure of the lungs of the porpoise (Tursiops truncatus). Amer. Journ. Anat., vol. 44, no. 1, pp. 47-77.

WORTMAN, J. L. 1921. On some hitherto unrecognized reptilian characters in the skull of the Insectivora and other mammals. Proc. U. S. Nat. Mus., vol. 57, pp. 1-52.

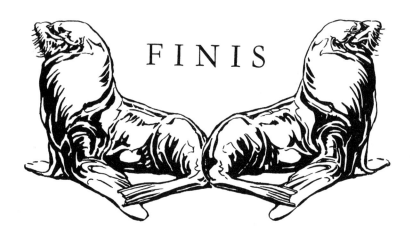

FINIS

Index

INDEX

whale,
  sei, *see* Balaenoptera
  sperm, *see* Physeter
  sulfurbottom, *see* Sibbaldus
  toothed, *see* Odontoceti
  whalebone, *see* Mysticeti
  white, *see* Delphinapterus
  ziphioid, 85, 117, 177

xiphoid, 173

yapok, *see* Chironectes

Zalophus, 107, 142, 194, 234, 292
Zarhachis, 56, 113
zeuglodont, 41, 114, 139, 198
Ziphiidae, 80, 117, 148, 177

[338]

A CATALOGUE OF SELECTED DOVER BOOKS
IN ALL FIELDS OF INTEREST

A CATALOGUE OF SELECTED DOVER BOOKS
IN ALL FIELDS OF INTEREST

WHAT IS SCIENCE?, *N. Campbell*
The role of experiment and measurement, the function of mathematics, the nature of scientific laws, the difference between laws and theories, the limitations of science, and many similarly provocative topics are treated clearly and without technicalities by an eminent scientist. "Still an excellent introduction to scientific philosophy," H. Margenau in *Physics Today.* "A first-rate primer . . . deserves a wide audience," *Scientific American.* 192pp. 5⅜ x 8.
60043-2 Paperbound $1.25

THE NATURE OF LIGHT AND COLOUR IN THE OPEN AIR, *M. Minnaert*
Why are shadows sometimes blue, sometimes green, or other colors depending on the light and surroundings? What causes mirages? Why do multiple suns and moons appear in the sky? Professor Minnaert explains these unusual phenomena and hundreds of others in simple, easy-to-understand terms based on optical laws and the properties of light and color. No mathematics is required but artists, scientists, students, and everyone fascinated by these "tricks" of nature will find thousands of useful and amazing pieces of information. Hundreds of observational experiments are suggested which require no special equipment. 200 illustrations; 42 photos. xvi + 362pp. 5⅜ x 8.
20196-1 Paperbound $2.75

THE STRANGE STORY OF THE QUANTUM, AN ACCOUNT FOR THE GENERAL READER OF THE GROWTH OF IDEAS UNDERLYING OUR PRESENT ATOMIC KNOWLEDGE, *B. Hoffmann*
Presents lucidly and expertly, with barest amount of mathematics, the problems and theories which led to modern quantum physics. Dr. Hoffmann begins with the closing years of the 19th century, when certain trifling discrepancies were noticed, and with illuminating analogies and examples takes you through the brilliant concepts of Planck, Einstein, Pauli, Broglie, Bohr, Schroedinger, Heisenberg, Dirac, Sommerfeld, Feynman, etc. This edition includes a new, long postscript carrying the story through 1958. "Of the books attempting an account of the history and contents of our modern atomic physics which have come to my attention, this is the best," H. Margenau, Yale University, in *American Journal of Physics.* 32 tables and line illustrations. Index. 275pp. 5⅜ x 8.
20518-5 Paperbound $2.00

GREAT IDEAS OF MODERN MATHEMATICS: THEIR NATURE AND USE, *Jagjit Singh*
Reader with only high school math will understand main mathematical ideas of modern physics, astronomy, genetics, psychology, evolution, etc. better than many who use them as tools, but comprehend little of their basic structure. Author uses his wide knowledge of non-mathematical fields in brilliant exposition of differential equations, matrices, group theory, logic, statistics, problems of mathematical foundations, imaginary numbers, vectors, etc. Original publication. 2 appendixes. 2 indexes. 65 ills. 322pp. 5⅜ x 8.
20587-8 Paperbound $2.50

THE MUSIC OF THE SPHERES: THE MATERIAL UNIVERSE — FROM ATOM
TO QUASAR, SIMPLY EXPLAINED, *Guy Murchie*
Vast compendium of fact, modern concept and theory, observed and calculated
data, historical background guides intelligent layman through the material
universe. Brilliant exposition of earth's construction, explanations for moon's
craters, atmospheric components of Venus and Mars (with data from recent
fly-by's), sun spots, sequences of star birth and death, neighboring galaxies,
contributions of Galileo, Tycho Brahe, Kepler, etc.; and (Vol. 2) construction
of the atom (describing newly discovered sigma and xi subatomic particles),
theories of sound, color and light, space and time, including relativity theory,
quantum theory, wave theory, probability theory, work of Newton, Maxwell,
Faraday, Einstein, de Broglie, etc. "Best presentation yet offered to the intelligent general reader," *Saturday Review*. Revised (1967). Index. 319 illustrations by the author. Total of xx + 644pp. 5⅜ x 8½.
21809-0, 21810-4   Two volume set, paperbound $5.00

FOUR LECTURES ON RELATIVITY AND SPACE, *Charles Proteus Steinmetz*
Lecture series, given by great mathematician and electrical engineer, generally
considered one of the best popular-level expositions of special and general
relativity theories and related questions. Steinmetz translates complex mathematical reasoning into language accessible to laymen through analogy, example
and comparison. Among topics covered are relativity of motion, location, time;
of mass; acceleration; 4-dimensional time-space; geometry of the gravitational
field; curvature and bending of space; non-Euclidean geometry. Index. 40
illustrations. x + 142pp. 5⅜ x 8½.   61771-8 Paperbound $1.50

HOW TO KNOW THE WILD FLOWERS, *Mrs. William Starr Dana*
Classic nature book that has introduced thousands to wonders of American
wild flowers. Color-season principle of organization is easy to use, even by
those with no botanical training, and the genial, refreshing discussions of
history, folklore, uses of over 1,000 native and escape flowers, foliage plants
are informative as well as fun to read. Over 170 full-page plates, collected from
several editions, may be colored in to make permanent records of finds. Revised
to conform with 1950 edition of Gray's Manual of Botany. xlii + 438pp.
5⅜ x 8½.   20332-8 Paperbound $2.50

MANUAL OF THE TREES OF NORTH AMERICA, *Charles Sprague Sargent*
Still unsurpassed as most comprehensive, reliable study of North American
tree characteristics, precise locations and distribution. By dean of American
dendrologists. Every tree native to U.S., Canada, Alaska; 185 genera, 717 species,
described in detail—leaves, flowers, fruit, winterbuds, bark, wood, growth
habits, etc. plus discussion of varieties and local variants, immaturity variations.
Over 100 keys, including unusual 11-page analytical key to genera, aid in
identification. 783 clear illustrations of flowers, fruit, leaves. An unmatched
permanent reference work for all nature lovers. Second enlarged (1926) edition.
Synopsis of families. Analytical key to genera. Glossary of technical terms.
Index. 783 illustrations, 1 map. Total of 982pp. 5⅜ x 8.
20277-1, 20278-X Two volume set, paperbound $6.00

IT'S FUN TO MAKE THINGS FROM SCRAP MATERIALS,
Evelyn Glantz Hershoff
What use are empty spools, tin cans, bottle tops? What can be made from rubber bands, clothes pins, paper clips, and buttons? This book provides simply worded instructions and large diagrams showing you how to make cookie cutters, toy trucks, paper turkeys, Halloween masks, telephone sets, aprons, linoleum block- and spatter prints — in all 399 projects! Many are easy enough for young children to figure out for themselves; some challenging enough to entertain adults; all are remarkably ingenious ways to make things from materials that cost pennies or less! Formerly "Scrap Fun for Everyone." Index. 214 illustrations. 373pp. 5⅜ x 8½.   21251-3 Paperbound $2.00

SYMBOLIC LOGIC and THE GAME OF LOGIC, *Lewis Carroll*
"Symbolic Logic" is not concerned with modern symbolic logic, but is instead a collection of over 380 problems posed with charm and imagination, using the syllogism and a fascinating diagrammatic method of drawing conclusions. In "The Game of Logic" Carroll's whimsical imagination devises a logical game played with 2 diagrams and counters (included) to manipulate hundreds of tricky syllogisms. The final section, "Hit or Miss" is a lagniappe of 101 additional puzzles in the delightful Carroll manner. Until this reprint edition, both of these books were rarities costing up to $15 each. Symbolic Logic: Index. xxxi + 199pp. The Game of Logic: 96pp. 2 vols. bound as one. 5⅜ x 8.
20492-8 Paperbound $2.50

MATHEMATICAL PUZZLES OF SAM LOYD, PART I
selected and edited by M. Gardner
Choice puzzles by the greatest American puzzle creator and innovator. Selected from his famous collection, "Cyclopedia of Puzzles," they retain the unique style and historical flavor of the originals. There are posers based on arithmetic, algebra, probability, game theory, route tracing, topology, counter and sliding block, operations research, geometrical dissection. Includes the famous "14-15" puzzle which was a national craze, and his "Horse of a Different Color" which sold millions of copies. 117 of his most ingenious puzzles in all. 120 line drawings and diagrams. Solutions. Selected references. xx + 167pp. 5⅜ x 8.
20498-7 Paperbound $1.35

STRING FIGURES AND HOW TO MAKE THEM, *Caroline Furness Jayne*
107 string figures plus variations selected from the best primitive and modern examples developed by Navajo, Apache, pygmies of Africa, Eskimo, in Europe, Australia, China, etc. The most readily understandable, easy-to-follow book in English on perennially popular recreation. Crystal-clear exposition; step-by-step diagrams. Everyone from kindergarten children to adults looking for unusual diversion will be endlessly amused. Index. Bibliography. Introduction by A. C. Haddon. 17 full-page plates, 960 illustrations. xxiii + 401pp. 5⅜ x 8½.
20152-X Paperbound $2.50

PAPER FOLDING FOR BEGINNERS, *W. D. Murray and F. J. Rigney*
A delightful introduction to the varied and entertaining Japanese art of origami (paper folding), with a full, crystal-clear text that anticipates every difficulty; over 275 clearly labeled diagrams of all important stages in creation. You get results at each stage, since complex figures are logically developed from simpler ones. 43 different pieces are explained: sailboats, frogs, roosters, etc. 6 photographic plates. 279 diagrams. 95pp. 5⅝ x 8⅜.
20713-7 Paperbound $1.00

PRINCIPLES OF ART HISTORY,
H. Wölfflin
Analyzing such terms as "baroque," "classic," "neoclassic," "primitive,"
"picturesque," and 164 different works by artists like Botticelli, van Cleve,
Dürer, Hobbema, Holbein, Hals, Rembrandt, Titian, Brueghel, Vermeer, and
many others, the author establishes the classifications of art history and style
on a firm, concrete basis. This classic of art criticism shows what really
occurred between the 14th-century primitives and the sophistication of the
18th century in terms of basic attitudes and philosophies. "A remarkable
lesson in the art of seeing," *Sat. Rev. of Literature*. Translated from the 7th
German edition. 150 illustrations. 254pp. 6⅛ x 9¼. 20276-3 Paperbound $2.50

PRIMITIVE ART,
Franz Boas
This authoritative and exhaustive work by a great American anthropologist
covers the entire gamut of primitive art. Pottery, leatherwork, metal work,
stone work, wood, basketry, are treated in detail. Theories of primitive art,
historical depth in art history, technical virtuosity, unconscious levels of pat-
terning, symbolism, styles, literature, music, dance, etc. A must book for the
interested layman, the anthropologist, artist, handicrafter (hundreds of unusual motifs), and the historian. Over 900 illustrations (50 ceramic vessels,
12 totem poles, etc.). 376pp. 5⅜ x 8. 20025-6 Paperbound $2.50

THE GENTLEMAN AND CABINET MAKER'S DIRECTOR,
Thomas Chippendale
A reprint of the 1762 catalogue of furniture designs that went on to influence
generations of English and Colonial and Early Republic American furniture
makers. The 200 plates, most of them full-page sized, show Chippendale's
designs for French (Louis XV), Gothic, and Chinese-manner chairs, sofas,
canopy and dome beds, cornices, chamber organs, cabinets, shaving tables,
commodes, picture frames, frets, candle stands, chimney pieces, decorations, etc.
The drawings are all elegant and highly detailed; many include construction
diagrams and elevations. A supplement of 24 photographs shows surviving
pieces of original and Chippendale-style pieces of furniture. Brief biography
of Chippendale by N. I. Bienenstock, editor of *Furniture World*. Reproduced
from the 1762 edition. 200 plates, plus 19 photographic plates. vi + 249pp.
9⅛ x 12¼. 21601-2 Paperbound $4.00

AMERICAN ANTIQUE FURNITURE: A BOOK FOR AMATEURS,
Edgar G. Miller, Jr.
Standard introduction and practical guide to identification of valuable
American antique furniture. 2115 illustrations, mostly photographs taken by
the author in 148 private homes, are arranged in chronological order in extensive chapters on chairs, sofas, chests, desks, bedsteads, mirrors, tables, clocks,
and other articles. Focus is on furniture accessible to the collector, including
simpler pieces and a larger than usual coverage of Empire style. Introductory
chapters identify structural elements, characteristics of various styles, how to
avoid fakes, etc. "We are frequently asked to name some book on American
furniture that will meet the requirements of the novice collector, the beginning dealer, and . . . the general public. . . . We believe Mr. Miller's two
volumes more completely satisfy this specification than any other work,"
Antiques. Appendix. Index. Total of vi + 1106pp. 7⅞ x 10¾.
21599-7, 21600-4 Two volume set, paperbound $10.00

THE BAD CHILD'S BOOK OF BEASTS, MORE BEASTS FOR WORSE CHILDREN, and A MORAL ALPHABET, H. Belloc
Hardly and anthology of humorous verse has appeared in the last 50 years without at least a couple of these famous nonsense verses. But one must see the entire volumes — with all the delightful original illustrations by Sir Basil Blackwood — to appreciate fully Belloc's charming and witty verses that play so subacidly on the platitudes of life and morals that beset his day — and ours. A great humor classic. Three books in one. Total of 157pp. 5⅜ x 8.
20749-8 Paperbound $1.25

THE DEVIL'S DICTIONARY, Ambrose Bierce
Sardonic and irreverent barbs puncturing the pomposities and absurdities of American politics, business, religion, literature, and arts, by the country's greatest satirist in the classic tradition. Epigrammatic as Shaw, piercing as Swift, American as Mark Twain, Will Rogers, and Fred Allen, Bierce will always remain the favorite of a small coterie of enthusiasts, and of writers and speakers whom he supplies with "some of the most gorgeous witticisms of the English language" (H. L. Mencken). Over 1000 entries in alphabetical order. 144pp. 5⅜ x 8.
20487-1 Paperbound $1.25

THE COMPLETE NONSENSE OF EDWARD LEAR.
This is the only complete edition of this master of gentle madness available at a popular price. A Book of Nonsense, Nonsense Songs, More Nonsense Songs and Stories in their entirety with all the old favorites that have delighted children and adults for years. The Dong With A Luminous Nose, The Jumblies, The Owl and the Pussycat, and hundreds of other bits of wonderful nonsense. 214 limericks, 3 sets of Nonsense Botany, 5 Nonsense Alphabets, 546 drawings by Lear himself, and much more. 320pp. 5⅜ x 8. 20167-8 Paperbound $1.75

THE WIT AND HUMOR OF OSCAR WILDE, ed. by Alvin Redman
Wilde at his most brilliant, in 1000 epigrams exposing weaknesses and hypocrisies of "civilized" society. Divided into 49 categories—sin, wealth, women, America, etc.—to aid writers, speakers. Includes excerpts from his trials, books, plays, criticism. Formerly "The Epigrams of Oscar Wilde." Introduction by Vyvyan Holland, Wilde's only living son. Introductory essay by editor. 260pp. 5⅜ x 8.
20602-5 Paperbound $1.50

A CHILD'S PRIMER OF NATURAL HISTORY, Oliver Herford
Scarcely an anthology of whimsy and humor has appeared in the last 50 years without a contribution from Oliver Herford. Yet the works from which these examples are drawn have been almost impossible to obtain! Here at last are Herford's improbable definitions of a menagerie of familiar and weird animals, each verse illustrated by the author's own drawings. 24 drawings in 2 colors; 24 additional drawings. vii + 95pp. 6½ x 6. 21647-0 Paperbound $1.00

THE BROWNIES: THEIR BOOK, Palmer Cox
The book that made the Brownies a household word. Generations of readers have enjoyed the antics, predicaments and adventures of these jovial sprites, who emerge from the forest at night to play or to come to the aid of a deserving human. Delightful illustrations by the author decorate nearly every page. 24 short verse tales with 266 illustrations. 155pp. 6⅝ x 9¼.
21265-3 Paperbound $1.50

THE PRINCIPLES OF PSYCHOLOGY,
William James
The full long-course, unabridged, of one of the great classics of Western literature and science. Wonderfully lucid descriptions of human mental activity, the stream of thought, consciousness, time perception, memory, imagination, emotions, reason, abnormal phenomena, and similar topics. Original contributions are integrated with the work of such men as Berkeley, Binet, Mills, Darwin, Hume, Kant, Royce, Schopenhauer, Spinoza, Locke, Descartes, Galton, Wundt, Lotze, Herbart, Fechner, and scores of others. All contrasting interpretations of mental phenomena are examined in detail—introspective analysis, philosophical interpretation, and experimental research. "A classic," *Journal of Consulting Psychology.* "The main lines are as valid as ever," *Psychoanalytical Quarterly.* "Standard reading ... a classic of interpretation," *Psychiatric Quarterly.* 94 illustrations. 1408pp. 5⅜ x 8.
20381-6, 20382-4 Two volume set, paperbound $6.00

VISUAL ILLUSIONS: THEIR CAUSES, CHARACTERISTICS AND APPLICATIONS,
M. Luckiesh
"Seeing is deceiving," asserts the author of this introduction to virtually every type of optical illusion known. The text both describes and explains the principles involved in color illusions, figure-ground, distance illusions, etc. 100 photographs, drawings and diagrams prove how easy it is to fool the sense: circles that aren't round, parallel lines that seem to bend, stationary figures that seem to move as you stare at them — illustration after illustration strains our credulity at what we see. Fascinating book from many points of view, from applications for artists, in camouflage, etc. to the psychology of vision. New introduction by William Ittleson, Dept. of Psychology, Queens College. Index. Bibliography. xxi + 252pp. 5⅜ x 8½.   21530-X Paperbound $1.75

FADS AND FALLACIES IN THE NAME OF SCIENCE,
Martin Gardner
This is the standard account of various cults, quack systems, and delusions which have masqueraded as science: hollow earth fanatics. Reich and orgone sex energy, dianetics, Atlantis, multiple moons, Forteanism, flying saucers, medical fallacies like iridiagnosis, zone therapy, etc. A new chapter has been added on Bridey Murphy, psionics, and other recent manifestations in this field. This is a fair, reasoned appraisal of eccentric theory which provides excellent inoculation against cleverly masked nonsense. "Should be read by everyone, scientist and non-scientist alike," R. T. Birge, Prof. Emeritus of Physics, Univ. of California; Former President, American Physical Society. Index. x + 365pp. 5⅜ x 8.   20394-8 Paperbound $2.00

ILLUSIONS AND DELUSIONS OF THE SUPERNATURAL AND THE OCCULT,
D. H. Rawcliffe
Holds up to rational examination hundreds of persistent delusions including crystal gazing, automatic writing, table turning, mediumistic trances, mental healing, stigmata, lycanthropy, live burial, the Indian Rope Trick, spiritualism, dowsing, telepathy, clairvoyance, ghosts, ESP, etc. The author explains and exposes the mental and physical deceptions involved, making this not only an exposé of supernatural phenomena, but a valuable exposition of characteristic types of abnormal psychology. Originally titled "The Psychology of the Occult." 14 illustrations. Index. 551pp. 5⅜ x 8. 20503-7 Paperbound $3.50

FAIRY TALE COLLECTIONS, *edited by Andrew Lang*
Andrew Lang's fairy tale collections make up the richest shelf-full of traditional children's stories anywhere available. Lang supervised the translation of stories from all over the world—familiar European tales collected by Grimm, animal stories from Negro Africa, myths of primitive Australia, stories from Russia, Hungary, Iceland, Japan, and many other countries. Lang's selection of translations are unusually high; many authorities consider that the most familiar tales find their best versions in these volumes. All collections are richly decorated and illustrated by H. J. Ford and other artists.

THE BLUE FAIRY BOOK. 37 stories. 138 illustrations. ix + 390pp. 5⅜ x 8½.
21437-0 Paperbound $1.95

THE GREEN FAIRY BOOK. 42 stories. 100 illustrations. xiii + 366pp. 5⅜ x 8½.
21439-7 Paperbound $2.00

THE BROWN FAIRY BOOK. 32 stories. 50 illustrations, 8 in color. xii + 350pp. 5⅜ x 8½.
21438-9 Paperbound $1.95

THE BEST TALES OF HOFFMANN, *edited by E. F. Bleiler*
10 stories by E. T. A. Hoffmann, one of the greatest of all writers of fantasy. The tales include "The Golden Flower Pot," "Automata," "A New Year's Eve Adventure," "Nutcracker and the King of Mice," "Sand-Man," and others. Vigorous characterizations of highly eccentric personalities, remarkably imaginative situations, and intensely fast pacing has made these tales popular all over the world for 150 years. Editor's introduction. 7 drawings by Hoffmann. xxxiii + 419pp. 5⅜ x 8½.
21793-0 Paperbound $2.25

GHOST AND HORROR STORIES OF AMBROSE BIERCE,
edited by E. F. Bleiler
Morbid, eerie, horrifying tales of possessed poets, shabby aristocrats, revived corpses, and haunted malefactors. Widely acknowledged as the best of their kind between Poe and the moderns, reflecting their author's inner torment and bitter view of life. Includes "Damned Thing," "The Middle Toe of the Right Foot," "The Eyes of the Panther," "Visions of the Night," "Moxon's Master," and over a dozen others. Editor's introduction. xxii + 199pp. 5⅜ x 8½.
20767-6 Paperbound $1.50

THREE GOTHIC NOVELS, *edited by E. F. Bleiler*
Originators of the still popular Gothic novel form, influential in ushering in early 19th-century Romanticism. Horace Walpole's *Castle of Otranto*, William Beckford's *Vathek*, John Polidori's *The Vampyre*, and a *Fragment* by Lord Byron are enjoyable as exciting reading or as documents in the history of English literature. Editor's introduction. xi + 291pp. 5⅜ x 8½.
21232-7 Paperbound $2.00

BEST GHOST STORIES OF LEFANU, *edited by E. F. Bleiler*
Though admired by such critics as V. S. Pritchett, Charles Dickens and Henry James, ghost stories by the Irish novelist Joseph Sheridan LeFanu have never become as widely known as his detective fiction. About half of the 16 stories in this collection have never before been available in America. Collection includes "Carmilla" (perhaps the best vampire story ever written), "The Haunted Baronet," "The Fortunes of Sir Robert Ardagh," and the classic "Green Tea." Editor's introduction. 7 contemporary illustrations. Portrait of LeFanu. xii + 467pp. 5⅜ x 8.
20415-4 Paperbound $2.50

EASY-TO-DO ENTERTAINMENTS AND DIVERSIONS WITH COINS, CARDS, STRING, PAPER AND MATCHES, *R. M. Abraham*
Over 300 tricks, games and puzzles will provide young readers with absorbing fun. Sections on card games; paper-folding; tricks with coins, matches and pieces of string; games for the agile; toy-making from common household objects; mathematical recreations; and 50 miscellaneous pastimes. Anyone in charge of groups of youngsters, including hard-pressed parents, and in need of suggestions on how to keep children sensibly amused and quietly content will find this book indispensable. Clear, simple text, copious number of delightful line drawings and illustrative diagrams. Originally titled "Winter Nights' Entertainments." Introduction by Lord Baden Powell. 329 illustrations. v + 186pp. 5⅜ x 8½.   20921-0 Paperbound $1.25

AN INTRODUCTION TO CHESS MOVES AND TACTICS SIMPLY EXPLAINED, *Leonard Barden*
Beginner's introduction to the royal game. Names, possible moves of the pieces, definitions of essential terms, how games are won, etc. explained in 30-odd pages. With this background you'll be able to sit right down and play. Balance of book teaches strategy — openings, middle game, typical endgame play, and suggestions for improving your game. A sample game is fully analyzed. True middle-level introduction, teaching you all the essentials without oversimplifying or losing you in a maze of detail. 58 figures. 102pp. 5⅜ x 8½.   21210-6 Paperbound $1.25

LASKER'S MANUAL OF CHESS, *Dr. Emanuel Lasker*
Probably the greatest chess player of modern times, Dr. Emanuel Lasker held the world championship 28 years, independent of passing schools or fashions. This unmatched study of the game, chiefly for intermediate to skilled players, analyzes basic methods, combinations, position play, the aesthetics of chess, dozens of different openings, etc., with constant reference to great modern games. Contains a brilliant exposition of Steinitz's important theories. Introduction by Fred Reinfeld. Tables of Lasker's tournament record. 3 indices. 308 diagrams. 1 photograph. xxx + 349pp. 5⅜ x 8.20640-8 Paperbound $2.50

COMBINATIONS: THE HEART OF CHESS, *Irving Chernev*
Step-by-step from simple combinations to complex, this book, by a well-known chess writer, shows you the intricacies of pins, counter-pins, knight forks, and smothered mates. Other chapters show alternate lines of play to those taken in actual championship games; boomerang combinations; classic examples of brilliant combination play by Nimzovich, Rubinstein, Tarrasch, Botvinnik, Alekhine and Capablanca. Index. 356 diagrams. ix + 245pp. 5⅜ x 8½.   21744-2 Paperbound $2.00

HOW TO SOLVE CHESS PROBLEMS, *K. S. Howard*
Full of practical suggestions for the fan or the beginner — who knows only the moves of the chessmen. Contains preliminary section and 58 two-move, 46 three-move, and 8 four-move problems composed by 27 outstanding American problem creators in the last 30 years. Explanation of all terms and exhaustive index. "Just what is wanted for the student," Brian Harley. 112 problems, solutions. vi + 171pp. 5⅜ x 8.   20748-X Paperbound $1.50

SOCIAL THOUGHT FROM LORE TO SCIENCE,
H. E. Barnes and H. Becker
An immense survey of sociological thought and ways of viewing, studying, planning, and reforming society from earliest times to the present. Includes thought on society of preliterate peoples, ancient non-Western cultures, and every great movement in Europe, America, and modern Japan. Analyzes hundreds of great thinkers: Plato, Augustine, Bodin, Vico, Montesquieu, Herder, Comte, Marx, etc. Weighs the contributions of utopians, sophists, fascists and communists; economists, jurists, philosophers, ecclesiastics, and every 19th and 20th century school of scientific sociology, anthropology, and social psychology throughout the world. Combines topical, chronological, and regional approaches, treating the evolution of social thought as a process rather than as a series of mere topics. "Impressive accuracy, competence, and discrimination . . . easily the best single survey," *Nation.* Thoroughly revised, with new material up to 1960. 2 indexes. Over 2200 bibliographical notes. Three volume set. Total of 1586pp. 5⅜ x 8.
20901-6, 20902-4, 20903-2 Three volume set, paperbound $10.50

A HISTORY OF HISTORICAL WRITING, *Harry Elmer Barnes*
Virtually the only adequate survey of the whole course of historical writing in a single volume. Surveys developments from the beginnings of historiography in the ancient Near East and the Classical World, up through the Cold War. Covers major historians in detail, shows interrelationship with cultural background, makes clear individual contributions, evaluates and estimates importance; also enormously rich upon minor authors and thinkers who are usually passed over. Packed with scholarship and learning, clear, easily written. Indispensable to every student of history. Revised and enlarged up to 1961. Index and bibliography. xv + 442pp. 5⅜ x 8½.
20104-X Paperbound $3.00

JOHANN SEBASTIAN BACH, *Philipp Spitta*
The complete and unabridged text of the definitive study of Bach. Written some 70 years ago, it is still unsurpassed for its coverage of nearly all aspects of Bach's life and work. There could hardly be a finer non-technical introduction to Bach's music than the detailed, lucid analyses which Spitta provides for hundreds of individual pieces. 26 solid pages are devoted to the B minor mass, for example, and 30 pages to the glorious St. Matthew Passion. This monumental set also includes a major analysis of the music of the 18th century: Buxtehude, Pachelbel, etc. "Unchallenged as the last word on one of the supreme geniuses of music," John Barkham, *Saturday Review Syndicate.* Total of 1819pp. Heavy cloth binding. 5⅜ x 8.
22278-0, 22279-9 Two volume set, clothbound $15.00

BEETHOVEN AND HIS NINE SYMPHONIES, *George Grove*
In this modern middle-level classic of musicology Grove not only analyzes all nine of Beethoven's symphonies very thoroughly in terms of their musical structure, but also discusses the circumstances under which they were written, Beethoven's stylistic development, and much other background material. This is an extremely rich book, yet very easily followed; it is highly recommended to anyone seriously interested in music. Over 250 musical passages. Index. viii + 407pp. 5⅜ x 8.
20334-4 Paperbound $2.50

THE TIME STREAM
John Taine
Acknowledged by many as the best SF writer of the 1920's, Taine (under the name Eric Temple Bell) was also a Professor of Mathematics of considerable renown. Reprinted here are *The Time Stream,* generally considered Taine's best, *The Greatest Game,* a biological-fiction novel, and *The Purple Sapphire,* involving a supercivilization of the past. Taine's stories tie fantastic narratives to frameworks of original and logical scientific concepts. Speculation is often profound on such questions as the nature of time, concept of entropy, cyclical universes, etc. 4 contemporary illustrations. v + 532pp. 5⅜ x 8⅜.
21180-0 Paperbound $3.00

SEVEN SCIENCE FICTION NOVELS,
H. G. Wells
Full unabridged texts of 7 science-fiction novels of the master. Ranging from biology, physics, chemistry, astronomy, to sociology and other studies, Mr. Wells extrapolates whole worlds of strange and intriguing character. "One will have to go far to match this for entertainment, excitement, and sheer pleasure . . ."*New York Times.* Contents: The Time Machine, The Island of Dr. Moreau, The First Men in the Moon, The Invisible Man, The War of the Worlds, The Food of the Gods, In The Days of the Comet. 1015pp. 5⅜ x 8.
20264-X Clothbound $5.00

28 SCIENCE FICTION STORIES OF H. G. WELLS.
Two full, unabridged novels, *Men Like Gods* and *Star Begotten,* plus 26 short stories by the master science-fiction writer of all time! Stories of space, time, invention, exploration, futuristic adventure. Partial contents: *The Country of the Blind, In the Abyss, The Crystal Egg, The Man Who Could Work Miracles, A Story of Days to Come, The Empire of the Ants, The Magic Shop, The Valley of the Spiders, A Story of the Stone Age, Under the Knife, Sea Raiders,* etc. An indispensable collection for the library of anyone interested in science fiction adventure. 928pp. 5⅜ x 8.
20265-8 Clothbound $5.00

THREE MARTIAN NOVELS,
Edgar Rice Burroughs
Complete, unabridged reprinting, in one volume, of Thuvia, Maid of Mars; Chessmen of Mars; The Master Mind of Mars. Hours of science-fiction adventure by a modern master storyteller. Reset in large clear type for easy reading. 16 illustrations by J. Allen St. John. vi + 499pp. 5⅜ x 8½.
20039-6 .Paperbound $2.50

AN INTELLECTUAL AND CULTURAL HISTORY OF THE WESTERN WORLD,
Harry Elmer Barnes
Monumental 3-volume survey of intellectual development of Europe from primitive cultures to the present day. Every significant product of human intellect traced through history: art, literature, mathematics, physical sciences, medicine, music, technology, social sciences, religions, jurisprudence, education, etc. Presentation is lucid and specific, analyzing in detail specific discoveries, theories, literary works, and so on. Revised (1965) by recognized scholars in specialized fields under the direction of Prof. Barnes. Revised bibliography. Indexes. 24 illustrations. Total of xxix + 1318pp.
21275-0, 21276-9, 21277-7 Three volume set, paperbound $7.75

HEAR ME TALKIN' TO YA, *edited by Nat Shapiro and Nat Hentoff*
In their own words, Louis Armstrong, King Oliver, Fletcher Henderson, Bunk Johnson, Bix Beiderbecke, Billy Holiday, Fats Waller, Jelly Roll Morton, Duke Ellington, and many others comment on the origins of jazz in New Orleans and its growth in Chicago's South Side, Kansas City's jam sessions, Depression Harlem, and the modernism of the West Coast schools. Taken from taped conversations, letters, magazine articles, other first-hand sources. Editors' introduction. xvi + 429pp. 5⅜ x 8½. 21726-4 Paperbound $2.50

THE JOURNAL OF HENRY D. THOREAU
A 25-year record by the great American observer and critic, as complete a record of a great man's inner life as is anywhere available. Thoreau's Journals served him as raw material for his formal pieces, as a place where he could develop his ideas, as an outlet for his interests in wild life and plants, in writing as an art, in classics of literature, Walt Whitman and other contemporaries, in politics, slavery, individual's relation to the State, etc. The Journals present a portrait of a remarkable man, and are an observant social history. Unabridged republication of 1906 edition, Bradford Torrey and Francis H. Allen, editors. Illustrations. Total of 1888pp. 8⅜ x 12¼.
20312-3, 20313-1 Two volume set, clothbound $30.00

A SHAKESPEARIAN GRAMMAR, *E. A. Abbott*
Basic reference to Shakespeare and his contemporaries, explaining through thousands of quotations from Shakespeare, Jonson, Beaumont and Fletcher, North's *Plutarch* and other sources the grammatical usage differing from the modern. First published in 1870 and written by a scholar who spent much of his life isolating principles of Elizabethan language, the book is unlikely ever to be superseded. Indexes. xxiv + 511pp. 5⅜ x 8½. 21582-2 Paperbound $3.00

FOLK-LORE OF SHAKESPEARE, *T. F. Thistelton Dyer*
Classic study, drawing from Shakespeare a large body of references to supernatural beliefs, terminology of falconry and hunting, games and sports, good luck charms, marriage customs, folk medicines, superstitions about plants, animals, birds, argot of the underworld, sexual slang of London, proverbs, drinking customs, weather lore, and much else. From full compilation comes a mirror of the 17th-century popular mind. Index. ix + 526pp. 5⅜ x 8½.
21614-4 Paperbound $3.25

THE NEW VARIORUM SHAKESPEARE, *edited by H. H. Furness*
By far the richest editions of the plays ever produced in any country or language. Each volume contains complete text (usually First Folio) of the play, all variants in Quarto and other Folio texts, editorial changes by every major editor to Furness's own time (1900), footnotes to obscure references or language, extensive quotes from literature of Shakespearian criticism, essays on plot sources (often reprinting sources in full), and much more.

HAMLET, *edited by H. H. Furness*
Total of xxvi + 905pp. 5⅜ x 8½.
21004-9, 21005-7 Two volume set, paperbound $5.50
TWELFTH NIGHT, *edited by H. H. Furness*
Index. xxii + 434pp. 5⅜ x 8½. 21189-4 Paperbound $2.75

LA BOHEME BY GIACOMO PUCCINI,
translated and introduced by Ellen H. Bleiler
Complete handbook for the operagoer, with everything needed for full enjoy-
ment except the musical score itself. Complete Italian libretto, with new,
modern English line-by-line translation—the only libretto printing all repeats;
biography of Puccini; the librettists; background to the opera, Murger's La
Boheme, etc.; circumstances of composition and performances; plot summary;
and pictorial section of 73 illustrations showing Puccini, famous singers and
performances, etc. Large clear type for easy reading. 124pp. 5⅜ x 8½.
20404-9 Paperbound $1.50

ANTONIO STRADIVARI: HIS LIFE AND WORK (1644-1737),
W. Henry Hill, Arthur F. Hill, and Alfred E. Hill
Still the only book that really delves into life and art of the incomparable
Italian craftsman, maker of the finest musical instruments in the world today.
The authors, expert violin-makers themselves, discuss Stradivari's ancestry, his
construction and finishing techniques, distinguished characteristics of many
of his instruments and their locations. Included, too, is story of introduction
of his instruments into France, England, first revelation of their supreme
merit, and information on his labels, number of instruments made, prices,
mystery of ingredients of his varnish, tone of pre-1684 Stradivari violin and
changes between 1684 and 1690. An extremely interesting, informative account
for all music lovers, from craftsman to concert-goer. Republication of original
(1902) edition. New introduction by Sydney Beck, Head of Rare Book and
Manuscript Collections, Music Division, New York Public Library. Analytical
index by Rembert Wurlitzer. Appendixes. 68 illustrations. 30 full-page plates.
4 in color. xxvi + 315pp. 5⅜ x 8½. 20425-1 Paperbound $3.00

MUSICAL AUTOGRAPHS FROM MONTEVERDI TO HINDEMITH,
Emanuel Winternitz
For beauty, for intrinsic interest, for perspective on the composer's personality,
for subtleties of phrasing, shading, emphasis indicated in the autograph but
suppressed in the printed score, the mss. of musical composition are fascinating
documents which repay close study in many different ways. This 2-volume
work reprints facsimiles of mss. by virtually every major composer, and many
minor figures—196 examples in all. A full text points out what can be learned
from mss., analyzes each sample. Index. Bibliography. 18 figures. 196 plates.
Total of 170pp. of text. 7⅞ x 10¾.
21312-9, 21313-7 Two volume set, paperbound $5.00

J. S. BACH,
Albert Schweitzer
One of the few great full-length studies of Bach's life and work, and the
study upon which Schweitzer's renown as a musicologist rests. On first appearance (1911), revolutionized Bach performance. The only writer on Bach to
be musicologist, performing musician, and student of history, theology and
philosophy, Schweitzer contributes particularly full sections on history of German Protestant church music, theories on motivic pictorial representations
in vocal music, and practical suggestions for performance. Translated by
Ernest Newman. Indexes. 5 illustrations. 650 musical examples. Total of xix
+ 928pp. 5⅜ x 8½. 21631-4, 21632-2 Two volume set, paperbound $5.00

THE METHODS OF ETHICS, *Henry Sidgwick*
Propounding no organized system of its own, study subjects every major
methodological approach to ethics to rigorous, objective analysis. Study discusses and relates ethical thought of Plato, Aristotle, Bentham, Clarke, Butler,
Hobbes, Hume, Mill, Spencer, Kant, and dozens of others. Sidgwick retains
conclusions from each system which follow from ethical premises, rejecting
the faulty. Considered by many in the field to be among the most important
treatises on ethical philosophy. Appendix. Index. xlvii + 528pp. 5⅜ x 8½.
21608-X Paperbound $3.00

TEUTONIC MYTHOLOGY, *Jakob Grimm*
A milestone in Western culture; the work which established on a modern
basis the study of history of religions and comparative religions. 4-volume
work assembles and interprets everything available on religious and folk-
loristic beliefs of Germanic people (including Scandinavians, Anglo-Saxons,
etc.). Assembling material from such sources as Tacitus, surviving Old Norse
and Icelandic texts, archeological remains, folktales, surviving superstitions,
comparative traditions, linguistic analysis, etc. Grimm explores pagan deities,
heroes, folklore of nature, religious practices, and every other area of pagan
German belief. To this day, the unrivaled, definitive, exhaustive study. Translated by J. S. Stallybrass from 4th (1883) German edition. Indexes. Total of
lxxvii + 1887pp. 5⅜ x 8½.
21602-0, 21603-9, 21604-7, 21605-5 Four volume set, paperbound $12.00

THE I CHING, *translated by James Legge*
Called "The Book of Changes" in English, this is one of the Five Classics
edited by Confucius, basic and central to Chinese thought. Explains perhaps
the most complex system of divination known, founded on the theory that all
things happening at any one time have characteristic features which can be
isolated and related. Significant in Oriental studies, in history of religions and
philosophy, and also to Jungian psychoanalysis and other areas of modern
European thought. Index. Appendixes. 6 plates. xxi + 448pp. 5⅜ x 8½.
21062-6 Paperbound $2.75

HISTORY OF ANCIENT PHILOSOPHY, *W. Windelband*
One of the clearest, most accurate comprehensive surveys of Greek and Roman
philosophy. Discusses ancient philosophy in general, intellectual life in Greece
in the 7th and 6th centuries B.C., Thales, Anaximander, Anaximenes, Herac-
litus, the Eleatics, Empedocles, Anaxagoras, Leucippus, the Pythagoreans, the
Sophists, Socrates, Democritus (20 pages), Plato (50 pages), Aristotle (70 pages),
the Peripatetics, Stoics, Epicureans, Sceptics, Neo-platonists, Christian Apolo-
gists, etc. 2nd German edition translated by H. E. Cushman. xv + 393pp.
5⅜ x 8.
20357-3 Paperbound $3.00

THE PALACE OF PLEASURE, *William Painter*
Elizabethan versions of Italian and French novels from *The Decameron*,
Cinthio, Straparola, Queen Margaret of Navarre, and other continental sources
— the very work that provided Shakespeare and dozens of his contemporaries
with many of their plots and sub-plots and, therefore, justly considered one of
the most influential books in all English literature. It is also a book that any
reader will still enjoy. Total of cviii + 1,224pp.
21691-8, 21692-6, 21693-4 Three volume set, paperbound $8.25

THE WONDERFUL WIZARD OF OZ, *L. F. Baum*
All the original W. W. Denslow illustrations in full color—as much a part of
"The Wizard" as Tenniel's drawings are of "Alice in Wonderland." "The
Wizard" is still America's best-loved fairy tale, in which, as the author expresses
it, "The wonderment and joy are retained and the heartaches and nightmares
left out." Now today's young readers can enjoy every word and wonderful picture of the original book. New introduction by Martin Gardner. A Baum
bibliography. 23 full-page color plates. viii + 268pp. 5⅜ x 8.
20691-2 Paperbound $1.95

THE MARVELOUS LAND OF OZ, *L. F. Baum*
This is the equally enchanting sequel to the "Wizard," continuing the adventures of the Scarecrow and the Tin Woodman. The hero this time is a little
boy named Tip, and all the delightful Oz magic is still present. This is the
Oz book with the Animated Saw-Horse, the Woggle-Bug, and Jack Pumpkin-
head. All the original John R. Neill illustrations, 10 in full color. 287pp.
5⅜ x 8.
20692-0 Paperbound $1.75

ALICE'S ADVENTURES UNDER GROUND, *Lewis Carroll*
The original *Alice in Wonderland*, hand-lettered and illustrated by Carroll
himself, and originally presented as a Christmas gift to a child-friend. Adults
as well as children will enjoy this charming volume, reproduced faithfully
in this Dover edition. While the story is essentially the same, there are slight
changes, and Carroll's spritely drawings present an intriguing alternative to
the famous Tenniel illustrations. One of the most popular books in Dover's
catalogue. Introduction by Martin Gardner. 38 illustrations. 128pp. 5⅜ x 8½.
21482-6 Paperbound $1.00

THE NURSERY "ALICE," *Lewis Carroll*
While most of us consider *Alice in Wonderland* a story for children of all
ages, Carroll himself felt it was beyond younger children. He therefore provided this simplified version, illustrated with the famous Tenniel drawings
enlarged and colored in delicate tints, for children aged "from Nought to
Five." Dover's edition of this now rare classic is a faithful copy of the 1889
printing, including 20 illustrations by Tenniel, and front and back covers
reproduced in full color. Introduction by Martin Gardner. xxiii + 67pp.
6⅛ x 9¼.
21610-1 Paperbound $1.75

THE STORY OF KING ARTHUR AND HIS KNIGHTS, *Howard Pyle*
A fast-paced, exciting retelling of the best known Arthurian legends for young
readers by one of America's best story tellers and illustrators. The sword
Excalibur, wooing of Guinevere, Merlin and his downfall, adventures of Sir
Pellias and Gawaine, and others. The pen and ink illustrations are vividly
imagined and wonderfully drawn. 41 illustrations. xviii + 313pp. 6⅛ x 9¼.
21445-1 Paperbound $2.00

Prices subject to change without notice.

Available at your book dealer or write for free catalogue to Dept. Adsci,
Dover Publications, Inc., 180 Varick St., N.Y., N.Y. 10014. Dover publishes more
than 150 books each year on science, elementary and advanced mathematics,
biology, music, art, literary history, social sciences and other areas.

Date Due

N. L. Bailey				
FE 3 70				
DE				
DE 18 71				
FE 2				
NO 26 74				
DE 9 74				
NOV 28 91				
JUN 28 1995				
APR				